Assessments of Regional and Global Environmental Risks

Designing Processes for the Effective Use of Science in Decisionmaking

EDITED BY
Alexander E. Farrell
Jill Jäger

RESOURCES FOR THE FUTURE
Washington, DC, USA

An RFF Press book
Published by Resources for the Future
1616 P Street NW
Washington, DC 20036–1400
USA
www.rffpress.org

Library of Congress Cataloging-in-Publication Data

Assessments of regional and global environmental risks : designing processes for the effective use of science in decisionmaking / edited by Alexander E. Farrell, Jill Jäger.
    p. cm.
  ISBN 1-933115-04-1 (hardcover : alk. paper) -- ISBN 1-933115-05-X (pbk. : alk. paper)
  1. Environmental impact analysis. 2. Decision making. I. Farrell, Alexander E. II. Jäger, Jill.
TD194.6.T875 2005
333.71'4--dc22   2005019552

The paper in this book meets the guidelines for permanence and durability of the Committee on Production Guidelines for Book Longevity of the Council on Library Resources. This book was typeset by Peter Lindeman. It was copyedited by Bonnie Nevel. Cover image, "Conceptual Earth Matrix," ©2004, Brand X Pictures. The cover was designed by Maggie Powell.

ISBN 1-933115-04-1 (cloth)   ISBN 1-933115-05-X (paper)

About Resources for the Future *and* RFF Press

RESOURCES FOR THE FUTURE (RFF) improves environmental and natural resource policymaking worldwide through independent social science research of the highest caliber. Founded in 1952, RFF pioneered the application of economics as a tool for developing more effective policy about the use and conservation of natural resources. Its scholars continue to employ social science methods to analyze critical issues concerning pollution control, energy policy, land and water use, hazardous waste, climate change, biodiversity, and the environmental challenges of developing countries.

RFF PRESS supports the mission of RFF by publishing book-length works that present a broad range of approaches to the study of natural resources and the environment. Its authors and editors include RFF staff, researchers from the larger academic and policy communities, and journalists. Audiences for publications by RFF Press include all of the participants in the policymaking process—scholars, the media, advocacy groups, NGOs, professionals in business and government, and the public.

Dedication

The editors dedicate their efforts to their families: Alice and Edward Farrell, and Fredy, Marc, and Andreas Jäger.

Contents

Foreword

THIS BOOK EMERGES FROM the Global Environmental Assessment (GEA) Project. We launched the project in 1995 as an international, interdisciplinary effort directed at understanding the role of organized efforts to bring scientific information to bear in shaping social responses to large-scale environmental change. The focus of the project was the growing number of such efforts—ranging from the periodic reports of the Intergovernmental Panel on Climate Change to the Global Biodiversity Assessment and the Millennium Ecosystem Assessment—that have been conducted in support of international policymaking over the past 25 years. Our central concern was to understand the impacts of environmental assessments on large-scale interactions between nature and society, and how changes in the conduct of those assessments could alter their impacts. We attempted to advance a common understanding of what it might mean to say that one effort to mobilize scientific information is more "effective" than another. We tried to view such issues from the perspectives of both decisionmakers at multiple scales and the experts who provide scientific advice. We attempted to embed our research approaches and interpretation of findings in contemporary theoretical frameworks of science studies, policy studies, and international relations. At the same time, we tried to keep our efforts grounded in reality through a series of workshops that engaged practitioners, users, and scholars of assessment in an off-the-record dialog that allowed them to compare insights and experiences.

Over its five year study, the GEA Project engaged a group of more than 50 senior scholars, post-doctoral fellows, and students drawn from the natural, social, and policy sciences in an intensive program of training and research. Our series of workshops with assessment practitioners and managers engaged another 50 or so individuals. The project produced more than 40 working papers, many of which subsequently appeared in peer-reviewed literature. Three synthesis volumes have emerged from the GEA Project: *Earthly Politics: Local and Global in Environmental Governance,* edited by Sheila Jasanoff and Marybeth Long Martello, analyzes a variety of approaches to environmental governance that balance the local and the global to encourage new, more flexible frameworks of global governance. *Global Environmental Assessments: Information and Influence,* edited by Ronald Mitchell, William Clark, David Cash, and Nancy Dickson, addresses the community of scholars seeking to understand the interactions of information and institutions in structuring international affairs. Finally, this volume, *Assessments of Regional and Global Environmental Risks: Designing Processes for the Effective Use of Science in Decisionmaking,* edited by Alexander E. Farrell and Jill Jäger, is directed to assessment practitioners and summarizes GEA Project findings on how the practices of global environmental assessment can be reformed to improve their utility to decisionmakers.

We acknowledge with gratitude the numerous groups that have provided financial and institutional support for the GEA Project. Initial support for the project was provided by a core grant from the National Science Foundation (Award No. SBR-9521910) to the "Global Environmental Assessment Team." Supplemental support to the team was provided by the National Oceanic and Atmospheric Administration, the U.S. Department of Energy, the National Aeronautics and Space Administration, the National Science Foundation, and the National Institute for Global Environmental Change. Additional support has been provided by the Department of Energy (Award No. DE-FG02-95ER62122) for the project, "Assessment Strategies for Global Environmental Change," the National Institute for Global Environmental Change (Awards No. 901214-HAR, LWT 62-123-06518) for the project "Towards Useful Integrated Assessments," the Center for Integrated Study of the Human Dimensions of Global Integrated Assessment at Carnegie Mellon University (NSF Award No. SBR-9521914) for the project "The Use of Global Environmental Assessments," the European Environment Agency, the Belfer Center for Science and International Affairs at Harvard University's Kennedy School of Government, the International Human Dimensions Programme on Global Environmental Change, Harvard University's Weatherhead Center for International Affairs, Harvard University's Environmental Information Center, the International Institute for Applied Systems Analysis, the German Academic Exchange Service, the Heinrich Böll Foundation in Germany, the Massachusetts Institute of Technology's Center for Environmental Initiatives, the Heinz Family Foundation, the Heinz Center for Science, Economics and the Environment, and the National Center for Environmental Decisionmaking Research.

A number of individuals who worked in the background of the GEA Project also deserve our thanks. Finally, we extend our deepest thanks and admiration to J. Michael Hall, director of the Office of Global Programs at the National Oceanic and Atmospheric Administration, whose vision, wisdom, and commitment has sustained us through the GEA effort.

William Clark
Nancy Dickson
Jill Jäger
Sheila Jasanoff
James J. McCarthy

Contributors

ALEXANDER E. FARRELL is an assistant professor in the Energy and Resources Group of the University of California, Berkeley. His research focuses on energy and environmental technology, economics, and policy. More specifically, he is interested in the use of technical (scientific and engineering) information in policymaking, market-based environmental regulation (e.g., emission trading), the environmental impacts of energy, the application of sustainability in decision-making, security in energy systems, and alternative transportation fuels.

ALASTAIR ILES is a visiting scholar at the Energy and Resources Group at the University of California, Berkeley. Originally trained in Australia and at Harvard University, where he was a pre-doctoral member of the Global Environmental Assessment Project, he works on sustainable industry and policy, learning tools, environmental health, and technology and justice. Under a National Science Foundation award, he has studied the policy and science of green chemistry. He has recently published in *Science and Public Policy, Global Environmental Politics*, and *Environmental Values*.

JILL JÄGER is a member of the Steering Group of the Initiative on Science and Technology for Sustainability. Dr. Jäger has worked as a consultant on energy, environment, and climate for numerous national and international organizations, including the International Institute for Applied Systems Analysis and the International Human Dimensions Programme on Global Environmental Change. Her main field of interest is in the linkages between science and policy in the development of responses to global environmental issues.

BERND KASEMIR is a founding director of sustainserv™, an international management consulting firm that focuses on linking sustainability and business. He is also a visiting scientist with the Swiss Paul Scherrer Institute. At the time of the study on sustainability and institutional investment reported here, he was a research fellow at Harvard University's Kennedy School of Government.

TERRY J. KEATING has been a participant-observer of environmental assessment processes at the local, state, national, and international scales for almost 20 years. He is a senior environmental scientist with the Office of Air and Radiation at the U.S. Environmental Protection Agency, where he advises senior management on scientific issues related to air quality management at the national and international level. Dr. Keating joined EPA after spending two years working within the agency on a fellowship from the American Association for the Advancement of Science. Prior to the fellowship, he was a post-doctoral fellow with the Global Environmental Assessment Project at Harvard University. He previously worked as an air quality consultant in Los Angeles for both public and private sector clients on a variety of regulatory issues.

MOJDEH KEYKHAH has advised firms in the oil and gas sector and the insurance industry, and served as project coordinator at the International Institute for Applied Systems Analysis. With the Global Environmental Assessment Project, she researched the use of expertise by the insurance industry in dealing with catastrophic risk issues.

DAVID C. LUND is currently a doctoral student at the Massachusetts Institute of Technology-Woods Hole Joint Program in Oceanography, studying long-term variations in the strength of the Gulf Stream. Mr. Lund spent a year at the National Oceanic and Atmospheric Administration's Office of Global Programs as a Sea Grant Fellow involved in the dissemination and application of El Niño forecast information in Latin America and California. At NOAA, he became interested in societal resiliency to abrupt climate change, which in turn led him to a secondment to the Global Environmental Assessment Project and the subject of his chapter in this book.

MARYBETH LONG MARTELLO is a research associate at Harvard University's Kennedy School of Government. Her research examines the ways in which science shapes and is shaped by democratic governance. Her research projects have examined the evolution of twentieth century intergovernmental efforts to address dryland degradation, local knowledge as an increasingly important organizing concept in environmental science and policy, social and political dimensions of vulnerability analysis, and the meanings and practices attached to sustainability in corporate settings. Ms. Martello co-edited (with Sheila Jasanoff) the book *Earthly Politics: Local and Global in Environmental Governance*.

CLARK A. MILLER is assistant professor in the Robert M. La Follette School of Public Affairs and the Robert and Jean Holtz Center for Science and Technology Studies at the University of Wisconsin-Madison. His research explores the politics of expertise in global governance. He co-edited (with Paul Edwards) the book *Changing the Atmosphere: Expert Knowledge and Environmental Governance*.

OLADELE OGUNSEITAN is a professor of environmental health, science, and policy at the University of California, Irvine, where he also directs the multidisciplinary Program in Industrial Ecology in the School of Social Ecology. His research has been funded by the National Science Foundation (Biocomplexity in the Environment Program) and the Global Forum for Health Research, Switzerland. He was a Macy Foundation fellow at the Marine Biological Laboratory in Woods Hole, Massachusetts, and an AT&T Foundation Industrial Ecology faculty fellow. His research has appeared in *Global Environmental Change* and *Environmental Pollution*.

EDWARD A. PARSON is a professor of law and associate professor of natural resources and environment at the University of Michigan, where he conducts research on international environmental policy, negotiations, and the role of science and technology in public policy. His most recent book, *Protecting the Ozone Layer: Science and Strategy*, won the 2004 Harold and Margaret Sprout Award of the International Studies Association. Dr. Parson has served on the Committee on Human Dimensions of Global Change of the National Academy of Sciences, on the Synthesis Team for the U.S. National Assessment of Climate Impacts, and as an advisor to the White House Office of Science and Technology Policy.

ANTHONY PATT is an assistant professor in the Geography Department and Center for Energy and Environmental Studies at Boston University. His research focuses on the analysis, communication, and use of information about risk and uncertainty. Much of his work focuses on the use of such information in developing countries to facilitate adaptation to climate change and climate variability. He is a leading researcher in using experimental economic methods to identify important behavioral decisionmaking biases in developing-country field locations.

RAPHAEL SCHAUB is a consultant with sustainserv™, a management consulting company focused on linking sustainability and business, with offices in Boston and Zurich. At the time of the study discussed in his chapter, he was a research assistant at the Swiss Federal Institute for Environmental Science and Technology.

NOELLE ECKLEY SELIN researches the science and policy of hazardous chemicals and metals. She was a fellow of the Global Environmental Assessment Project in 1999–2000. During 2000–2001, she was a Fulbright Fellow at the European Environment Agency in Copenhagen, Denmark, where her research focused on European chemicals management. She is currently a Ph.D. candidate in Harvard University's Department of Earth and Planetary Sciences, Atmospheric Chemistry Modeling Group, where she is developing a model of the global transport of mercury.

BERND SIEBENHÜNER is an assistant professor of ecological economics and head of the GELENA-research group on social learning and sustainability at Carl von Ossietzky University in Oldenburg, Germany. He is also deputy leader of the Global Governance Project, and has been working as a post-doctoral research fellow at Harvard University's Kennedy School of Government.

ANDREA SÜESS is currently a Ph.D. candidate at the Swiss Federal Institute for Forest, Snow and Landscape Research, where her research focuses on economic instruments in land use planning. At the time of the study reported in this book, she was a research fellow at the Swiss Federal Institute for Environmental Science and Technology, working on issues of sustainable pension fund investment.

STACY D. VANDEVEER is the 2003–2006 Ronald H. O'Neal Associate Professor in the University of New Hampshire's Department of Political Science. His research interests include international environmental policymaking and its domestic impacts, the connections between environmental and security issues, and the role of expertise in policymaking. Before taking a faculty position, he spent two years as a post-doctoral research fellow in the Belfer Center for Science and International Affairs at Harvard University's Kennedy School of Government.

Assessments of Regional and Global Environmental Risks

CHAPTER I

Overview

Understanding Design Choices

Alexander E. Farrell, Jill Jäger, and Stacy D. VanDeveer

*E*NVIRONMENTAL ASSESSMENTS ARE increasingly important and complex endeavors designed to harness scientifically grounded information to inform decisionmaking for businesses, local and national governments, and international arenas. For the purposes of this book, *environmental assessment* refers to the entire social process by which expert knowledge related to a policy problem is organized, evaluated, integrated, and presented in documents to inform policy choices or other decisionmaking (Farrell et al. 2001). Although one might think first of assessments in terms of the reports they often produce, the research presented in this book reveals that the implications of scientific assessment are better understood by viewing assessments as a communication process, rather than just as a report. Assessment, therefore, is a process that bridges expert knowledge to policy and seeks to inform policymakers and the scientific community.

Many people and organizations now participate in environmental assessments, partly because many of the world's more than 200 multilateral environmental agreements require periodic assessments to support their implementation and revision. These practitioners, as well as most scientists conducting the research to be used in an assessment, generally want their efforts to increase knowledge and improve environmental policy (Bolin 1994; Davies 1993). Decisionmakers in business and government want their decisions to be firmly grounded in scientifically supported data and analysis so that they improve environmental quality (Bronk 1974; Carnegie Commission on Science 1991). In part, this desire reflects a belief that scientific and engineering research produces knowledge more likely to be effective than personal opinion, political ideology, or other sources of information. And in part, it reflects efforts by leaders to show that their policy positions are not merely pursuits of self-interest but constitute an objectively defensible means for achieving agreed upon public ends (Ezrahi 1990). All sides, it seems, have an interest in effective environmental assessments.

Environmental assessments are likely to become more important (or at least more common) in the future for a number of reasons. Increased population and increased industrialization will place greater demands on resources (e.g., natural habitats, an adequate supply of clean water, and less-polluted air), while at the same time these effects will increase the social demand for environmental quality (Anderson and Cavendish 2001; Arrow et al. 1995; Dasgupta et al. 2002). Further, increased economic and cultural linkages (i.e., globalization) will tend to make international environmental policy more important, all of which will raise the importance of environmental assessments.

Thus, environmental assessment processes are significant (and often expensive) undertakings that may influence public policy, strategic decisions for firms, and, ultimately, the quality of life for people. They would merit study even if they generally worked well. However, scientific assessments often do *not* demonstrate significant influence on decisions affecting environmental quality (Social Learning Group 2001a, 2001b; Zehr 1994). This book addresses a central question: How can environmental assessment processes be designed such that scientific and engineering knowledge will most likely influence decisionmaking? It explores the latter observation further to identify design elements that can help lead to more effective assessments. This chapter introduces some of the major issues that arise when considering the effectiveness of environmental assessments in linking the scientific realm to the policymaking realm. These issues were derived through extensive study of a wide variety of assessment processes, interviews with assessment practitioners and users, workshops, and study of relevant literature.

The authors of the subsequent chapters have been fortunate in being able to meet and talk with participants in many assessment activities around the world, from climate modelers in universities and research centers, to government officials in states and cities faced with tough pollution management choices, to industry leaders contemplating the implications of environmental challenges for their businesses. Like the authors, many of these practitioners were keenly interested in the questions implicit in the use of science in policymaking, but they were usually deep in the details of the problems at hand and unable to devote the time and energy to reflective thinking on the process. A key goal of this book is to aid practitioners of environmental assessments: those who may be asked to participate in an assessment or to manage one, those who think an environmental assessment is needed and want to understand how to start one up, and those who want to use the results of an assessment. To aid these individuals and groups, this book focuses on the design of assessments in terms of the choices that practitioners make in the organization, operation, and follow-up of an assessment and the possible consequences of those choices.

Background

As environmental issues began to receive increasing attention in both national and international political circles, assessment processes that connect scientific research and policy became increasingly common and important. The earliest and most common form is project-based assessment, often called the Environmental Impact Statement (EIS). The use of the EIS is now institutionalized in more than 100 countries and in international organizations such as the World Bank (Sadler 1996). An international treaty on project-level transboundary environmental assessment was signed in 1991, although it has not yet entered into force (United Nations Economic Commission for Europe 1991). This book deals with assessments that examine large-scale environmental phenomena beyond the scope of a single project, sometimes called "ecosystem-level" or "strategic" environmental assessments (Merkle and Kaupenjohann 2000; Partidário and Clark 2000). A key feature of the assessments discussed here is that they examine environmental phenomena that cross important political boundaries, usually international boundaries but sometimes jurisdictional boundaries within a single nation (e.g., state boundaries within the United States).

There are two basic points about assessments that must be understood before an attempt is made to analyze them. First, assessments are fundamentally communication processes, not simply reports, and, second, assessment processes share many important features irrespective of topic or discipline, making generalizations possible (Farrell et al. 2001; Miller et al. 1997). That is, while the *product* of an environmental assessment has obvious value, to really understand effectiveness it is critical to understand the *process* that was used. A process–oriented approach is needed to understand what effectiveness means for environmental assessments, and what design features yield effectiveness. Furthermore, understanding the process facilitates the observation that the similarities across different cases highlights the many commonalities across assessments of different topical areas in terms of the challenges and opportunities presented to organizers, participants, and users.

Environmental assessment organizers, participants, and users have a large array of concerns at the outset of the process. The usual list includes such things as recruiting appropriate technical experts to participate, obtaining funding and other resources, and determining timelines. However, a small but growing body of empirical research has shown that a number of other, more subtle features are also important.

Environmental Assessment Processes

Assessments are often organized as a way to inform decisionmakers about issues that are controversial or new in the policy realm. Assessment processes are embedded in a variety of institutional settings within which scientists, decisionmakers, and other stakeholders communicate to define relevant questions for analysis, mobilize certain kinds of experts and expertise, and interpret findings in particular ways.

Approaches to Environmental Assessments

The study of environmental assessments has been carried out by numerous researchers, most of whom look at one or a few specific assessments on a single environmental issue, rather than across many assessments and multiple topic areas (e.g., Boehmer-Christiansen 1994a, 1994b; Castells and Ravetz 2001; Christoffersen et al. 2000; Cohen 1997; Cowling 1995; Elzinga 1997; NAPAP Oversight Review Board 1991; Rubin 1991; Tuinstra et al. 1999; Wettestad 1997; Winstanley et al. 1998). Other researchers concentrate on approaches for effective use of technical advice in government in general (Carnegie Commission on Science 1991; Collingridge and Reeve 1986; Golden 1988). Still others address even broader questions, such as how various parts of government are structured and operate to receive technical advice (White 1948), how productivity and integrity can be maintained in government-funded research (Guston 1999), and, most fundamentally, the meaning of and the proper place for scientific endeavor in a democracy (Ezrahi 1990; Price 1965; Jasanoff 1990). Recent debate has also focused on assessment practices such as participatory technology assessment processes (Joss and Bellucci 2002).

Unlike previous studies, this book presents systematic research efforts undertaken to develop general observations and recommendations about environmental assessment processes. This approach implies that assessment is a separate, independent activity from either scientific research or political choice, containing its own features, norms of behavior, and limitations. The evidence presented in this book is meant to support this claim and to extend the insights that have been made so far on the basis of evaluations of single assessments. In particular, we hope to show the great variety in how an assessment can be designed and some of the choices made in design that can have implications for effectiveness.

The Paths to Policy Outcomes

Implicit in much of the research and common wisdom in environmental policy is the assumption that the primary influence of technical information on organizations is through formal legislative and regulatory approaches, including international agreements and treaties. But recent research in the environmental realm (Social Learning Group 2001a, 2001b), including our own initial findings (Global Environmental Assessment Project 1997), emphasizes that this is not necessarily the case. Instead, these studies suggest that technical information and the institutions that shape it may influence factors ranging from who participates in debates about significant problems, to what they believe those problems are, and to how they strategize about what to do. Moreover, these influences can create substantial changes over the long term (Clark et al. forthcoming).

But changes in what? Previous studies typically focused on "policy outcomes," that is, authoritative decisions by government (or by a firm's executives) on a course of action. Policy outcomes include rules enforced by governments, but these outcomes also encompass other actions by other actors and changes in the cognitive framing of problems that may well precede and shape action. We therefore need to examine the "issue domain"—the broad range of actors, institutions, behaviors, and impacts associated with global and regional environmental risks. In particular, we are concerned with the factors that lead to changes in the issue domain.

A principal reason for selecting the issue domain as our unit of analysis is that it allows us to focus not just on policy outcomes but also upon a much richer set of factors that earlier studies have suggested may affect long-term issue development. Moreover, this term can be linked to earlier useful concepts in the study of environmental policy such as the issue-attention cycle and advocacy coalitions, which were helpful in the development of the issue domain approach (Sabatier and Jenkins-Smith 1993; Schreurs et al. 2001).

We characterize an issue domain at any particular moment in terms of four interrelated elements:

- the actors who participate in the issue domain, including their interests, the resources they control, their beliefs about the issue, and their strategies for using their resources to advance their interests;

- the institutional settings within which actors interact with one another inside the issue domain and with the world outside the issue domain;

- the behaviors (decisions, policies, agreements) that emerge from those interactions; and

- the impacts of those behaviors on the world (e.g., improvements in environmental quality).

Change in the issue domain (i.e., issue development) is a continuous process and can be characterized in terms of changes in the individual components and their relationship with one another. In general, this change will be shaped by internal interactions among elements of the issue domain (including attempts to recruit new participants, negotiations among those already inside, and the discovery of new technical information through ongoing research and practice); and external events impinging on the issue domain, including changes in the physical environment.

Importantly, the individuals and organizations that are part of an issue domain regularly interact over long periods and identify themselves in some way with the issue. The members of an issue domain will typically disagree on some (possibly many) things, including how to understand the issue and what (if anything) ought to be done about it.

These basic features of the issue development framework are shown in Figure 1-1. This extremely simplified view is useful for addressing the questions at hand, and it is not meant to be a universal or exclusive view. For instance, a wide array of activities is subsumed under the term "internal processes," including attempts to recruit new participants into the issue domain, various negotiations among those already inside, and the discovery of new technical information through ongoing research and practice. Change in issue domains is a continuous process, as indicated by the arrow labeled "$t + n$" on the right-hand side of the figure. Over time (decades), disagreements about how best to characterize the issue tend to be resolved, generally followed by agreements on what to do about it.

Within this framework, assessments can be viewed in terms of the strategic actions taken by some participants to advance their interests by collecting and organizing scientific and engineering knowledge to change the issue domain. By definition, an assessment must come from within the issue domain, because any individual or group that would be motivated to ask for one (or promote government to ask for one) is in the issue domain. Knowledge changes an issue domain

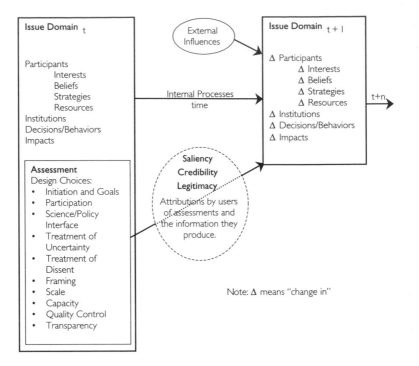

Figure 1-1. *A Simplified Framework for Understanding How Assessments Affect Policy*

primarily by changing the beliefs of other participants, which over time leads to changes in other parts. Some members of an issue domain will participate in assessments while others will not, possibly being users instead. Further, new organizations may be recruited into the issue domain first by participating in an assessment.

Recent research conducted as part of the larger project of which this book was a part (Mitchell et al. forthcoming) carefully evaluated how, and under what conditions different types of assessments influenced issue domains and political and economic decisionmaking at the local, national, and international levels. To assess the influence of environmental assessments, the authors conducted case studies of a wide range of environmental assessments. The authors of the case studies (Mitchell et al. forthcoming) relied on counterfactuals, and focused comparisons within and across cases, process tracing, and evaluation of alternative hypotheses. These techniques allowed the authors to examine cases in which an assessment took place but factors other than the assessment caused observed changes in the issue domains, and distinguished them from cases in which the assessment could be accurately attributed as the cause of issue domain change. Equally important, the authors of the larger project carefully tracked both the empirical evidence and the causal logic to determine what features of assessments or conditions in which those assessments were performed explained their

Table 1-1. *Potential Goals of Participants in an Environmental Assessment*

- Obtain funding for research, some of which may be part of a pre-existing research agenda
- Solidify or stabilize one's own beliefs, especially by accumulating new evidence or analysis
- Collect data, develop and test theories, and otherwise improve scientific understanding of an issue
- Change the perception of knowledge needs and thus identify new priorities for research and development
- Build an argument for increased resources for further research
- Identify one's own interests and agenda, as well as those of others
- Change or discredit the beliefs of others
- Identify various options for action and evaluate the relative merits of each
- Legitimize a policy preference by showing that it serves a public purpose, rather than just private interests
- Demonstrate competence and leadership to enhance personal or institutional prestige and credibility
- Increase the awareness of the issue among individuals and in organizations outside the issue domain
- Recruit participants into the issue domain
- Change the framing and perceptions of issues
- Speed changes in the behavior of key actors relevant to the physical environment (e.g., decreasing emissions from large polluters)
- Delay actions that would change the status quo

ability to change the issue domains into which they introduced information. Many authors' findings are based on comparing the same assessments in different contexts or comparing influential assessments with noninfluential assessments in the same context. In many cases, those authors found a particular assessment had considerable influence in one country but not in another, at one point in time but not at another, or on one economic sector but not on another. In all these cases, assessments appeared to be more influential when either the assessment itself or the social context in which it was produced led the assessment to be more salient, credible, or legitimate with a particular audience.

The Many Meanings of Effectiveness

The idea that assessments should be effective makes intuitive sense; however, attempts to parsimoniously and rigorously define what "effective" means quickly fail—the goals of different participants in an assessment often diverge, and may even be in conflict. A simple definition that fits most situations is that more effective assessments are more likely to have significant influences on the corresponding issue domain and its development. However, this straightforward idea can hide many differences among participants. Table 1-1 lists some of the goals that users and assessors often have for an assessment. An effective assessment may accomplish several of these changes.

Determinants of Effectiveness

The research described in the subsequent chapters (and in Clark et al. forthcoming) has shown that some assessments have failed to achieve some of the goals of the participants for one or more reasons. By mapping the pitfalls that have commonly led to ineffective assessments, three major attributes emerge that seem to make it more likely that the knowledge contained in an assessment would induce change in the issue domain: *salience, legitimacy,* and *credibility.*[1] An assessment that is viewed as more salient, legitimate, and credible to a particular assessment participant or user, therefore, is more likely to change his or her beliefs and thus be effective.

Importantly, these three determinants of effectiveness are attributions made by users, not factors the assessment participants can include directly. Thus to be effective in issue development, assessment designers and participants must understand how to encourage users to label the assessment as salient, legitimate, and credible.

Salience,[2] or relevance, is intended to reflect the ability of an assessment and its results to address the particular concerns of a user. An assessment is salient to a user if that user is aware of the assessment and if that user deems the assessment to be relevant to current policy and behavioral decisions. One example of an assessment that lacks salience is a process that simply produces a report that remains on a shelf in perpetuity, never referred to again. Another example of an assessment that lacks salience is one that asks questions in which a particular user is not interested. A user might, for example, commission an assessment about acidification to inform a decision about regulating electricity generation; if the resulting assessment focuses on the mechanisms of pollutant transport, it will likely not be salient to that user.

One major element of making assessments salient involves ensuring that participants in the assessment context are drawn from organizations that contain the users who might find the assessment salient, if only they knew about it. Our research has consistently uncovered assessments lacking influence over a given issue domain simply because they did not include participants from organizations they hoped to influence.

Even when an assessment is salient, it is unlikely that all elements of the assessment will be equally salient. Findings considered central by the assessments' creators may be ignored by certain participants in a given issue domain, while an assumption given little thought can become a major focus of debate. The salience of an assessment (or a particular element of an assessment) depends to some degree on coincidence with external events. Over time, the salience of assessments of a particular issue can rise and fall.

Legitimacy is a measure of the political acceptability or perceived fairness of an assessment to a user. A legitimate assessment process is one in which the process was fair and conducted in a manner that allows users to be satisfied that their interests were taken into account. An example of a lack of legitimacy occurs when a "global" assessment is questioned by industrializing nations because they feel their input was not included or their interests were ignored. Participants and users must believe that their interests, concerns, views, and perspectives were included and given appropriate weight and consideration in an assessment if they are to grant the assessment legitimacy. A key observation from the research presented in this book is that an assessment cannot promote knowledge regarding facts and causal beliefs without simultaneously, if often implicitly, promoting certain goals and values over others. For this reason, individuals and organizations in policy debates are quick to evaluate how any assessment affects their interests.

This does not necessarily mean that the assessment must strictly support the existing interests of all participants and users. An assessment that runs counter to the interests of an organization may still be accepted as legitimate by that organization if they believe that their interests and concerns were accurately represented in the assessment context. Such a situation will be difficult of course, and the credibility of the assessment may be examined even harder. At the extreme, where the outcome of an assessment yields a direct threat to key interests, an assessment user (or participant) may refuse to accept the outcome, possibly citing uncertainty as the reason. However, such a situation may induce the organization to re-evaluate its interests or its strategies to accommodate the knowledge gained through the assessment.

Frequently, legitimacy is an issue when an assessment is perceived as recommending behavioral changes by one group of actors that would disproportionately benefit some other group of actors. Indeed, an assessment's legitimacy is rarely questioned by those whose interests would be furthered by the policy implications of the assessment. Even assessments coincident with a participant's interests in a given policy context can be questioned if they are produced by actors viewed by the participant as opposed to their interests. Increasing the perception of an assessment as legitimate for particular participants in an issue domain often can be accomplished by engaging representatives whom those participants believe voiced their views, goals, and concerns or by allowing those decisionmakers to participate in the assessment.

Credibility refers to the assessment's scientific and technical believability to a defined user of that assessment, often in the scientific community. If an assessment captures the attention of some participants in a given issue domain, the assessment's influence still depends on whether and which participants consider that information to be credible. An important component of credibility is captured by the notion that the potential user of the information must be convinced that the facts and causal beliefs promoted in the assessment correspond to those that the user would have arrived at had they conducted the assessment. More credible assessments do better at ensuring this sort of technical adequacy. Assessments gain credibility from several bases.

First, an important criterion involves the conformance of the new information to competing sources of information. New information that is consistent with existing information, particularly well-established facts and causal beliefs, will be accepted as credible more quickly than information that refutes existing facts and theories. Second, assessments are often deemed credible based on the process by which they were created. Assessments can seek to build credibility by ensuring that the assessment "passes muster" with respect to standards of scientific rigor and process, such that those decisionmakers who cannot assess the validity of the findings directly will be willing to view the information as credible based on such process criteria. Third, assessments can also be deemed credible based on the credentials of the participants.[3] Although the credentials that lead to acceptance of information will vary from participant to participant, they will typically include education, source of financial support, and prior research record (especially publication in the peer-reviewed literature). These credentials help document that the assessment participants are both expert and trustworthy.

Closely associated with credentials are reputations, which can be based on factors such as publication record or personal interactions. Fourth, credibility also is a function of the degree of consensus on an issue and the correspondence between the information being evaluated and such consensus as exists. When an assessment makes claims regarding an arena in which considerable uncertainty and variation in scientific opinion exists, either about facts or causal relationships, the credibility of an assessment may prove hard to establish.

Adding to the challenge, these three determinants are often in tension, because the easiest ways of enhancing any single attribute almost invariably cause declines in another. Thus, a crucial job for those who design and manage assessment processes is to balance efforts to enhance salience, legitimacy, and credibility. For instance, credibility can be maximized by addressing only questions in which scientific certainty is high or by allowing only the most renowned scientists to participate, regardless of the nation or sector they represent. The problem is that in the former case, the assessment risks losing salience by failing to ask what decisionmakers want to know, and in the latter case it risks losing legitimacy by failing to take the interests of potential users into account. Similarly, decisionmakers may be more likely to receive advice salient to their interests by ordering an in-house assessment, but the results are not likely to have much legitimacy with others in the issue domain. Box 1-1 summarizes typical "fatal flaws" that lead to ineffective assessments.

To make matters more difficult, the relative importance of salience, credibility, and legitimacy in making an assessment influential changes as an issue develops. Tradeoffs change as an issue progresses from early stages in which policy-relevant actors attempt to push new issues onto the international agenda, toward later stages in which those actors attempt to get national policy agendas to engage issues already being addressed at the international level, toward still later stages in which those actors attempt to induce local decisionmakers to take action in response to international agreements (van Eijndhoven et al. 2001).

The discussion above points to a direct rationale for the common use of large-scale, peer-review efforts in environmental assessments. Well-designed peer reviews can combine several of these bases for credibility and thus help improve the effectiveness of the assessment. Some potential problems should also be clear. An assessment that lacks credibility, for example, might be challenged by scientists for being based on shoddy methods, ignoring important empirical evidence, or for drawing inappropriate conclusions from scientific data. The conclusions of an assessment might be questioned because a user of the assessment believes that a laboratory measurement was in error, a crucial process was omitted in an atmospheric model, or an inappropriate analytical methodology was employed.

Assessment Design Features

If assessments become effective by being credible, salient, and legitimate to the users, what factors can the designers of assessments use to promote these properties? We use the term "design features" because they are the product of choices made by individuals and organizations participating in the assessment (sometimes

---

Box 1-1. Fatal Flaws for Environmental Assessments

Several single factors can, by themselves, make assessments ineffective. So-called fatal flaws are strongly connected to the attributes of credibility, salience, and legitimacy.

- Lack of scientific credibility is the most frequent fatal flaw for assessments. Credibility can be lost through inadequate quality control over technical arguments, unresolved disagreements over what constitutes appropriate standards of evidence and argument, or the appearance of substantive discrepancies between the body of an assessment report and its executive summary.

- Assessments that are credible can be rendered ineffective from lack of salience due to a number of issues. Most importantly, salience can be lost by assuming that the questions important to the scientific community are the same as those important to the policy community, when in fact they are not.

- Salience can be lost by adopting a "one size fits all" approach to policy questions, instead of tailoring assessments to specific users. Salience can also be lost by delivering assessments too slowly to play a meaningful role in rapidly evolving policy processes.

- Failure to secure political legitimacy in the view of relevant stakeholders can also doom an assessment to ineffectiveness. For international environmental assessments, this most classically occurs when poorer countries perceive that they have been left out of a process largely run by a few wealthy nations.

---

unintentionally). The rest of this introduction describes what these design features are.

Although some very pragmatic assessment design issues are not addressed here, it is largely because they are the more routine and well-understood design issues to which alert managers are already attuned. These include such things as the time budget available for the assessment and the scope of work and staff needs for completion.

A critical task when thinking about the design of an assessment is to keep the intended users of the assessment in mind. Our research has identified three particular characteristics of assessment users that stand out as important in this regard: interest, capacity, and openness. The user or audience must be interested in or concerned with what the assessment has to say, and willing to listen, in order for an assessment to have an impact. The most effective assessment processes seem to devote a substantial amount of time and energy to negotiating with potential users the particular questions about which those users are most passionately and urgently interested. Assessment strategies that opt instead to address generic issues of presumed interest to generic users seem rarely to be effective. A user's capacity to "hear" and evaluate the messages communicated in the assessment also emerges as important. A user without the technical expertise or resources to engage in an assessment process is unlikely to find the assessment effective. Ensuring such capacity by capacity-building efforts and other mechanisms might well be part of an assessment strategy. Another characteristic of assessment users that our research has found to be significant is the openness of the user to hearing different messages. Decisionmakers are more likely to treat an assessment as credible if they hear similar findings and recommendations from a variety of sources. A user community that is more open to different channels of communication, therefore, especially where there are multiple assessments or multiple channels through which news media and advocacy groups can fine tune and amplify existing assessments, seems more likely to respond to those assessments, compared with a user community that received information from only a single assessment voice.

Initiation and Goals

In examining assessment processes, important influences on the characteristics and outcomes of assessment processes and the roles played by individuals and institutions can be traced to the origins of particular assessment processes. Who called for a particular assessment process and why? Do participants share an understanding of why assessment processes were initiated, or do they hold different views? What is the organizational context of the assessment process? Does it take place within a particular organization, such as an environmental policy bureaucracy, or does the assessment process cross numerous levels of jurisdiction and types of organizations? Frequently, the answers to these questions may be hard to find in an assessment document, because the designers of the assessment may not choose to present this information, especially if some of the participants would not want their goals openly acknowledged or if prior conflicts led to the assessment. For instance, behind some U.S. assessments described as "cooperative" lie lawsuits, court orders, or the threat of regulation. Nonetheless, users of assessments may be aware of these backgrounds.

Participation

Questions surrounding participation are central to any discussion of assessment process design, partly because it is such a strong determinant of legitimacy. By participation, we refer to which individuals and organizations are involved in an environmental assessment, and when and how they are involved in the process. We distinguish between nominal and engaged participation. Nominal participants are formally part of assessment processes, but they may or may not have much understanding of the issues at hand or much influence on outcomes. Nominal participation often results from a combination of resource constraints or a lack of real interest in the relevant issues but where the participants need to *appear* to be engaged. Engaged participation refers to active participation in meetings, attempts to influence decisions, contributions to writing and editing of reports, and so forth.

Participation in the various phases of an assessment depends on the assessment goals and design. Participation in an assessment can vary substantially, from developing the initial scope of work for the assessment, to the day-to-day conduct of the assessment, to the communication of its results. In transboundary assessments, the governments of the nations whose borders are involved are the principal participants. In some cases, special concessions are made to allow particular actor groups, such as firms or environmental nongovernmental organizations, to participate directly in the assessment process. Other assessments use more formal mechanisms such as public hearings.

There are various reasons for choosing different levels and types of participation in designing an assessment process. Most obviously, broad participation will increase diversity of the actors involved in an assessment process, which is valuable if the views or suggestions from different groups are desired. Broad participation is seen by some as a way of permitting power to be shared. It may ease implementation of measures, by ensuring that the interests of important groups are taken into account in designing those measures and thus making it easier to gain their agreement on the final decision. For these reasons, broad participation may be one of the best ways to increase the legitimacy and salience of assessments.

However, if participation is used to engage a wider array of groups in a dialogue, capacity constraints may limit the ability of some to participate and/or limit the ability of assessment organizers to include diverse voices and expertise. More generally, expanding participation in an assessment process can increase input (or appear to do so). Decisions about participation may also be stimulated by nontechnical considerations, such as the perceived need to build a political constituency, the attempt to influence the policy agenda, or the creation or removal of political cover.

The self-interest of assessment organizers is another important motivation for participation decisions, when assessors attempt to use participation to preempt criticisms. Participation can be expanded to generate new insights; to shape research agendas, practices and methods; and/or to build and expand issue networks and professional communities. There are also reasons to limit participation, especially if assessment organizers and designers feel it necessary to separate scientists and engineers from decisionmakers.

Lastly, there is an interplay between decisions regarding who participates and decisions about the rules and norms of participation in particular assessments. If, for example, assessment decisions are to be made by consensus, then incorporating critical (or opposition) perspectives becomes more problematic. Consensus decisionmaking, in effect, gives each participant a kind of veto power. As such, incorporating critical perspectives into consensual decisionmaking processes may lead to an inability to act or say anything specific.

Science/Policy Interface

The structure of interactions between the scientists and the policymakers within assessment processes can take on different forms. These structures can be thought of as falling along a spectrum, ranging from attempts to isolate scientists from the policy process on one end, to highly institutionalized collaboration between various groups on the other end. At this end of the spectrum, some participants may even be both scientist and decisionmaker. No matter where along the spectrum the science/policy interaction falls for any given assessment, each group must maintain its self-identity and protect its sources of legitimacy and credibility. Boundaries are therefore commonly negotiated, articulated, and maintained by assessment participants. It is important to note that these descriptions apply to the formal interactions between scientists and policymakers; there may be important informal interactions as well, as described below.

One view of where assessments *should* be located on that spectrum is articulated by Lee (1993), who argues: "Science and politics serve different purposes. Politics aims at the responsible use of power; in a democracy, 'responsible' means accountable, eventually to voters. Science aims at finding truths—results that withstand the scrutiny of one's fellow scientists." Lee uses an idea of occupational roles and social functions originally proposed by Price (1965) to identify distinctive contributions made by different groups to a technological society. Lee argues that the roles of individuals as politician, administrator, professional analyst, and scientist are separate and that a single person cannot play several different roles at once, at least without the risk of losing legitimacy. The cases discussed in this book suggest that this assertion is not always true—in several very effective assessments, participants serve more than one role, although doing so is often not easy.

An analytically useful approach treats environmental assessments as "boundary organizations." This concept emerges from the scholarship in the social studies of science, which shows how scientists assert and maintain their authority to speak definitively about the character of the world. Jasanoff (1990) extends this analysis by demonstrating the need for scientists and policymakers to negotiate boundaries between their domains in any given assessment or advisory process. These boundaries become agreements that outline the issues under each group's exclusive authority. That is, they are lines between concepts, not lines on a map. Jasanoff's work uses examples of standard-setting for pollutants and decisions about drug-use approval to illustrate the challenges inherent in the production of politically legitimate and relevant advice that is also scientifically credible. These challenges are resolved through ongoing "boundary negotiation," which involve agreements between regulatory bodies and expert advisory groups as to what issues each will deal with and what issues will be shared between them (Guston 1999).

Jasanoff (1990) concludes that the issue is not *whether* there are boundaries between science and policy but *where* they are and *how* and *why* they are located where they are. One of the key roles of expert advisory bodies is to provide a forum to define these boundaries. Allowing expert bodies to perform this function within given institutional contexts is crucial for obtaining the political and scientific acceptability of advice. When the process is successful (and scientific credibility is obtained) it is not possible for political adversaries to deconstruct the results or attack them as "bad science." Thus, Jasanoff notes that the problem of politically motivated bias in appointing members of expert committees is unavoidable and should be countered with administrative devices to limit it, rather than doomed attempts to banish it (244–245).[4]

Uncertainty

One of the key aspects of research in environmental science and technology is that it involves considerable uncertainty, which may be very difficult or impossible to resolve in the short term (Arrow and Fisher 1974; Hellstrom 1996; Hellstrom and Jacob 1996; vanAsselt et al. 1996). Thus, an important feature of environmental assessment is how uncertainty will be treated. Although sophisticated techniques for dealing with uncertainty have existed for some time (e.g., Klein-dorfer et al. 1993; Morgan and Henrion 1990), many scientific assessments fail to give adequate treatment to the extremes of the distribution and instead focus their attention on the central tendency. Approaches that are sometimes used include the use of scenarios, stochastic analysis (such as Monte Carlo techniques), and subjective elicitation of expert opinion.

Dissent

Many environmental assessment processes are carried out using some form of consensus rules so that the assessment products must achieve unanimous support (or at least no strong objections) before they are released. This is the approach used in the Intergovernmental Panel on Climate Change (IPCC), for instance. However, consensus is only one approach to dealing with differing opinions of the participants, which is labeled here as "dissent." Other approaches exist, such as defection, the establishment of competing assessment processes (possibly in conjunction with defection), and the inclusion of "minority reports." For instance, in both the European and U.S. air pollution assessment processes described by Farrell and Keating in this volume, several competing modeling approaches were established for a while. In contrast, minority reports were used by the German Parliamentary Enquete Commission on climate change in the early 1990s. Perhaps the most typical approach is for assessment processes to simply avoid areas in which great dissent exists, as Patt discusses in this volume. These choices in assessment design can have important implications for how the assessment is perceived and for its potential effectiveness.

Framing

One of the most fundamental design choices (on par with participation and the science/policy interface) is how the assessment is framed. The assessment framing essentially encompasses the overarching beliefs that define what an issue is about, including the basic worldviews or underlying assumptions that will be used in the assessment. Framing choices largely determine which features of an issue will be given more attention and which less, and these choices vary over time and among different groups (van Eijndhoven et al. 2001). Framing involves both the inclusion and exclusion of ideas as well as the way in which ideas are used, shaped, and interpreted in the context of the issue domain. The diverse goals, justifications, and institutional contexts present at the inception of assessment processes strongly determine the framing of an assessment. Framing is crucial for the everyday activities of practitioners of assessment, policymaking, and scientific research who routinely make (implicit and explicit) decisions within particular frames and sometimes make decisions when choosing among alternative frames. A major finding of our research team has been the influence assessments wield by shaping the information that is introduced into an issue domain and by molding the rhetoric of policy debates.

One of the key questions about framing is: How narrowly or broadly should assessments be focused? The idea of integration is increasingly emphasized as a goal in many environmental assessments and policies. Our research has shown that integrated assessment, however, is not the most effective strategy in all situations; there are certain contexts in which narrowly focused assessments are more likely to gain credibility, salience, and legitimacy.

The framing of assessments is an important determinant in the selection of people involved in the assessment process and the design of the process itself. Assessments rarely have one widely agreed-upon goal. In many cases, multiple actors want particular types of assessment processes for a variety of reasons. That many different actors and goals coexist within an assessment essentially requires a process view of assessments. The task for analysts, in part, is to parse out how certain strategic interests, frames, and patterns of participation arise or come to dominate particular assessment processes. Contextual factors may include the organizational bodies administering assessments, the kind of decisions they may be intended to inform, or the level of crisis perceived by participants.

Scale

Environmental phenomena and issue domains are increasingly understood to have implications for assessments that span multiple scales. Scale refers to any specific geographically or temporally bounded level at which a particular phenomenon is recognizable. Scale can also—and sometimes simultaneously—imply a level of organization or a functional unit. There is considerable disagreement about the precise extent or definition of any scale (e.g., determining the boundaries of something "local"), and there is rarely perfect congruence of, for example, a spatial and a functional unit identified at the same scale. In designing an environmental assessment, the organizers impose a definition of scale for each particular issue and for particular purposes. As such, scale is a heuristic employed by assessment practitioners and users to organize their understanding of the world and the relationships and interactions therein (e.g., ecologists find it useful to think of trees, forests, and biomes; politicians find it useful to think of cities, counties, states, and nations).

Environmental assessment and management increasingly recognize the importance of scale and cross-scale dynamics in understanding and addressing global environmental change. Scale is an important design feature of environmental assessments, but one that is too frequently given scant attention in the design of assessments, leading to significant problems for the effectiveness of assessments. One of the most important problems is that most issue domains have a multiscale nature of biogeophysical and human systems, with the interactions between them falling across scales.

Capacity

Assessment capacity refers to the ability of relevant groups, organizations, or political jurisdictions to meaningfully engage and participate in an assessment (i.e., to get past nominal participation) and to sustain that ability over time. Most obviously, capacity requires possessing the necessary linguistic, scientific, and technical skills (i.e., knowledgeable personnel), material capabilities (i.e., financial resources and equipment), and organizational support. Developing and maintaining technical and organizational aspects of assessment capacity requires resources—differences in wealth are therefore often an important cause of differences in assessment capacity among relevant and/or participating jurisdictions. Differences in resource allocation may also result from the fact that not all participants have the same level of internal (i.e., domestic) interest in the issues under assessment.

Quality Control

One of the most important determinants of credibility is the quality control that is used in an assessment process. Quality control is the process (or processes) used to ensure the substantive material contained in the assessment report agrees with underlying data and analysis, as agreed to by competent experts. Quality control is closely linked to tough questions concerning what makes up expert opinion, and who is an expert. Eventually, answers to these questions rely on a fundamental judgment. Sometimes expert opinion is defined as that which appears in peer-reviewed journals, although this excludes a vast amount of classified and proprietary information that national and business leaders rely on for their most important decisions. Other times, expert opinion is defined as the content of testimony, reports, or other communications created for the assessment by individuals invited to do so, sometimes using specific techniques or data. Frequently, an important quality control technique is to expose part of an assessment to repeated review by the participants of the assessment. Thus, quality control can be somewhat related to legitimacy as well.

Transparency

Transparency is an important way of establishing legitimacy and credibility, and it has been called for in a wide number of regulatory and assessment processes (Jasanoff 1998; Mayer and Stirling 2002; Von Damme 1998; Zobel et al. 2002). Generally speaking, transparency means that interested observers can readily see into an assessment process and judge for themselves the data, methods, and decisions used in the process. In practice, this means making a significant amount of information available and explaining decisions based on this information. The issue of transparency has been investigated extensively in the context of spent nuclear waste storage (Mohanty and Sagar 2002), which highlights two critical features to ensure transparency. First, because expertise among assessment users and interested parties varies significantly, the level of information required to assure transparency needs to vary as well. To meet the needs of all, assessments should make summaries and basic data available. Second, transparency will be easiest to achieve if standard procedures exist to make the necessary information available and if these procedures are institutionalized.

One of the more difficult aspects of ensuring transparency is in the area of computer models, which are often important features of environmental assessment processes (Ha-Duong 2001). The tension arises because of the rather unique and specialized nature of computer models, and because some of the most widely used models are proprietary. Even within the academic research community, there is resistance to fully disclosing models for public review. In either case, mastery of these models requires significant time and effort, and many models are moving targets in this respect, because they undergo regular upgrades. This evolving process makes it difficult to undertake comparative analyses or to attempt to reproduce results found by others. One solution, proposed by Ha-Duong, is to use an open-source software approach to model development. Another solution is the use of multimodel comparisons that establish an agreed-upon set of scenarios to drive different models, as has been practiced by Stanford's Energy Modeling Forum for several years.

Outline of the Book

The book contains this introduction, 12 case study chapters, and a concluding chapter. Most of the cases are related to international assessments associated with atmospheric phenomena, including global climate change and acid rain. Other cases are associated with the use of genetically modified organisms, pesticide regulation, natural resource management, catastrophic risk, and other environmental issues. Individually, these chapters provide a variety of concrete examples of the diversity observed in the design of environmental assessments. The chapters are organized into three groups: assessments associated with the Convention on Long-Range Transboundary Air Pollution (LRTAP; see Box 1-2 for a brief description of the convention's assessment framework), convention assessments associated with climate change (see Box 1-3 for an outline of the assessment process conducted through the Intergovernmental Panel on Climate Change), and assessments conducted largely (or solely) by private companies.

In the first group of chapters, Stacy VanDeveer (Chapter 2) examines varying national participation within the LRTAP assessment and negotiating bodies (mostly on the issue of acidification). He discusses participation patterns and their ramifications for international bodies involved in the construction of consensus positions around scientific, technical, and policy questions concerning transboundary air pollution in Europe. The chapter focuses on "peripheral" European states—many of them formerly communist "transition" states—in LRTAP assessment processes and organizations, and it describes the level and nature of these countries' participation in international environmental and scientific cooperation processes. Alexander Farrell and Terry Keating (Chapter 3) compare the LRTAP assessment process used for regional tropospheric ozone (or smog) pollution with an assessment conducted in the United States on the same phenomenon at about the same time. They highlight the treatment of uncertainty and dissent, the role of the private sector, capacity, and transparency for each case. They also examine the reasons behind design choices made. Noelle Eckley Selin (Chapter 4) compares how the LRTAP assessment process addressed two very different issues: sulfur pollution (or acid rain) and persistent organic pollutants. Design features facilitated scientists and policymakers to communicate repeatedly about assessment procedures and outcomes. Another key design element injected a sense of adaptability or dynamism in the assessment and negotiation processes, which allowed

Box 1-2. A Framework for Assessing Transboundary Air Pollutants

Since its adoption in 1979, the Convention on Long-Range Transboundary Air Pollution (LRTAP) has addressed several major environmental problems of the transatlantic and Eurasian region. This success has been achieved through overlapping processes of scientific assessment and policy negotiation between national representatives. The LRTAP agreement was signed partly to demonstrate and further East–West cooperation during the Cold War, although acidification remained a concern mostly for a few northern West European countries.

The LRTAP convention established a framework for international cooperation and information sharing around scientific, technical, and policy issues associated with transboundary air pollutants. The convention's eight individually negotiated protocols address continental-scale environmental monitoring; acidification; eutrophication; ground-level ozone; hazardous substance pollution; and the related reductions of emissions of sulfur, nitrogen oxides, volatile organic compounds, heavy metals, and persistent organic pollutants (for a complete list, see Table 2-1 in Chapter 2).

In addition to laying down the general principles of international cooperation for air pollution abatement, the convention provides an extensive institutional framework linking scientific and technical expertise to international and national policymaking. LRTAP's array of scientific and technical working groups, task forces, and international centers gather and analyze environmental data and assess ongoing scientific and technical research intended to support environmental policymaking by national representatives (see Figure 2-1 in Chapter 2).

The European Monitoring and Evaluation Programme (EMEP) is an international monitoring initiative and modeling effort developed in conjunction with LRTAP and later supported by the European Union. EMEP, begun in 1984, was initially concerned with acidification, but it now also looks at tropospheric ozone, heavy metals, and persistent organic pollutants. EMEP, and the Regional Air Pollution Information and Simulation (RAINS) model that is now closely associated with it, has involved many prominent research scientists and scientific organizations, but these were mostly organizations from pro-environmental protection nations. During LRTAP policy negotiations, EMEP/RAINS information is often the point of departure, and the model is frequently relied upon to forecast expected outcomes of possible emission-reduction agreements.

EMEP/RAINS is open to all European nations, but it is influenced strongly by countries interested in pollution control, partly because they devote so many resources to the issue and produce so much research. This group of "lead countries" includes Norway, the Netherlands, Germany, Sweden, and Austria. In the leading states, there is significant scientist–politician interaction at the domestic level, both formally and informally. This interaction is not the case in the other nations, creating an asymmetry in the way scientists and politicians interact at the international level as well.

EMEP/RAINS significantly increased European air pollution assessment capacity, mostly by installing many monitoring sites in countries that otherwise would not have done so. Assessment capacity in poorer European countries (i.e., those in the South and East) did not develop much due to EMEP/RAINS, although the European Union fostered capacity development in indirect ways.

As of 2004, the LRTAP convention had 49 member countries, with a small administrative secretariat located in Geneva at the United Nations Economic Commission for Europe.

Box 1-3. Climate Change Assessment through Intergovernmental Cooperation

The Intergovernmental Panel on Climate Change (IPCC) was established by the World Meteorological Organization (WMO) and the United Nations Environment Programme (UNEP) in 1988, in response to increasing concern about anthropogenic climatic change. The IPCC's mandate is to produce a state-of-the-art assessment of climate change, including the scientific basis of concern, possible impacts, and possible response options. To that end, the IPCC publishes state-of-the-art assessment reports, including those published in 1990, 1992, 1995, and 2001. A fourth full assessment report is due in 2007. Since the mid-1990s, special reports have been published on issues raised in the negotiations on the United Nations Framework Convention on Climate Change. The IPCC is an intergovernmental body, so governments nominate authors of the reports. Nominated authors work together to produce draft chapters that undergo expert review and government review, and these reviews are taken into account in the final draft chapters.

IPCC participants assess scientific, technical, and socioeconomic information relevant to understanding climate change. The IPCC does not carry out research, nor does it monitor climate related data or other relevant parameters. It bases its assessments mainly on peer reviewed and published scientific/technical literature. Over time, IPCC participants have developed detailed principles and rules for their operation, decisionmaking, and participation. A policymaker's summary of each report is drafted and approved line-by-line by governmental officials in a final plenary session.

The IPCC has three working groups and one task force. Working Group I assesses the scientific aspects of the climate system and climate change. Working Group II assesses the vulnerability of socioeconomic and natural systems to climate change, negative and positive consequences of climate change, and options for adapting to it. Working Group III assesses options for limiting greenhouse gas emissions and otherwise mitigating climate change. The Task Force on National Greenhouse Gas Inventories is responsible for the IPCC National Greenhouse Gas Inventories Programme. Each Working Group and the Task Force has a technical support unit. These support units are supported by the government of the developed country co-chair of that working group or task force and hosted by a research institution in that country. A number of other institutions provide in-kind support for IPCC activities.

The IPCC is managed by the IPCC Secretariat, which is hosted by WMO in Geneva and supported by the UNEP and WMO.

policymakers to make science-based decisions with confidence that they would be later revisited, resulting in scientific assessment processes that were more effective at moving negotiations forward.

The second group of chapters, focusing on climate change, starts with a chapter by Marybeth Long Martello and Alastair Iles (Chapter 5), who look at the changes in social and ecological systems that climate change is expected to induce. Based on comparison of assessments for coastal zones/small island states and agriculture between the 1970s and early 1990s, this chapter illustrates how the framing of the assessments influences their salience, credibility, and legitimacy. Next, Anthony Patt (Chapter 6) examines how uncertainty has been treated (or not treated) in a number of different climate assessments, and he develops a general framework for identifying the conditions under which uncertainty should be included in the design of an assessment process. Similarly, David Lund (Chapter 7) compares two assessments that used knowledge of an often-ignored phenomenon (abrupt climate change) to understand natural resource management, including water in the Colorado River Basin and salmon in the Columbia River Basin. He finds that neither assessment was particularly effective in affecting policy in the near term and describes how institutional factors were key to these limitations.

The chapter by Bernd Siebenhüner (Chapter 8) looks at the questions of if and how processes learn over time and the effects of learning on policy. To do so, he studies the case of strategic decisionmaking within IPCC since its establishment in 1988. Clark Miller (Chapter 9) examines the role of assessment within the Subsidiary Body for Scientific and Technological Advice of the United Nations Framework Convention on Climate Change (UNFCCC). He argues that the process of designing and managing assessments at the request of government is constrained by two important sets of design criteria. Not only must these assessments produce good science, but they must also meet standards of good governance—a highly complex and contested task in the rapidly changing milieu of global environmental regulatory institutions. The final chapter of this group is by Oladele Ogunseitan (Chapter 10), who sets out to better understand the design of environmental assessments in developing countries by focusing on the experience of the U.S. Country Studies Program, an effort to assist developing and transitional countries in meeting their obligations to the UNFCCC. Typically, such efforts emphasize "capacity development" without systematic analysis of the design issues relevant to sustaining capacity for environmental assessments in developing countries. Ogunseitan's analysis suggests initial remedies for this problem (Chapter 2 also addresses this issue.)

Although private firms play a role in some of the assessments in the first two sections of this book, the final group of case studies looks at assessments in which private actors were the primary (or sole) participants. First, Edward Parson (Chapter 11) discusses the assessment of technical options to solve an environmental problem, which is often the hardest part because the potentially regulated industry holds most of the necessary information and has little interest in helping regulators. He focuses on the Montreal Protocol Technical Assessment Panels, with the most conspicuous instance of this deadlock being overcome. Mojdeh Keykhah (Chapter 12) investigates the salience, credibility, and legitimacy of two forms of catastrophe assessments for Atlantic hurricane risk and evaluates their effectiveness in terms of reinsurance decision outcomes. She illustrates how Hurricane Andrew compelled the insurance industry to invest in hazard assessments and found that better management of chance events was possible through greater use of scientific expertise. In the last of the case studies, Bernd Kasemir, Andrea Süess, and Raphael Schaub (Chapter 13) look at the implications of design choices in sustainability assessments for pension fund managers seeking socially responsible investments. The formal products of existing sustainability assessments are found to be of little use, mostly because of questions of scope and framing.

The concluding chapter (Chapter 14) draws together the findings of the individual case studies with respect to the assessment attributes of salience, credibility, and legitimacy (which are addressed in most of the case studies), as well as the particular design elements referred to in this introductory chapter. This drawing together of results and conclusions also provides an opportunity to reflect on how they relate to other work in the field. Furthermore, the conclusions indicate a number of fruitful areas for future research.

Notes

1. These attributes map onto the criteria for evaluating efforts to link knowledge and action investigated extensively by the Social Learning Group (2001a,b).

2. From Oxford English Dictionary, 2nd edition: **salience** (noun) 2a: the fact, quality or condition of being salient...b: the quality or fact of being more prominent in a person's awareness or in his memory of past experience. **salient** (adjective): 5b: standing out from the rest, prominent, conspicuous.

3. A number of scientists who were interviewed or attended a workshop as part of this research indicated that within their own field, they could readily decide on whether an assessment was credible by asking who has led the effort.

4. Jasanoff's "constructivist" approach has been criticized by positivists such as Aaron Wildavsky (1992), who claimed in a review of Jasanoff's work that "the best thing for scientists to do when they are far apart is to speak the truth exactly as they understand it to the powers that be."

References

Anderson, D., and W. Cavendish. 2001. Dynamic Simulation and Environmental Policy Analysis: Beyond Comparative Statics and the Environmental Kuznets Curve. *Oxford Economic Papers—New Series* 53(4): 721–746.

Arrow, K., and A. Fisher. 1974. Environmental Preservation, Uncertainty, and Irreversibility. *Quarterly Journal of Economics* 88: 312–319.

Arrow, K., B. Bolin, R. Costanza, P. Dasgupta, C. Folke, C.S. Holling, B.-O. Jansson, S. Levin, K.-G. Maler, C. Perrins, and D. Pimentel. 1995. Economic Growth, Carrying Capacity, and The Environment. *Science* 268: 520–521.

Boehmer-Christiansen, S. 1994a. Global Climate Protection Policy: The Limits of Scientific Advice 1. *Global Environmental Change—Human and Policy Dimensions* 4(2): 140–159.

———. 1994b. Global Climate Protection Policy: The Limits of Scientific Advice 2. *Global Environmental Change—Human and Policy Dimensions* 4(3): 185–200.

Bolin, B. 1994. Science and Policy Making. *Ambio* 23(1): 25–29.

Bronk, D.W. 1974. Science Advice in the White House. *Science* 186(11): 116–121.

Carnegie Commission on Science. 1991. *Science, Technology, and Congress: Expert Advice and the Decision-Making Process.* New York: Carnegie Commission on Science, Technology, and Government. http://www.carnegie.org/sub/pubs/science_tech/cong-exp.txt (accessed January 31, 2005).

Castells, N., and J. Ravetz. 2001. Science and Policy in International Environmental Agreements: Lessons from the European Experience on Transboundary Air Pollution. *International Environmental Agreements: Politics, Law and Economics* 1: 405–425.

Christoffersen, L., N. Denisov, K. Folgen, C. Heberlein, and L. Hislop. 2000. *Impact of Information on Decision-Making Processes.* GRID-Arendal issues paper, May 11, 2000. http://www.grida.no/pub/impact.htm (accessed January 31, 2005).

Clark, W.C., R. Mitchell, and D. Cash. Forthcoming. Evaluating the Influence of Global Environmental Assessments. In *Global Environmental Assessments: Information and Influence,* edited by R. Mitchell, W.C. Clark, D. Cash, and N. Dickson. Cambridge, MA: MIT Press.

Cohen, S.J. 1997. Scientist-Stakeholder Collaboration in Integrated Assessment of Climate Change: Lessons from a Case Study of Northwest Canada. *Environmental Modeling and Assessment* 2: 281–293.

Collingridge, D., and C. Reeve. 1986. *Science Speaks to Power: The Role of Experts in Policy-Making.* New York: St. Marten's Press, 175.

Cowling, E. 1995. Lessons Learned in Acidification Research: Implications for Future Environmental Research and Assessment. In *Acid Rain Research: Do We Have Enough Answers?* edited by G.J. Heij and J.W. Erisman. Oxford, United Kingdom: Elsevier Science, 307–319.

Dasgupta, S., B. Laplante, H. Wang, and D. Wheeler. 2002. Confronting the Environmental Kuznets Curve. *Journal of Economic Perspectives* 16(1): 147–168.

Davies, J.C. 1993. Environmental Regulation and Technical Change: Overview and Observations. In *Keeping Pace with Science and Engineering: Case Studies in Environmental Regulation,* edited by M. Uman. Washington, DC: National Academy Press, 251–262.

Elzinga, A. 1997. From Arrhenius to Megascience: Interplay between Science and Public Decisionmaking. *Ambio* 26(1): 72–80.

Ezrahi, Y. 1990. The Descent of Icarus: Science and the Transformation of Contemporary Democracy. Cambridge, MA: Harvard University Press, 354.

Farrell, A., J. Jaeger, and S. VanDeveer. 2001. Environmental Assessments: Four Under-Appreciated Elements of Design. *Global Environmental Change—Human and Policy Dimensions* 11(4): 311–333.

Global Environmental Assessment Project. 1997. *A Critical Evaluation of Global Environmental Assessments: The Climate Experience.* Calverton, MD: Center for the Application of Research for the Environment (CARE), 171.

Golden, W.T. (ed.). 1988. *Science and Technology Advice to the President, Congress, and Judiciary.* Elmsford, NY: Pergamon Books, 523.

Guston, D.H. 1999. Stabilizing the Boundary between U.S. Politics and Science: The Role of the Office of Technology Transfer as a Boundary Organization. *Social Studies of Science* 29(1): 87–111.

Ha-Duong, M. 2001. Transparency and Control in Engineering Integrated Assessment Models. *Integrated Assessment* 2(4): 209–218.

Hellstrom, T. 1996. The Science–Policy Dialogue in Transformation: Model-Uncertainty and Environmental Policy. *Science and Public Policy* 23(2): 91–97.

Hellstrom, T., and M. Jacob. 1996. Uncertainty and Values: The Case of Environmental Impact Assessment. *Knowledge and Policy: The International Journal of Knowledge Transfer and Utilization* 9(1): 70–84.

Jasanoff, S. 1990. The Fifth Branch: Science Advisors as Policymakers. Cambridge, MA: Harvard University Press.

———. 1998. Harmonization: The Politics of Reasoning Together. In *The Politics of Chemical Risk,* edited by R. Bal and W. Halfmann. Boston: Kluwer Academic Publishers, 173–202.

Joss, S., and S. Bellucci. 2002. *Participatory Technology Assessment, European Perspectives.* London, United Kingdom: University of Westminster.

Kleindorfer, P., H. Kunreuther, and P. Schoemaker. 1993. *Decision Sciences: An Integrative Perspective.* New York: Cambridge University Press, 470.

Lee, K.N. 1993. *Compass and Gyroscope: Integrating Science and Politics for the Environment.* Washington, DC: Island Press.

Mayer, S., and A. Stirling. 2002. Finding a Precautionary Approach to Technological Developments—Lessons for the Evaluation of GM Crops. *Journal of Agricultural and Environmental Ethics* 15(1): 57–71.

Merkle, A., and M. Kaupenjohann. 2000. Derivation of Ecosystemic Effect Indicators—Method. *Ecological Modeling* 130(1–3): 39–46.

Miller, C., S. Jasanoff, M. Long, W.C. Clark, N. Dickson, A. Iles, and T. Parris. 1997. Global Environmental Assessment Project Working Group 2 Background Paper: Assessment as Communications Process. In *A Critical Evaluation of Global Environmental Assessments: The Climate Experience,* edited by Global Environmental Assessment Project. Calverton, MD: Center for the Application of Research for the Environment (CARE), 79–113.

Mitchell, R., W.C. Clark, D. Cash, and N. Dickson (eds.). Forthcoming. *Global Environmental Assessments: Information and Influence.* Cambridge, MA: MIT Press.

Mohanty, S., and B. Sagar. 2002. Importance of Transparency and Traceability in Building a Safety Case for High-Level Nuclear Waste Repositories. *Risk Analysis* 22(1): 7–15.

Morgan, M.G., and M. Henrion. 1990. *Uncertainty: A Guide to Dealing with Uncertainty in Quantitative Risk and Policy Analysis.* New York: Cambridge University Press, 332.

National Acid Precipitation Assessment Program (NAPAP) Oversight Review Board. 1991. *The Experience and Legacy of NAPAP.* Washington, DC: NAPAP.

Partidário, M.R., and R. Clark (eds.). 2000. *Perspectives on Strategic Environmental Assessment.* New York: Lewis Publishers, 287.

Price, D.K. 1965. *The Scientific Estate.* New York: Oxford University Press.

Rubin, E.S. 1991. Benefit–Cost Implications of Acid Rain Controls—An Evaluation of the NAPAP Integrated Assessment. *Journal of the Air and Waste Management Association* 41(7): 914–921.

Sabatier, P.A., and H.C. Jenkins-Smith (eds.). 1993. *Policy Change and Learning: An Advocacy Coalition Approach.* Boulder, CO: Westview Press.

Sadler, B. 1996. *International Study of the Effectiveness of Environmental Assessment.* Quebec, Canada: Canadian Environmental Assessment Agency.

Schreurs, M.A., W.C. Clark, N.M. Dickson, and J. Jaeger. 2001. Issue Attention, Framing, and Actors: An Analysis Across Arenas. In *Learning to Manage Global Environmental Risks: Volume 1, A Comparative History,* edited by The Social Learning Group. Cambridge, MA: MIT Press, 349–364.

Social Learning Group (ed.). 2001a. *Learning to Manage Global Environmental Risks: Volume I, A Comparative History.* Cambridge, MA: MIT Press, 361.

———. 2001b. *Learning to Manage Global Environmental Risks: Volume 2, A Functional Analysis of Social Responses to Climate Change, Ozone Depletion, and Acid Rain.* Cambridge, MA: MIT Press, 222.

Tuinstra, W., L. Hordijk, and M. Amman. 1999. Using Computer Models in International Negotiations: Acidification in Europe. *Environment* 41(9): 32–42.

United Nations Economic Commission for Europe. 1991. Convention on Environmental Impact Assessment in a Transboundary Context. Geneva: United Nations Economic Commission for Europe. http://www.unece.org/env/eia/ (accessed January 31, 2005).

van Eijndhoven, J., W.C. Clark, and J. Jaeger. 2001. The Long-Term Development of Global Environmental Risk Management: Conclusions and Implications for the Future. In *Learning to Manage Global Environmental Risks: Volume 2, A Functional Analysis,* edited by The Social Learning Group. Cambridge, MA: MIT Press, 181–197.

van Asselt, M.B.A., A.H.W. Beusen, and H.B.M. Hilderink. 1996. Uncertainty in Integrated Assessment: A Social Science Perspective. *Environmental Modeling and Assessment* 1:71–90.

Von Damme, K. 1998. Some Considerations on the European Union and the Politics of Chemical Risk for Worker's Health. In *The Politics of Chemical Risk,* edited by R. Bal and W. Halfmann. Boston: Kluwer, 131–157.

Wettestad, J. 1997. Acid Lessons? LRTAP Implementation and Effectiveness. *Global Environmental Change* 7(3): 235–249.

White, L.D. (ed.). 1948. *Introduction to the Study of Public Administration.* New York: The MacMillian Company, 612.

Wildavsky, A. 1992. Book Review: The Fifth Branch—Science Advisors as Policymakers. *Journal of Policy Analysis and Management* 11(3): 505–513.

Winstanley, D., R.T. Lackey, W.L. Warnick, and J. Malanchuk. 1998. Acid Rain: Science and Policy Making. *Environmental Science and Policy* 1(1): 51–57.

Zehr, S.C. 1994. The Centrality of Scientists and the Translation of Interests in the U.S. Acid-Rain Controversy. *Canadian Review of Sociology and Anthropology* 31(3): 325–353.

Zobel, T., C. Almroth, J. Bresky, and J.O. Burman. 2002. Identification and Assessment of Environmental Aspects in an EMS Context: An Approach to a New Reproducible Method Based on LCA Methodology. *Journal of Cleaner Production* 10(4): 381–396.

CHAPTER 2

European Politics with a Scientific Face

Framing, Asymmetrical Participation, and Capacity in LRTAP

Stacy D. VanDeveer

*T*HIS CHAPTER ADDRESSES three of the key design features identified in Chapter 1: framing (the overarching concepts that define the issue), participation (the admission of individuals and organizations into an environmental assessment and the definition of when and how they are involved in the process), and capacity (ability of relevant groups to meaningfully engage and participate in an assessment). It examines these factors as a means of understanding the role of "peripheral" European nations—many of them formerly communist "transition" countries—within one of the most important European air pollution activities, the Convention on Long Range Transboundary Air Pollution (LRTAP). Numerous forms of environmental assessment and formal modeling are used within LRTAP bodies. These assessment activities are often credited both by LRTAP analysts and practitioners as having encouraged national policymakers to take stronger actions in pursuit of environmental protection.

The empirical examination of participation patterns in LRTAP activities by representatives from Central and Eastern Europe (CEE) presented in this chapter reveals that CEE participation remains low—probably lower than before the post-communist transitions began. Thus, national participation within LRTAP assessment bodies is highly asymmetrical. This chapter shows how LRTAP cooperation has been influential in different ways for Northern and Western European "drivers" of LRTAP and for "followers" in the European periphery. These differences are examined in four ways: (1) the relative contribution to framing of multilateral scientific and technical consensus positions, (2) the use of multilateral assessment processes to assist in the formulation of foreign policy positions and assert policy positions to international representatives, (3) the existence of domestic institutions to link assessment processes to policymaking, and (4) the relative salience of LRTAP issues as compared with larger political and economic issues in the determination of environmental policy. These differences illustrate the important interactive effects between three assessment design parameters: participation patterns, issue framing, and participant capacities. They also help to explain why, in general, international assessment processes appear to have little discernible direct impact on national policy in the countries on the European periphery. Where some influence of assessment processes can be found, it remains indirect and occurs at different places in the policymaking process than is usually assumed and claimed by many international relations analysts.

Multilateral transboundary air pollution assessment efforts in Europe, like all assessment efforts, "take place in the context of existing knowledge, research communities, politics, institutions and history" (GEA 1997, 55). The political context of LRTAP assessments includes the overarching European political environment (e.g., the Cold War's East–West divisions and détente, the general growth in environmental awareness and concern among publics and elites, the construction of the European Union, the collapse of communist regimes, and the current drive toward deeper and more expansive pan-European political and economic integration). Participants in various forms of scientific and technical assessment are, ostensibly, organizing scientific and technical information. Yet they are doing so within much broader social and political contexts that they may be trying to influence at the same time. This chapter's examination of multilateral scientific and political air pollution cooperation demonstrates that such assessment efforts actually communicate and organize both scientific and political information (VanDeveer 2004). In other words, they help to produce scientific and political/social consensus (Jasanoff and Martello 2004; Jasanoff and Wynne 1998).

This chapter is based on fieldwork in eight European countries and more than fifty personal interviews with more than three dozen current and former LRTAP participants (governmental and nongovernmental) and CEE environmental policymakers (see Annex A). This fieldwork included observing LRTAP negotiations in Geneva. In addition, an extensive review of primary and secondary literature on interstate and transnational European air pollution cooperation was conducted, including the growing scholarly literature on LRTAP and United Nations Economic Commission for Europe (UNECE) documents and working papers, as well as internal reports in personal and library collections at the International Institute for Applied Systems Analysis (IIASA). LRTAP has spawned a large and growing body of literature on the international and domestic politics for transboundary air pollution control and acid deposition—most of them drawing attention to the roles of scientific and technical expertise in negotiations, policymaking, and assessment organizations (Castells and Funtowicz 1997; Churchill et al. 1995; Darst 2001; Levy 1993; McCormick 1985, 1989; Munton 1998; Munton et al. 1999; Selin 2000; Social Learning Group 2001; Soroos 1997; Tuinstra et al. 1999; Victor et al. 1998; Wettestad 1996, 1997, 2000).

The Origins of Cooperation: Participation, Framing, and Transboundary Pollution

Careful examination of the origins of international scientific and policymaker cooperation around acidification issues demonstrates the importance of participation patterns in the initial framing of transboundary environmental problems. Answers to questions associated with who participates in initial assessment activities and what their interests are prove centrally important in understanding how "long-range transboundary air pollution"—a relatively general term—became framed largely as a problem of acidification for more than two decades.

LRTAP's Origins

A small group of Scandinavians, particularly one Swedish researcher, began articulating claims about acid rain in Europe in the late 1960s (Backstrand and Selin 2000; Cowling 1982; Oden 1968). Once claims about acid rain were articulated, they entered the public realm, engendering great scientific and popular debate. By the early 1970s, Marc Levy notes, the central questions about acid rain could be desegregated into four distinct areas: "(1) Did sulfur dioxide travel long distances? (2) Did airborne deposition of sulfur dioxide harm rivers and lakes? (3) Did airborne sulfur dioxide harm forests and crops? (4) Would proposed domestic abatement measures bring comparable improvements in foreign environmental effects?" (Levy 1993, *80*).

The burning of fossil fuels releases sulfur dioxide and nitrogen oxides into the air, and these compounds can be deposited near their sources or travel hundreds of miles in atmospheric wind currents. Sulfur and nitrogen oxides form acids which, when deposited in wet form (e.g., in snow or rain) or in dry form may cause damage to terrestrial and aquatic ecosystems as well as to the built environment. At high levels—possible in local areas of air pollution—oxides of sulfur and nitrogen can have direct negative human health effects as a cause of respiratory illness. At lower levels—generally involved in atmospheric transport—human health impacts, when they exist, are generally indirect.

Over time, acid deposition has adverse impacts if it exhausts receptors' (such as lakes, forests, croplands, or buildings) abilities to withstand it. Thus, because of their composition, some more robust ecosystems or aspects of the built environment can continue to thrive at levels of acid deposition that would destroy a more vulnerable ecosystem. This variability has important ramifications for environmental policy, because deposition thresholds—levels below which ecosystems are not damaged by acid deposition—vary widely across ecosystems and hence across Europe.

To state the obvious, air pollution becomes transboundary only when it crosses political borders, which were constructed largely irrespective of ecosystem boundaries and meteorological phenomena. Thus, air pollution can travel only a few miles before becoming transboundary, as is common in much of Europe, or it can travel hundreds of miles and remain domestic, as in the United States of America. Thus, the political scale of air pollution issues is not determined by natural phenomenon, but by social institutions.

Examination of the early history of multilateral scientific and technical cooperation around air pollution issues and LRTAP development reveals that CEE representatives were absent or constituted a minor presence in four crucial international arenas of scientific and political cooperation and consensus building: (1) the Organisation for Economic Co-operation and Development (OECD)-sponsored Air Management Research Group (AMRG) in the late 1960s; (2) the 1972 United Nations Conference on the Human Environment (UNCHE) in Stockholm; (3) the data gathering, analysis, and scientific assessment activities over the course of the 1970s simultaneously done by the OECD and the Norwegian Interdisciplinary Research Programme (the so-called SNSF Project); and (4) the negotiation of the LRTAP convention in 1978–1979. The first three arenas contained little or no CEE participation, while the fourth involved them only as minor players doing the bidding of the (then) Soviet Union (VanDeveer in press).

The LRTAP convention's roots lie in Swedish concerns about increases in freshwater acidity during the 1950s and 1960s, which drove extensive European research on the issue (Bolin et al. 1972; Cowling 1982; Oden 1967; OECD 1979). The AMRG has been generally overlooked, but its activities, sponsored by the OECD at the urging of Swedish officials, provided the initial network building between air quality officials (and some researchers), and it explicitly worked to build a common set of terms and concepts for OECD involvement in multilateral efforts to assess and address air pollution issues throughout the 1970s. The AMRG surveyed OECD member states regarding air pollution policy interests and research needs, established international networks of state officials and experts, and began to survey the state of knowledge among participant states regarding air pollution measurement, management, and control technologies (see OECD 1968).

The AMRG constitutes an early effort at multilateral assessment around air pollution issues. It encouraged and facilitated international information sharing among OECD countries around air pollution issues. Surveys were conducted among air policy officials, international working groups, and lead countries; scientific and technical research agendas were established, and a multilingual glossary of air pollution terminology was prepared (OECD 1968). The glossary, apparently the first of its kind, was intended to begin to standardize terminology across borders and languages to facilitate data and information exchange. Linguistic standardization was necessary to facilitate communication and methodological standardization. As a demonstration of the conceptual and linguistic influence of the AMRG, the term "long-range transboundary air pollution" and the abbreviation "LRTAP" appear to have been coined and brought to prominence by the group. Thus AMRG participants established practices (linguistic, programmatic, and otherwise) around "boundaries" between policymaking and scientific and technical research. For example, AMRG participants grouped national officials with many interests (as gleaned from surveys) into three groups for additional assessment and priorities for research: measurement of air pollution, effects of air pollution, and control technologies and planning. In particular, they prioritized research and assessment of technologies to reduce sulfur emissions by establishing working groups for these subjects.

The boundaries around areas of inquiry established within the AMRG, illustrated by its organization, proved to be robust. Thirty years later, LRTAP incorporates separate groups and organizational structures in much the same way that the AMRG first organized its work. The AMRG's approach, as with the subsequent LRTAP institutions, frames air pollution issues primarily as management problems to be assessed and addressed via technical means (Backstrand 2001). This highly technocratic approach leaves the social institutions, practices, and values that underlie—some might say cause—transboundary air pollution off the table.

Swedish officials pushed air pollution and acidification issues at the 1972 UNCHE in Stockholm. The Swedish government, largely spurred by the work of Swedish University of Agricultural Sciences Professor Svante Oden and the work his research inspired, growing public environmental concern, and its interest in protecting valuable and potentially endangered fish stocks, commissioned a study that became "Sweden's Case Study for the United Nations Conference on the Human Environment: Air Pollution Across National Boundaries" (Bolin et al. 1972). Famously, at the UNCHE, Swedish participants pushed for international policy action to combat acid rain, citing the Swedish case study. Less well known is the fact that delegates from the Soviet Union and Eastern Europe participated in some preparatory meetings and drafting sessions prior to the conference. However, the Soviet Union led an East Bloc boycott of the 1972 Stockholm conference over the lack of Western formal diplomatic recognition of the German Democratic Republic (GDR). Of Eastern Bloc countries, only Romania sent a delegation to the UNCHE.

Two major scientific and technical assessment efforts had grown out of the increased scientific, governmental, and public awareness of acid rain issues by the early 1970s: Norway's SNSF Project entitled, "Acid Precipitation: Effects on Forests and Fish," and an OECD research program (from 1973 to 1975) focusing on the long-range transport of pollutants. Both initiatives excluded participants from CEE countries. The OECD study resulted in a publication (OECD 1977) that "confirmed the idea that pollutants are transported long distances and showed that the air quality in each European country is measurably affected from all other European countries" (Cowling 1982, 116A; OECD 1977). Many LRTAP participants and analysts agree that the OECD research constituted an important contribution to awareness raising among many Western European policymakers and publics. Importantly, it helped to de-legitimize flat denials of the occurrence of transboundary pollution transport, such as those previously voiced by British and West German officials (Wetstone and Rosencranz 1983). In this way, the OECD study altered the foreign policy of some opponents of air pollution cooperation, establishing the understanding that pollutants were being transported across borders and shifting the debate toward issues of assessing damages and policy proposals.

Following the OECD reports and growing public concern and media coverage of acid rain, a number of states began negotiating an international convention designed to address (or begin to address) problems associated with the long-range transboundary transport of pollutants in the air. Negotiations from 1977 to 1979 over the details of a new convention took place largely between Nordic states (Sweden, Norway, Denmark, and Finland) and other more reluctant Western Europeans, including West Germany and the United Kingdom (Chossudovsky n.d.). Generally speaking, CEE nations and the Soviet Union played a minor role in the negotiation process. The CEE nations were understood to be following the Soviet lead, taking no public or negotiating position not in accord with that of the Soviet Union. For their part, the Soviets signaled their willingness to sign a framework convention early in the LRTAP negotiating process. The United Nations Economic Commission for Europe (UNECE) was selected as the host of LRTAP negotiations—and for the new LRTAP secretariat—

because of the perception that it was the only existing organization with both environmental and economic interests that also included national members from both East and West. The notion that the UNECE operated on the basis of consensus was also attractive to many state officials generally protective of sovereign independence. Like most international secretariats, the LRTAP secretariat was given very limited duties and resources. As a result of these early LRTAP negotiating dynamics, LRTAP activities were framed largely by the debate between Western Europeans (over Northern European officials' environmental priorities), leaving CEE nations on the political margins from the beginning.

The resultant 1979 Convention on Long-Range Transboundary Air Pollution spells out no specific, binding pollution control or reduction commitments, leaving all specifics of multilateral environmental policy development for subsequent international agreements (see Box 1-2 in Chapter 1). Countries are obliged to cooperate in research and information sharing regarding environmental conditions, natural science, policy development, and control technologies, and convention delegates formally linked the LRTAP convention to the Cooperative Programme for Monitoring and Evaluation of the Long-Range Transmission of Air Pollutants in Europe (EMEP). Yet, Soviet and CEE officials remained unwilling to share emissions data, claiming that sensitive economic, security, and energy related information would be compromised. Instead, they agreed to report "transboundary fluxes" of pollutants—an estimate of the amount of sulfur crossing their borders in both directions, for example.

In addition, the acidification–related scientific, technical, and political activities in the West appear to have stimulated related scientific and technical research in CEE countries and the Soviet Union. Academic researchers and meteorological services, for example, picked up the trend toward greater interest in air pollution issues and dynamics and began to include these areas in their research. However, there was little linkage of domestic air pollution research to domestic or foreign policymaking in CEE countries. Nor were CEE environmental researchers given much latitude to pursue their interests absent ideological constraints from the government.

LRTAP Grows and Moves East

LRTAP's organizational structures, behavioral norms, framing of issues, and participation patterns generally developed around acidification concerns—framed as the initial challenge to European policymakers by the Scandinavian countries in the 1960s and 1970s. However, in the twenty-five years following the signing of the original LRTAP convention, a growing set of environmental problems have been subject to assessment and international treaty negotiations. Table 2-1 contains a brief explanation of the LRTAP convention and its eight protocols. The first decade of LRTAP activities focused almost exclusively on acidification-related activities, while the 1990s witnessed an expansion of issues addressed by LRTAP, including additional pollutants such as tropospheric ozone (e.g., photochemical smog from combustion of fossil fuels), heavy metals, and other pollutants, and additional effects such as those on human health. The recent changes

Table 2-1. *The LRTAP Convention and its Protocols (as of May 2004)*

1979	**LRTAP convention:** Framework convention; states agree to "endeavor to limit" and/or reduce air pollution using best available technologies and to share scientific, technical, and environmental policy information. Adopted in Geneva, Switzerland, November 13, 1979; entered into force March 16, 1983; 48 parties.
1984	**EMEP protocol:** Creates a multilateral trust fund for the long-term financial support of EMEP activities. Adopted in Geneva, Switzerland, September 28, 1984; entered into force January 28, 1988; 41 parties.
1985	**Sulfur protocol:** States agree to reduce sulfur emissions or their transboundary fluxes by 30 percent (from 1980 levels), by 1993. All parties in compliance by 1998. Adopted in Helsinki, Finland, July 8, 1985; entered into force September 2, 1987; 22 parties.
1988	**NOx protocol:** States commit to freezing NOx emissions (at 1987 or earlier levels) by the end of 1994, and to future co-operation to further reduce NOx emissions and to establish critical loads—18 of the protocol's 25 parties complied with the terms of the freeze, 12 West European states went further, aiming to reduce NOx emissions by 30 percent by 1998. By 1994, only 2–4 states appeared on track to achieve a 30 percent reduction on time. Adopted in Sophia, Bulgaria, October 31, 1988; entered into force February 14, 1991; 28 parties.
1991	**VOCs protocol:** States agree to reduce VOC emissions by 30 percent from a chosen baseline year between 1984 and 1990. Most countries chose 1988. Reliably assessing progress toward implementation remains complicated, if not impossible, by the lack of accepted emission data. Adopted in Geneva, Switzerland, November 18, 1991; entered into force September 29, 1997; 21 parties.
1994	**Second sulfur protocol:** Replaces the expired 1985 sulfur protocol. Retains 1980 levels as a baseline and uses an "effects based" approach setting "target loads" based on calculated critical loads; states agree to different emissions reductions by 2000 (toward their target loads), which represent a 60 percent reduction in the difference between existing deposition levels and critical loads. Adopted in Oslo, Norway, June 14, 1994; entered into force August 5, 1998; 25 parties.
1998	**Heavy metals protocol:** States commit to reduce emissions of lead, cadmium, and mercury below 1990 levels (or an alternate year between 1985 and 1995). Aims to cut emissions from industrial and combustion sources and from waste incineration. Sets limit values for stationary sources, suggests numerous BAT standards, and requires phase out of lead in petrol. Adopted in Aarhus, Denmark, June 24, 1998; entered into force December 29, 2003; 36 signatories and 21 parties.
1998	**POPs protocol:** States pledge to eliminate discharges, emissions, and losses of POPs; 16 POPS are covered. Various restrictions (e.g., bans, use criteria, and emission reduction targets) are applied to covered POPs. Adopted in Aarhus, Denmark, June 24, 1998; entered into force October 23, 2004; 36 signatories and 20 parties.
1999	**Multi-pollutant/multi-effects protocol:** States agree on emission ceilings for 2010 for sulfur, nitrogen, VOCs, and ammonia. When fully implemented, Europe's sulfur emissions will be cut by at least 63 percent, nitrogen emissions by 41 percent, VOC emissions by 40 percent, and ammonia emissions by 17 percent, compared to 1990 levels. Adopted in Göteborg, Sweden, November 30, 1999; 31 signatories and 9 parties.

Source: Adapted and expanded from McCormick (1997, 59) and UNECE (1995).

Note: Full names of the agreements are (respectively): Convention on Long-Range Transboundary Air Pollution; Protocol to the 1979 Convention on Long-Range Transboundary Air Pollution Long-Term Financing of the Cooperative Programme for Monitoring and Evaluation of the Long-Range Transmission of Air Pollution in Europe (EMEP); Protocol to the 1979 Convention on Long-Range Transboundary Air Pollution on the Reduction of Sulphur Emissions or Their Transboundary Fluxes by at Least 30 Percent; Protocol to the 1979 Convention on Long-Range Transboundary Air Pollution Concerning the Control of Emissions of Nitrogen Oxides or Their Transboundary Fluxes; Protocol to the 1979 Convention on Long-Range Transboundary Air Pollution Concerning the Control of Emissions of Volatile Organic Compounds or Their Transboundary Fluxes; Protocol to the 1979 Convention on Long-Range Transboundary Air Pollution on the Further Reduction of Sulphur Emissions; Protocol to the 1979 Convention on Long-Range Transboundary Air Pollution on Heavy Metals; Protocol to the 1979 Convention on Long-Range Transboundary Air Pollution on Persistent Organic Pollutants; and Protocol to the 1979 Convention on Long-Range Transboundary Air Pollution to Abate Acidification, Eutrophication and Ground-Level Ozone.

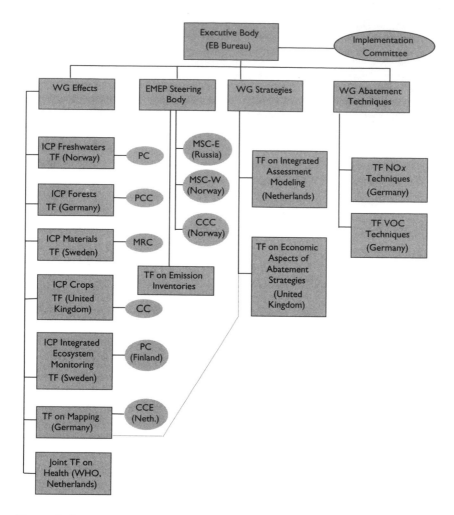

Figure 2-1. *LRTAP Organizational Structure*

Abbreviations appearing in chart: ICP = International Cooperative Program; PC = Programme Centre; PCC = Programme Coordination Centre; MRC = Main Research Centre; CCE = Coordination Centre for Effects; MSC-E = Meteorological Synthesizing Centre–East; MSC-W = Meteorological Synthesizing Centre–West; CCC = Chemical Coordinating Center; WHO = World Health Organization; TF = Task Force; VOC = Volatile Organic Compounds. Unless otherwise indicated, centers are located in the ICP host country. *Source*: VanDeveer 1997

in LRTAP's agenda present LRTAP participants with a host of challenges given that the effects of different environmental challenges vary across Europe.

The organizational structure of LRTAP-related international activities has grown in complexity and scope over time, while retaining key features established in the OECD work in the 1960s. As illustrated in Figure 2-1, the Executive Body (EB) sits atop the LRTAP organizational hierarchy. The EB consists of one representative from each party to LRTAP (i.e., from each country that has ratified LRTAP). In general, it operates on the basis of consensus, delegating some decisions and administrative tasks to the EB Bureau—a subset of EB members. The bureau always includes at least one CEE representative. The next level (down), the "working group level," is made up of the Working Group on Effects (WGE), Working Group on Strategies and Review (WGS), and Working Group on Abatement Techniques (WGAT), as well as the EMEP Steering Body. All countries are invited to send participants to bodies at the working group level. As discussed below, however, participation varies significantly across states and working groups.

Members of the WGE, WGAT, and the EMEP Steering Body administer their respective areas of scientific and technical assessment, data gathering, and information distribution. The WGS is LRTAP's formal negotiating body. Here delegates officially represent their countries, occasionally performing other tasks on a voluntary basis in service to cooperation within the group or to the secretariat. The Implementation Committee was recently established by members of the EB to assess implementation among the parties, and it reports to the EB with recommendations. Below the working group level are the many international cooperative programmes (ICPs) and task forces. As Figure 2-1 illustrates, LRTAP's major national actors (Germany, Netherlands, Norway, Sweden, and the United Kingdom) host all these activities, with the exception of the Meteorological Synthesizing Centre-East in Moscow. Thus, participation by individuals from these major national actors is likely to be higher. As Figure 2-1 illustrates, the organizational structure of LRTAP's scientific and technical assessment and monitoring bodies has grown quite complex (and large) since the convention's signing in 1979.

Table 2-2 shows which countries have signed or ratified the various LRTAP protocols. Ratification rates are considerably lower for CEE states as compared with Western and Northern European States. The typical transition country, for example, is not a party to either of the more stringent protocols on volatile organic compounds (VOCs) or the Second Sulfur protocols. From the environmental policy standpoint, the general lack of CEE ratification is a serious challenge to LRTAP effectiveness, because transition countries tend to be significant sources of the transboundary pollutant emissions that LRTAP protocols attempt to reduce.

Interviews conducted for this research indicate that CEE officials used the following process when considering ratification of LRTAP protocols: The CEE environmental policy officials and the consultants they hired used information generated within LRTAP (and EMEP) assessment processes to assess national prospects for compliance with LRTAP protocols. They report focusing on two basic questions: (1) What, if any, domestic legal and regulatory changes would be required to achieve compliance if a given protocol is ratified? and (2) How much (and to whom) would such changes cost? In cases where the answers to the first question constitute little or no change to existing policies and the costs of these small changes are perceived to be small, CEE governments tend to support ratification. When this is not the case, prospects for ratification dim and/or environmental officials search for other justifications in support of joining a specific protocol. Such justifications include the perceived requirements for accession to

Table 2-2. *LRTAP Parties and Signatories and Ratifications of Protocols (May 2004)*

LRTAP convention party	EMEP	First sulfur	Nitrogen	VOC	Second sulfur	Heavy metals	POPs	Multi-multi
Armenia						S	S	S
Austria	R	R	R	R	R	R	R	S
Azerbaijan								
Belarus	R	R	R					
Belgium	R	R	R	R	R	S	S	S
Bosnia and Herzegovina	R							
Bulgaria	R	R	R	R	S	R	R	S
Canada	R	R	R	S	R	R	R	S
Croatia	R				R	S	S	S
Cyprus	R					S	S	
Czech Republic	R	R	R	R	R	R	R	S
Denmark	R	R	R	R	R	R	R	R
Estonia	R	R	R	R				
Finland	R	R	R	R	R	R	R	R
France	R	R	R	R	R	R	R	S
Georgia								
Germany	R	R	R	R	R	R	R	S
Greece	R		R	S	R	S	S	S
Hungary	R	R	R	R	R	S	R	S
Iceland						S	R	
Ireland	R		R		R	S	S	S
Italy	R	R	R	R	R	S	S	S
Kazakhstan								
Kyrgyzstan								
Latvia	R					S	S	S
Liechtenstein		R	R	R	R	R	R	S
Lithuania	R					S	S	R
Luxembourg	R	R	R	R	R	R	R	R

membership in the European Union, a justification that became increasingly important in the late 1990s through the 2004 accession of eight CEE countries to EU membership. A priority concern of CEE environmental policymakers was the extent to which EU and LRTAP standards are mutually compatible. The key issue compelling CEE participation in LRTAP assessment processes is that these transition country officials are essentially recipients and users of assessment information. In general, assessment information does not shape their negotiating positions during international negotiations, because they tend to use the information only after international agreements have been reached.

For example, in 1998, two protocols concerning the long-rang transport of certain heavy metals and persistent organic pollutants (POPs) were completed. In general, CEE negotiators and domestic environmental policy officials reported using information and contacts associated with LRTAP's assessment processes to attempt to follow what Western states were negotiating. CEE and Russian negotiators expressed strong preferences in negotiating sessions only

Table 2-2. *LRTAP Parties and Signatories and Ratifications of Protocols (May 2004) (continued)*

LRTAP convention party	EMEP	First sulfur	Nitrogen	VOC	Second sulfur	Heavy metals	POPs	Multi-multi
Malta	R							
Monaco	R			R	R	R		
Netherlands	R	R	R	R	R	R	R	R
Norway	R	R	R	R	R	R	R	R
Poland	R				S	S	S	S
Portugal	R			S		S	S	S
Republic of Moldova						R	R	S
Romania	R					R	R	R
Russian Federation	R	R	R		S			
San Marino								
Slovakia	R	R	R	R	R	R	R	R
Slovenia	R				R	R	S	S
Spain	R		R	R	R	S	S	S
Sweden	R	R	R	R	R	R	R	R
Switzerland	R	R	R	R	R	R	R	S
Macedonia, FYR								
Turkey	R							
Ukraine	R	R	R	S	S	S	S	
United Kingdom	R		R	R	R	S	S	S
United States	R		R	S		R	S	S
Yugoslavia	R							
European Community	R		R	S	R	R	R	R

Note: Holy See has signed the LRTAP convention but not ratified the agreement and is therefore not included in this table.

Source: Regularly updated information about the status of various LRTAP agreements is provided on the UNECE website, at http://www.unece.org.

over regulatory provisions they believed to be too costly (or impossible) to achieve in CEE and former Soviet Union (FSU) countries in the time frames proposed. In most cases, "countries with economies in transition" were granted broad, sometimes indefinite and unconstrained, exemptions to the agreements' binding regulatory commitments. CEE officials reported their intentions to commission detailed national-level assessments of the prospects for (and costs of) implementation of these two new protocols only *after* the texts were finalized. They saw developments in both LRTAP political negotiating bodies and assessment activities in a similar way—as indications of where Western environmental policy officials (particularly EU policymakers and officials from "green leader" countries) were taking European environmental policy. In effect, for CEE officials LRTAP assessment processes communicate the direction of policy desired for the whole of Europe by the countries and societies that dominate the processes. In fact, CEE officials reported on LRTAP activities and developments to domestic individuals and groups preparing for and participating in EU accession negotiations. The subject of what LRTAP assessment processes communicate to CEE scientific and technical researchers is addressed below.

LRTAP's Agenda: Big Players and Big Science

The general lack of influence of LRTAP assessment information on CEE policymaking demonstrates that such information does not automatically leap from assessment practitioners or glossy reports into policy discussions and policymaking organizations. Without accepted practices to connect such information to policy, assessments remain much less likely to be influential in policymaking. The history of acidification research demonstrates that, at the international level, it has been "science for policy" from the beginning (Cowling 1982). In other words, major conferences, data gathering exercises, and research projects generally have been supported by governments with the intention of enhancing understanding of the physical environment and the ramifications of human activities in ways that afford public officials opportunities to change—and hopefully improve—public policy. This science for policy model did not apply well in the political and institutional contexts of communist states and societies. In the East Bloc, public officials generally subjected scientific and technical research topics and agendas to overt scrutiny. For researchers, one was ill advised to demonstrate that public policy was environmentally destructive—particularly in light of the fact that the official position of communist regimes, from the 1960s through the early 1980s, held that capitalism, not socialism, produced the kinds of environmental destruction common in the West. In other words, one of the notions underlying various forms of environmental assessment—namely that scientific and technical research about the environment ought to be used to improve or alter policy—did not apply in East Bloc societies. Institutions connecting environmental research to environmental policymaking were largely absent or, at most, were informal and ad hoc. Furthermore, such institutions cannot be built overnight. The struggles associated with the arrival of more democratic governance and greater freedoms of expression and association, as witnessed in CEE countries since 1990, cannot magically create such institutions.

Attention to three common themes often cited in analyses of LRTAP's successful cooperation helps to explain the relative lack of LRTAP influence in most transition countries: (1) the influential role of a small number of West European states and the importance for international policy outcomes of the interactions between these actors, (2) the important role of scientific research and multilateral data collection and analysis in creating common areas of understanding (and figures) around which to negotiate, and (3) the substantial emissions reductions in some substances (particularly sulfur) in many of the participant states.

LRTAP assessment processes and multilateral negotiations were originally driven by downwind Nordic states—later joined in support of multilateral action by Germany (following discovery of forest death/damage, or Waldsterben, and the growth of Green Party influence in German politics). The United Kingdom, following its reluctance to agree to a convention and its refusal to sign the first Sulfur Protocol is commonly identified as the premier European environmental laggard. The United Kingdom has subsequently engaged LRTAP institutions more fully. Thus, the "big players" include Sweden, the Netherlands, Norway, the United Kingdom, and Germany—with the United Kingdom often playing the role of skeptic. As Figure 2-1 illustrates, these countries are major sponsors of LRTAP-related multilateral bodies and programs—usually with state funds. Furthermore, they have frequently been host to related large scientific, technical, and political conferences and sponsors of related research institutes. As a result, nationals from these five states can be found in virtually all LRTAP bodies. These countries and the research produced within them constitute significant proportions of that which is assessed within LRTAP ICPs and task forces. Of course, LRTAP big players have not driven the agenda alone. The green leaders' political allies during international negotiations regularly included Austria, Denmark, Finland, and Switzerland.[1] The United Kingdom frequently relied on Southern European (and at times North American) support for its less environmentally ambitious policy positions. States on Europe's eastern and southern "peripheries" tended to remain in a reactive position vis-à-vis LRTAP, participating and/or expressing a policy preference only when they thought that actions proposed by others might harm their interests.

Regarding the influence of scientific and technical information, EMEP programs, particularly those associated with the provision of internationally legitimate deposition and transport data based on regular ambient air quality monitoring, are frequently cited in the literature on LRTAP (and by many participants in the process) as keys to successful international cooperation with LRTAP. The scientific and technical research (and the later critical loads modeling and mapping) driving growing awareness of adverse environmental effects was conducted almost entirely in Western Europe. The important role of researchers at IIASA in legitimizing and refining research, modeling, and visual representations of transboundary air pollution across Europe is a reoccurring theme within LRTAP literature and among long-standing participants. IIASA is a product of détente. Located in officially neutral Austria, its staff and programs seek to enhance multinational cooperation, especially across the East–West divide, in applied scientific and technical research on common solutions to common problems such as air pollution, forestry management, and transportation.

IIASA has a long history of interest in air pollution issues. The institute's researchers and model developers also have been strategic in selling their Regional Air Pollution Information and Simulation (RAINS) models of atmospheric transport and deposition of pollution in Europe to policymakers across the continent.[2] They held demonstrations and workshops involving policymakers and non-IIASA researchers and modelers in Geneva at the UNECE secretariat, at IIASA, and in West and East European capitals. For example, when U.S.-driven export bans constrained East European access to modern personal computers, the RAINS developers at IIASA kept the program accessible to less-advanced computers that were available in CEE countries and the Soviet Union. IIASA took the model to the GDR, Hungary, Poland, Czechoslovakia, and the Soviet Union to demonstrate it for policymakers and researchers and answer questions (Hordijk 1998). Leen Hordijk, the first leader of IIASA's Transboundary Air Pollution Project, which developed the RAINS model, asserts that these trips and demonstrations were centrally important for the Soviet policymakers' eventual acceptance of the RAINS model and the critical loads concept within the LRTAP framework. In addition to his work at IIASA, Hordijk also served as chair of LRTAP's Integrated Assessment Modeling Task Force. Numerous IIASA alumni, using personal and professional networks established there, were later central in the development of critical loads concepts, models, and maps.[3] These individuals are at the center of a kind of post-normal science issue network made up of people doing applied, policy-related scientific and technical work around long-range air pollution. The network contains many densely connected "nodes" among Scandinavian, Austrian, German, and UK researchers with far fewer participants on the European periphery.

Many scientific and technical participants in LRTAP assessment bodies interviewed for this research believed IIASA's association with the RAINS model added extra legitimacy to the work that other national models could not match. LRTAP WGS and EB members often voiced skepticism of transport models produced in national labs (such as those from the Netherlands and the United Kingdom). IIASA's multinational character, particularly its East–West membership seems to have allayed some officials' fears regarding the national character of other research. Leen Hordijk argues that a national model could never have achieved the status within LRTAP assessment and negotiating activities garnered by RAINS. IIASA's involvement in RAINS development and the IIASA modelers' campaign to sell the model contributed to capacity building around acidification research in a number of CEE countries, including Czechoslovakia, Hungary, Poland, and the Soviet Union. However, this capacity appears to have remained largely within national technical communities, not spilling over into the policymaking bureaucracy. Research found nothing that suggests that domestic policy was affected by this work, although greater understanding of the model likely contributed to CEE countries' acceptance of the RAINS model for use within LRTAP task forces and negotiations. The big national players in LRTAP politics are the same as the big players in LRTAP-related scientific research and modeling.

Regarding emissions reductions, debate over the relative impact of LRTAP activities on state policies and national emissions levels of various pollutants centers around the extent to which international activities are responsible for trends in emissions (see Connolly 1997, for example). In general, sulfur emissions are declining across Europe (UNECE 1995). For many countries, estimates of national emissions have declined by more than 50 percent, compared with 1980 levels. LRTAP participants and analysts also agree that emissions must decline further in most countries if European ecosystems are to be protected from acidification. Scholarly work about LRTAP cooperation often attributes regional political cooperation with more responsibility for the large sulfur emission cuts across much of Europe than it likely deserves. Yet many, perhaps most, participants in LRTAP activities and constituent bodies believe that international cooperation, particularly scientific and technical cooperation, has "raised the bar" and improved environmental protection relative to a counterfactual scenario in which no LRTAP convention exists (see, e.g., Levy 1993; McCormick 1997). LRTAP's existence as an arena for discourse and the continual production of scientific and technological assessment information may have had great influence on European air pollution policies. However, most of the analytical tools in the social sciences to examine policy development are not designed to measure such indirect and/or diffuse influences.

For example, rather than a direct effect of LRTAP cooperation, compliance with early emission reduction commitments by many states—some Western and some transition countries—might best be called accidental compliance. Such compliance results from the restructuring of energy use and economic production as well as from declines in industrial output, rather than from the introduction of stringent environmental regulation. In particular, one is hard pressed to assign credit to LRTAP for significant emissions reductions in the transition states. Reduced national sulfur emissions in transition countries are generally understood to be a result of steep declines in industrial output—and related energy use—in virtually all transition countries after the collapse of communism in 1989. Some contribution was also made by fuel switching away from dirty coal and lignite, mainly to address local pollution concerns. Few analysts have pointed out that future scenarios used in modeling have failed to account for or predict accidental compliance.

A somewhat different picture of LRTAP's influence emerges when one looks at pollutants other than sulfur (McCormick 1997, 1998). Regarding nitrogen oxides (NO_x) and VOCs, LRTAP offers few binding and/or stringent emissions reduction requirements. Far fewer states appear to be on the way to compliance (UNECE 1995) and far less accidental compliance is taking place. Furthermore, LRTAP documents and secretariat officials complain that reporting by states remains sporadic and vague. Peripheral states fail to submit reports and respond to information requests by the secretariat more frequently than do the wealthier, Northern European countries. Yet, the latter states also fail to submit reports. Relatively poor performance on these nonsulfur issues was in large part the impetus for the LRTAP regime's recent establishment of its Implementation Committee. In sum, existing descriptions and analysis of LRTAP politics, while contributing greatly to contemporary understanding of international politics and science among the big players, frequently leave unanswered questions about the rest of the member countries. In fact, analysis of LRTAP has often left questions about CEE countries unasked, as well as unanswered.

LRTAP Participation Patterns: Finding the Other Europeans

An empirical examination of participation patterns in LRTAP activities by CEE nations reveals that CEE participation remains low—probably lower than before the post-communism transitions began. International support for such participation, sometimes called "capacity building," is discussed below. There remains little (and even that is declining) support for broad participation in LRTAP-related international meetings, workshops, conferences, and working groups. Virtually no international financial resources are offered for research or other forms of support for participants to be prepared/informed for meetings, even if they are able to attend. LRTAP officials use and distribute information, including tables and charts, suggesting broad national participation, yet some of this information suggests greater equality of national participation than actually exists in practice. The data in this section do not indicate a complete lack of relevant, domestic scientific and technical expertise in the East. In fact, the level of domestic expertise varies widely in the region. Rather the data suggest that Easterners are generally not present at the international level. They were not, and are not, participating significantly in the construction of international scientific and technical consensus.

What role do "other" countries and domestic scientific communities (those not identified as big players) play in LRTAP processes? Prior to the dramatic geopolitical changes in 1989–1991, European scientists, technical experts, and policymakers—like most Europeans—inhabited institutional realms divided along East–West lines. Subsequently, many institutional differences between East and West have proven much harder to tear down than the Berlin Wall (see Yoder 1999). CEE participation and interest in LRTAP activities are generally explained by noting that the Soviet leadership and the policymaking officials in a few East European states were committed to LRTAP activities because multilateral environmental cooperation improved relations with Western Europe. Having identified international environmental and scientific cooperation as a fruitful area for détente-era politics, Eastern Bloc countries were committed to multilateral cooperation during the negotiations of the 1979 LRTAP convention. Environmental cooperation improved communist states' international reputations in the spirit of co-existence and détente. Furthermore, Soviet dominance of the Eastern Bloc significantly influenced these states' policy positions. Because Soviet officials identified international environmental cooperation in general, and LRTAP in particular, as priorities for détente-style cooperation, CEE officials were expected to follow along.

At the level of individual researchers, there was scientific and technical interest in air pollution and acidification issues in the Eastern Bloc from the late 1960s and early 1970s. A small number of personally committed researchers, mainly from Czechoslovakia, Hungary, Poland, and the Soviet Union, studied local air pollution problems and were interested in access to accurate and reliable data. These researchers, even into the 1980s, tended to focus on areas of locally severe air pollution damages and on human health impacts. As in the West, easily identifiable and observable effects produced early scientific interest. Unlike in the West, CEE researchers had little access to mass media or domestic public policymaking. Transition countries did not lack individuals with scientific and technical training and expertise regarding the physical/natural environment. Nor were the energy and production technologies that created most atmospheric pollutants absent. However, the proportion of individuals able to engage in environmental research, although the precise figure is not possible to calculate, likely remained lower than education figures suggest. It is, by now, not a surprising finding to report that financial and human resources dedicated to scientific and technical research in post-socialist countries have fallen substantially since 1989. Put simply, institutions of higher education and academic research remain in financial crisis. Resources for costly technical and field research have declined—so too has money for international travel for things like conferences and workshops. While some international support is available, it does not make up for the decline in available domestic resources.

In the Czech Republic, Hungary, and Poland, expertise regarding technical and environmental assessment is available to environmental policymakers, although it generally does not exist within government ministries. As with individuals possessing environmental law expertise and experience, environmental assessment expertise remains a scarce commodity in transition countries. Frankly, the public sector cannot pay such individuals enough to acquire or retain them. While public sector employment offers opportunities to acquire international technical training, people often leave the public sector once they acquire these marketable skills. Where scarce forms of expertise exist, they tend to be located in private (or quasi-privatized) consulting firms or research institutes that function like Western consulting firms. Many of these enterprises are located in close physical proximity to governmental ministries or publicly supported scientific organizations, such as national academies of science. In Poland, for example, a leading environmental assessment firm, Energysys, is physically located within the National Academy of Sciences and staffed by current and former academy members and employees.

Because it was rarely the intention of state-socialist policymakers to significantly improve environmental protection, assessment processes as used in the West remained largely absent until the late 1980s (at the earliest). This explains in part why CEE domestic officials rarely initiated assessments. It also suggests that their reasons for initially agreeing to international assessment often differed from those of the green leaders. Formal technical and scientific assessment processes have become somewhat more common only in the mid-1990s. To say that the domestic use of environmental assessment for policy was rare does not mean that the skills required for assessment did not exist in the East. Certainly, domestic scientific communities attempted to assess the state of their knowledge about environmental questions and processes—and could have done so for policymakers, if asked. However, domestic policymaker demand for assessment and institutions that connected assessment processes to policymaking were missing. In addition, some areas of expertise, such as cost–benefit analysis and cost optimization techniques, remained quite rare in the scientific community. Early in the transition period, the use of Western consultants for various kinds of formal assessments became common in response to domestic and (Western) international demand for rationalized information about the state of knowledge in specific areas. Gradually, growth in domestic scientific and technical assessment capacity is occurring. Yet, such skills exist mainly in capital cities, with chronic shortages in most other domestic regions.

Participation Patterns in LRTAP and EMEP

Patterns of national participation from two major international conferences on acidification research and policies—held in 1986 and 1991 and sponsored by the Dutch government in cooperation with UNECE—illustrate the typical low levels of CEE participation in international scientific activities related to acidification (Schneider 1986, 1992). At the 1986 conference, of the 232 recorded participants from 26 countries, 10 (or 4.3 percent) hailed from 5 CEE countries: Czechoslovakia (1 participant), Hungary (2 participants), Poland (3 participants),

Soviet Union (2 participants), and Yugoslavia (2 participants). In 1991, of 106 participants from 16 states, 3 (or 2.8 percent) of conference participants were from the East (1 each from Czechoslovakia, Hungary, and Poland). In contrast, conference delegates from the five big players in LRTAP (Germany, the Netherlands, Norway, Sweden, and the United Kingdom) made up 60 percent of attendees in 1986 and 85 percent in 1991.

Papers and reports presented at these conferences demonstrate vast differences in the general level of data gathering, monitoring, and acidification research between the heavily engaged Western countries (Canada, Finland, Germany, the Netherlands, Norway, Sweden, the United Kingdom, and the United States) and all others. In the heavily engaged countries, literally hundreds of people were involved in the monitoring, data gathering and analysis, and research by the early 1990s. They have produced large data sets; models of transport, deposition, and effects; maps of emissions, deposition, and effects; and innumerable recommendations for national and international policy. In short, these countries have developed overlapping domestic scientific, technical, and policymaking communities with institutionalized links between environmental expertise and policymaking and across state boundaries.

The size of national contingents at international conferences has implications for the construction of scientific consensus as well. Groups from countries represented by a small number of individuals (e.g., one to three persons) rarely include environmental scientists. Small delegations tend to be made up of people from state environmental bureaucracies and administrators from meteorological services (see, e.g., Schneider 1986, 1992). Only rarely do they include a technical advisor from outside an official government body. As such, scientific assessors and policymakers may be the same individuals. Large delegations, such as those from the big player countries, include administrators and policymakers in addition to individual scientists and directors of research programs. The larger (national) groups of experts bring more than a greater number of voices and resources to international conferences. They possess greater capacity to direct discourse and to shape research agendas and the nature of scientific and technical consensus at the international level. Networks like the Meetings of Acidification Research Coordinators, a group of west European research coordinators who began meeting in 1984 (Levy 1993, 90), enhance this capacity, where those developing and working with new concepts, ideas, and research have more interaction with one another and have increased opportunity to shape larger transnational research and assessment agendas.

Generally speaking, most states send representatives to meetings of LRTAP's "high level bodies," including the EB, WGS, and EMEP Steering Body. Table 2-3 presents data on national participation at the working group level from 1994 to 1998. Of the 16 states with perfect attendance for mid-1990s working group meetings, only 2 are transition states: Poland and Russia. The Central European states of Slovenia, the Czech Republic, and Slovakia have also regularly sent representatives to meetings of these bodies. In recent years, Bulgarian and Ukrainian attendance has also increased. Ukrainian participation has recently improved in both quantity and quality, such that Ukraine is now seen as a regular and informed participant in EB and WGS sessions. Other transition states' attendance has been

Table 2-3. *National Participation in LRTAP at the Working Group Level*

Country	Executive Body (4 meetings 12/94–1/98)	WG Strategies (9 meetings 5/96–2/98)	EMEP Steering Body (6 meetings 9/93–6/97)	Total
Austria	4	9	6	19
Canada	4	9	6	19
Denmark	4	9	6	19
Finland	4	9	6	19
France	4	9	6	19
Germany	4	9	6	19
Hungary	4	9	6	19
Italy	4	9	6	19
Netherlands	4	9	6	19
Norway	4	9	6	19
Poland	4	9	6	19
Russia	4	9	6	19
Spain	4	9	6	19
Sweden	4	9	6	19
Switzerland	4	9	6	19
United Kingdom	4	9	6	19
Slovenia	4	9	5	18
United States	4	9	5	18
European Commission	4	8	6	18
Czech Republic	4	9	4	17
Belgium	4	9	3	16
Slovakia	4	9	3	16
Bulgaria	2	9	4	15
Ukraine	4	8	2	14
Portugal	4	3	6	13
Croatia	1	4	5	10
Cyprus	3	1	4	8
Greece	1	4	1	6
Iceland*	1	5	0	6
Ireland	3	9	0	6
Armenia*	1	3	1	5
Romania*	1	4	0	5
Malta	1	3	0	4
Belarus	1	1	0	2
Latvia	0	2	0	2
Lithuania*	1	0	0	1
Luxembourg	0	1	0	1
Turkey	1	1	0	1
Yugoslavia	0	0	1	1
Bosnia-Herzegovenia	0	0	0	0
Holy See*	0	0	0	0
Liechtenstein	0	0	0	0
Moldova*	0	0	0	0
San Marino*	0	0	0	0
Macedonia*	0	0	0	0

Note: * = not a party to the EMEP protocol.
Source: UNECE documents 1992–1998.

Table 2-4. *National Participation in LRTAP Research Programs (as of October 1990)*

Country	Task force memberships	Research sites/labs	EMEP emissions reports 1980–1989	Coordination and synthesis centers	Total
FRG	6	26	12	2	46
Norway	6	12	14	4	40
Sweden	6	15	11	1	34
Finland	6	12	11	1	31
Netherlands	6	9	11	1	28
USSR	5	23	15	1	45
United Kingdom	5	14	16	2	39
Czechoslovakia	5	10	10	2	27
Denmark	5	7	13		25
Italy	5	10	9		24
Belgium	4	4	13		21
Poland	4	5	12		21
France	4	4	12		20
Spain	4	8	6		18
Ireland	4	3	10		17
Hungary	4	3	8		15
Portugal	4	4	4		12
Yugoslavia	3	7	10		20
Austria	3	3	9		15
Switzerland	3	3	9		15
GDR	3	3	7		13
Bulgaria	3		7		10
Greece	3	2	2		7
Turkey	2		3		5
Luxembourg	2		2		4
Liechtenstein	2				2
Iceland	1	2	4		7
Holy See	1				1
Romania	0	6	1		7
San Marino					0

Notes: The total for each country includes the number of task forces to which countries send representatives, the number of identified research sites and labs in the LRTAP network, the number of national emissions reports within the EMEP process, and a doubling of the number of coordinating and synthesizing centers hosted. Thus, hosting two centers is counted as four in the total column. This doubling reflects the higher level of national participation and commitment of human and material resources entailed in serving as a host country. Countries are ranked by task force membership and total.

Sources: Levy (1993, 112), Table 3.5. Levy's work did not count and double the coordinating and synthesizing centers.

generally less frequent and more sporadic. This has been the case with states such as Armenia, Croatia, and Romania. Romanian attendance picked up as a result of the state's interest in eventual EU membership (Mihu 1998). The CEE and FSU states of Belarus, Bosnia-Herzegovina, Latvia, Lithuania, Macedonia, Moldova, and Yugoslavia rarely attend any working group level meetings.

Transition country meeting attendance and participation in LRTAP ICPs, task forces, ad hoc study groups, and informal research networks is lower than in the WGS, EB, and EMEP Steering Body. It is in these groups that the technical work—the various forms of assessment—is designed and carried out. Tables 2-4 and 2-5 contain quantitative data on the nominal national participation of individuals in LRTAP's programmatic activities in the 1980s and 1990s (nominal participation is a rough measure of participation, capturing the named quantity of participants). Membership in a task force "requires that research be conducted at a national level and that this research be harmonized and shared with other participants" (Levy 1993). Most, but not all, participants meet these requirements most of the time. Thus, the data may suggest higher participation than actually occurred. Such data are gathered and distributed by the LRTAP secretariat to demonstrate broad participation in LRTAP activities and to encourage state officials to attempt to raise participation in areas where it is low. Serving as a host for a coordinating or synthesizing center requires sustained commitment of resources and staff and illustrates national commitment and a high degree of participation. Thus, hosting a center is double-counted in each table in the total columns. Of the eight countries host to a coordinating or synthesizing center, two are transition states (the Czech Republic and Russia). These two transition states host three of the eight LRTAP-related subcenters. No program coordinating centers are in CEE countries.

Table 2-6 contains rankings, by country, of the participation data in Tables 2-4 and 2-5. This rough measure of participation contains a number of measurement and comparison problems. For example, and as an illustration of the broad changes in European politics in the early 1990s, at least 12 states on the 1998 list did not exist in 1990, and 2 states on the 1990 list had ceased to exist by 1998 (i.e., the GDR and the Soviet Union). By any definition, the many newly independent states of Europe are "new entrants" to the LRTAP process. Some of the individuals who represent these new states at the working group and EB levels and serve at the ICP and task force level have some previous experience in LRTAP bodies, having been "inherited" from former states like the Soviet Union and the Czech Republic. The figures in Tables 2-4, 2-5, and 2-6 show the dominance of big players, but they also show nominal participation from four transition states—Czechoslovakia/Czech Republic, Hungary, Poland, and Soviet Union/Russia—to be almost as high as in the big players. Note that these are the transition countries most affected by acidification. Furthermore, these CEE countries have nominal participation levels equal to or higher than those of many non-big-player West European states.

Qualitative Dimensions: Delegation Support and Size

Asymmetries in national participation between groups from the big player countries and those from elsewhere get larger as additional quantitative and qualitative aspects of participation are added to the discussion of nominal participation. For example, the size of delegations to WGS sessions varies greatly. Delegations from the big player countries (as well as from Canada and the United States) generally bring at least two to four individuals. Other countries are represented by one person, if they are represented at all. CEE states, for example, are never represented by more than one individual at the working group level. Much of the

Table 2-5. *National Participation in LRTAP Assessment Activities (as of January 1998)*

| Country | EMEP | ICP Forests | | ICP Waters | | ICP Materials | | ICP Crops | | ICP Ecol. Mon. | | TF Mapping | | TF IAM | Total |
		T	D	T	D	T	D	T	D	T	D	T	D	T	
Germany	I	P	I	I	I	S	I	I	I	I	I	I	I	I	16
United Kingdom	I	I	I	I	I	S	I	P	I	I	I	I	I	I	16
Finland	I	I	I	I	I	I	I	I	I	P	I	I	I	I	15
Netherlands	I	I	I	I	I	I	I	I	I	I	I	I	C	I	15
Sweden	I	I	I	I	I	P	I	I	I	I	I	I	I	I	15
Austria	I	I	I		I	S	I	I	I	I	I	I	I	I	14
Czech Republic	I	S	I	I	I	S	I			I	I	I	I	I	14
Italy	I	I	I	I	I	I	I	I	I	I	I	I	I	I	14
Norway	I	I	I	P	I	S	I			I	I	I	I	I	14
Russia	I	I	I	I	I	I	I		I	I	I	S	I	I	14
Switzerland	I	I	I	I	I	I	I	I	I	I	I	I	I	I	14
Spain	I	I	I	I	I	I	I	I	I	I	I	I	I	I	13
France	I	I	I	I	I	I	I	I	I			I	I	I	12
Poland	I	I	I	I	I			I	I	I	I	I	I	I	12
Denmark	I	I	I	I	I			I	I	I	I	I	I	I	12
Hungary	I	I	I	I	I					I	I	I	I	I	11
Canada	I	I	I	I	I	I	I			I	I			I	10
Estonia*	I	I	I	I	I	I	I			I	I	I	*		9
Belgium	I	I	I	I	I			I	I					I	8
Ireland	I	I	I	I	I							I	I	I	8
Latvia	I	I	I	I	I					I	I	I			8
United States	I	I	I	I	I	I	I							I	8
Greece	I	I	I			I	I	I	I						7
Portugal	I	I	I			I	I			I	I				7
Slovakia	I	I	I	I	I							I	I		7
Bulgaria	I	I	I	I								I	I		6
Lithuania	I	I	I	I						I	I	I			6
Belarus	I	I								I	I			I	5
Slovenia	I	I	I					I	I						5
Ukraine	I	I	I							I	I				5
Croatia	I	I	I										I		4
Romania	I	I	I	I											4
European Commission	I	I	I											I	4

detailed negotiating, deal making, and compromising occurs outside of plenary sessions in small "breakout groups" of delegates. Such work is reviewed, discussed, and sometimes changed when it is brought back into plenary. Only delegations with more than one member can work in the breakout groups, shaping the language and details of compromises and side agreements prior to reintroduction in plenary.

Other important indicators of participation, which are more difficult to measure and assess than is delegation size, include factors such as the preparation of state representatives and individual task force members for LRTAP meetings and activities and the influence of coordinating and synthesizing centers. For example, working group level delegates from peripheral European states are generally and widely regarded as highly overburdened with responsibilities. Such delegates report having grossly inadequate budgets and little or no support staff. The time available to these individuals is further constrained by their many other responsi-

Table 2-5. *National Participation in LRTAP Assessment Activities (as of January 1998) (continued)*

Country	EMEP	ICP Forests T	ICP Forests D	ICP Waters T	ICP Waters D	ICP Materials T	ICP Materials D	ICP Crops T	ICP Crops D	ICP Ecol. Mon. T	ICP Ecol. Mon. D	TF Mapping T	TF Mapping D	TF IAM T	Total
Israel*		I				I	I								3
Luxembourg	I	I		I											3
Yugoslavia	I	I		I											3
Cyprus	I			I											2
Liechtenstein	I	I													2
Moldova		I											I		2
Turkey	I	I													2
Bosnia-Herzegovina	I														1
Armenia															0
Holy See															0
Iceland															0
Malta															0
San Marino															0
Macedonia															0

Notes: Notes on abbreviations and symbols: T = Task force; D = Data provision; P = Programme coordinating centre; C = Coordinating centre for effects; S = Subcentre; * = Nonparty to the convention. Notes on Columns: EMAP = EMEP party/participant; ICP Forests = International Cooperative Programme (ICP) Assessment and Monitoring of Air Pollution Effects on Forests, established 1985; ICP Waters = ICP Assessment and Monitoring of Acidification of Rivers and Lakes, established in 1985; ICP Materials = ICP for Effects on Materials, including Historic Cultural Monuments, established 1985; ICP Crops = ICP on Effects of Air Pollution and Other Stressors on Crops and Non-wood Plants, established 1987; ICP Ecol. Mon. = ICP on Integrated Monitoring of Air Pollution Effects on Ecosystems, established 1987 (as a pilot program); TF Mapping = Task Force (TF) on Modeling of Critical Levels and Loads, established 1988; TF IAM = TF on Integrated Assessment Modeling, established 1987 (numbers indicate that a national representative attended at least one meeting of the TF; and Total = Totals for each country include a doubling of the number of coordinating (P, C) and synthesizing (S) centers hosted). Thus, hosting two centers is counted as 4 in the "Total" column. This doubling reflects the higher level of national participation and commitment of human and material resources that serving as host country entails. Countries are ranked by total.
Sources: Adapted and expanded from information provided by R. Chrast, LRTAP Secretariat (UNECE 1995); and Leen Hordijk (1998) for "TF IAM" column information.

bilities. It is not uncommon for LRTAP delegates to be responsible for large areas of domestic and/or international environmental policy (like "air pollution") and to represent their states in as many as a dozen international fora—all with multiple meetings for which they must prepare without support staff. Some delegates spend 50–75 percent of their time outside of their countries, severely limiting time for communicating with domestic colleagues (e.g., officials, media).

Furthermore, working group and EB level delegates from big player states report regular informal contact with their country's members on task forces and with coordinating and synthesizing centers, particularly prior to working group level meetings. Working group level delegates and task force members from these states tend to serve for several years, building personal relationships over time. Here again, serving as a national host for a coordinating or synthesizing center is likely to increase expert–policymaker interactions. No national delegate from any country reported informally consulting non-national experts on scientific

Table 2-6. *National Participation Ranked by Country*

Country	FT Membership 1990 rank (number)	TF Membership 1998 rank (number)	Total Participation from Table 4 rank (total)	Total Participation from Table 5 rank (total)	Average Rank
FRG	1 (6)	1 (7)	1 (46)	1 (16)	1
Sweden	1 (6)	1 (7)	5 (34)	3 (15)	2.5
Finland	1 (6)	1 (7)	6 (31)	3 (15)	2.75
Netherlands	1 (6)	1 (7)	7 (28)	3 (15)	3
United Kingdom	6 (5)	1 (7)	4 (39)	1 (16)	3
USSR	6 (5)	-	2 (45)	-	4
Norway	1 (6)	9 (6)	3 (40)	6 (14)	4.75
Italy	6 (5)	1 (7)	10 (24)	6 (14)	5.75
Czechoslovakia	6 (5)	-	7 (28)	-	6.5
Czech Republic	-	9 (6)	-	6 (14)	7.5
Russia	-	9 (6)	-	6 (14)	7.5
Denmark	6 (5)	9 (6)	9 (25)	15 (11)	9.75
Spain	11 (4)	1 (7)	15 (18)	12 (13)	9.75
Switzerland	18 (3)	1 (7)	17 (15)	6 (14)	10.5
Poland	11 (4)	9 (6)	11 (21)	13 (12)	11
France	11 (4)	9 (6)	13 (20)	13 (12)	11.5
Austria	18 (3)	9 (6)	17 (15)	6 (14)	12.5
Hungary	11 (4)	16 (5)	17 (15)	15 (11)	14.75
Belgium	11 (4)	21 (3)	11 (21)	18 (8)	15.25
Ireland	11 (4)	18 (4)	16 (17)	18 (8)	15.75
Estonia*	-	16 (5)	-	17 (9)	16.5
Latvia	-	18 (4)	-	18 (8)	18
Portugal	11 (4)	21 (3)	21 (12)	21 (7)	18.5
German Democratic Republic	18 (3)	-	20 (13)	-	19
Lithuania	-	18 (4)	-	23 (6)	20.5
Greece	18 (3)	21 (3)	23 (7)	21 (7)	20.75
Bulgaria	18 (3)	21 (3)	22 (10)	23 (6)	21
Slovakia	-	21 (3)	-	23 (6)	22

and technical questions, with the exception of those at IIASA. Because CEE countries, in general, have fewer nationals as task force members, they have fewer people to consult informally. Furthermore, because individuals from these countries are overburdened and often serve for shorter periods of time, they have less time to consult with one another and build personal relationships. The result of these dynamics is that many CEE delegates are less prepared for LRTAP work than are their Northern European colleagues and less likely to be viewed by other LRTAP participants as experts or authorities on LRTAP activities and cooperation.

Participation and Capacity Building

Available funds from both UNECE/LRTAP and domestic task force sponsors to cover the costs of CEE representative attendance at meetings are in decline—at least they appear to be declining, although no exact figures are available. Financial support for attendance to LRTAP EB and working group level meetings, the

Table 2-6. *National Participation Ranked by Country (continued)*

Country	FT Membership 1990 rank (number)	TF Membership 1998 rank (number)	Total Participation from Table 4 rank (total)	Total Participation from Table 5 rank (total)	Average Rank
Yugoslavia	18 (3)	30 (1)	13 (20)	31 (3)	23
Slovenia	-	26 (2)	-	26 (5)	26
Ukraine	-	26 (2)	-	26 (5)	26
Romania	29 (0)	26 (2)	23 (7)	29 (4)	26.75
Belarus	-	29 (1)	-	26 (5)	27.5
Luxembourg	24 (2)	29 (1)	27 (4)	31 (3)	27.75
Turkey	24 (2)	29 (1)	26 (5)	34 (2)	28.25
Iceland	27 (1)	26 (2)	23 (7)	39 (0)	28.75
Liechtenstein	24 (2)	29 (1)	28 (2)	34 (2)	28.75
Croatia	-	29 (1)	-	29 (4)	29
Israel**	-	29 (1)	-	31 (3)	30
Moldova	-	29 (1)	-	34 (2)	31.5
Holy See	27 (1)	36 (0)	29 (1)	39 (0)	32.75
San Marino	29 (0)	36 (0)	30 (0)	39 (0)	33.5
Cyprus	-	36 (0)	-	34 (2)	35
Bosnia-Herzegovina	-	36 (0)	-	38 (1)	37
Armenia	-	36 (0)	-	39 (0)	37.5
Malta	-	36 (0)	-	39 (0)	37.5
Macedonia	-	36 (0)	-	39 (0)	37.5

Notes: Because of the way in which national data are compiled and reported, both with LRTAP institutions and by scholars, partially overlapping country names are all included in the table (e.g. Czechoslovakia and the Czech Republic, USSR and Russia).

The numbers in parentheses refer to the number of memberships in each respective period (columns 2 and 3) and the total nominal participation numbers from tables 2-4 and 2-5 (columns 4 and 5).

The rank numbers refer to the rank order of the participation numbers among the countries. Thus, a rank of 1 means that the country had the highest participation in the group.

* = nonparty to the LRTAP convention; - = no data and/or not applicable.

only meetings for which assistance is available from the secretariat, is drawn from the general UNECE fund. The UNECE offers LRTAP-related financial support only for attending meetings, not for related research in transition countries. West European states and the European Union provide only small amounts, most premised on building networks across the old East–West divisions. None explicitly offer funds for things like staff support for preparation for LRTAP meetings. However, one LRTAP-related capacity-building effort bears noting: EMEP officials run regular training sessions on issues related to EMEP methodologies, equipment, and research—many targeted at the perceived needs of CEE technicians. These programs are generally aimed at improving quality control through the standardization of practices among EMEP participants.

In addition to the funding necessary to attend international meetings, transition states lack other types of capacity to participate equally in LRTAP. At the top of the resource-related challenges associated with public sector organizations like environmental ministries and universities discussed above, most peripheral states have little in the way of ongoing LRTAP-related research. In particular, effects research and modeling work remain in short supply in peripheral countries. Repeatedly, LRTAP participants and analysts cite the important influence of effects research, magnified by mass media, in driving demand for more stringent environmental policies in lead states like Germany, Netherlands, Norway, and Sweden. EU foreign assistance programs help to build national assessment and policymaking and analysis capacity in CEE countries, but these are short term, aimed at facilitating EU accession, and do not support scientific research per se (see Botcheva 1998).

Capacity building for environmental policy, research, and scientific and technical advice are intensely complex endeavors.[4] Alone, programs that offer incentives to individuals to attend meetings or to receive short-term technical training cannot bridge the expertise, experiential, and institutional gaps across states and societies (see Grindle 1997; Miller 1998). This does not mean they have no impact, however. They can improve some aspects of domestic-level policy and policymaking processes and link principles and policy norms promulgated in international fora to particular domestic-level actors and spheres (VanDeveer 1997, 2000). Nevertheless, increasing attendance and the technical skills of a few actors are unlikely to create institutions that demand scientific and technical information for policymaking.

Assessment Lessons from the Periphery

Although CEE participation in LRTAP is low and these countries remain minor voices in the construction of international consensus, national participation levels in LRTAP scientific and technical bodies still matter for policymaking at the international and domestic levels in several ways. National delegates to WGS negotiating sessions are generally more informed about the scientific and technical issues at hand when they have nationals on task forces. WGS delegates from countries with active members in task forces report frequent formal and informal contact with their national task force representatives prior to WGS meetings. Such exchanges generally give them greater capability to engage aspects of the technical debates in "political" sessions within LRTAP. Yet, practitioners (and analysts) often treat information from the multilateral assessment bodies such as LRTAP task forces as symmetric information. Simply because such bodies issue publicly available reports does not mean that the process by which the group reaches consensus involves symmetric participation. Nor are various individuals and states equally equipped to process information produced by assessments.

Impacts of Assessment on the Periphery

What impacts of LRTAP assessment activities can be identified in CEE countries and what aspects of these accomplishments might be generalized to other peripheral countries? First, in the days of communist rule, the international LRTAP regime offered officially sanctioned data for use by domestic researchers and nascent environmental groups. Prior to the inception of LRTAP reporting, information sharing, and assessment processes, such data were largely unavailable domestically in CEE countries. The existence of the data—and the fact that they bore the imprimatur of the state—helped spur domestic scientific and technical interest in air pollution issues. Once such data were available for most European countries, and once research using and evaluating it became common across Europe, researchers began working on ways to improve its accuracy to build transnational professional networks (Hordijk 1998; Agren 1998). As such, the quality of available data became, in itself, an acceptable area of research work.

Second, the results of LRTAP assessment processes frequently has added greater understanding and specificity to generally accepted consensus positions. As one observer said, "We knew better what we already knew." For example, it was known prior to the inception of LRTAP and EMEP that air pollution was traveling across the borders between Czechoslovakia, Poland, and the GDR (one need only go to some parts of these national borders and look around.) LRTAP and EMEP work offered numerical estimates—long contested but consistently cited in policy debates—of pollution imports and exports around which to negotiate, conduct research, and argue. As in the West, this type of assessment work delegitimized claims that nothing was wrong or that insignificant amounts of pollution were being transported.

The third impact of LRTAP assessment processes has been a kind of "legacy of limited information" around air pollution issues in the post-communist period. In comparison to many other environmental issues in which effects play out mostly in localized areas (e.g., waste disposal, ground water pollution), LRTAP assessments and the domestic research it has helped to spur in some CEE countries has led to air quality data gathering and analysis. Thus while in many issue areas, transition-state environmental policymakers began the post-communist era with almost no data, this was not the case in the area of air quality.

Research suggests that the single most important function of LRTAP's assessment activities—in relation to both its political and scientific and technical bodies—is the capacity to communicate and justify the direction in which Europe's most significant political actors want to take environmental policy in Europe. In this sense, the most important thing LRTAP assessment communicates to the periphery is the direction of future policy debates. The mandates and requests of assessment groups by EB and WGS members for such things as assessment techniques for an effects-based approach to POPs and heavy metals regulation, as well as the push to establish critical loads and models for the multipollutant/multi-effects approach for the next LRTAP protocol, sends signals to the periphery about where the central players want to take policy debates. CEE officials understand that LRTAP's green leader states are also the environmental policy leaders in the European Union. The growing membership and influence of the European Union in setting pan-European environmental standards, and the role of most of LRTAP big player states in driving EU policy enhances this communication function. As such, CEE officials participating in LRTAP consistently report that they use LRTAP assessment information to help prepare them to react to ongoing and future debates and proposals. Of the ten European countries with the highest participation rankings in Table 2-6, seven were EU members before 2004, one (Czech Republic) joined the European Union in mid-2004, and one (Norway) coordinates its environmental policy with the European

Union. Of the top ten, only Russia remains largely outside the EU sphere of influence regarding environmental policy.

In sum, in Central and Eastern Europe LRTAP assessment accomplishes four main points: (1) it communicates to CEE officials which scientific, technical, and regulatory ideas are becoming influential in the big player countries; (2) it gives big player and peripheral countries a common discourse in which to discuss and debate regulatory approaches and scientific and technical ideas; (3) it teaches CEE officials the "rules of the game" driven by the big player states; and (4) it organizes and helps to transfer new knowledge among national scientific and technical communities.

The Uses of Assessment on the Periphery

How do policymakers in transition states use the processes and products of LRTAP assessment? Officials in environmental ministries in transition countries and such states' delegates to LRTAP agree that, generally speaking, the information and reports generated by the task forces are useful in preparation for LRTAP meetings and multilateral discussions. The reports or "products" of the task forces, however, are not always adequate in their coverage and presentation of ongoing technical debates. Where participating technical experts lack consensus, the differences are noted. In these situations, task forces tend to state that "the group lacked consensus." LRTAP task force reports generally reveal little detailed information regarding the processes by which consensus positions are reached—nor do they record many details about what types of issues caused lack of consensus. To gain detailed information about such things, one must participate in the task force or communicate with those who do so. In the absence of such contacts, delegates and policymakers can be left with little to go on besides a lack of consensus.

Task force information is sometimes used to constrain Western European policy (or negotiating) positions that appear "too green" from CEE perspectives. CEE officials sometimes use the scientific or technical evidence, or a lack of evidence or consensus, to argue that current knowledge does not warrant action. Such arguments are frequently supported by, or made in alliance with, Southern European countries and the United Kingdom.

In LRTAP, CEE state policymakers generally use internationally produced assessment information *after* international protocols are finalized. In a sense, national officials "nationalize" the international assessment products produced within LRTAP processes, attempting to adapt them to special circumstances at the national and CEE regional levels. This re-nationalization lends the assessment information greater domestic utility in CEE countries. Botcheva's (1998) analysis also suggests that such national level "learning by doing" assessment adds credibility to assessment conclusions as well.

In CEE states, assessment reports are consulted during domestic discussions of the prospects and costs of ratifying and complying with existing protocols. Perhaps the most direct impact of internationally generated assessment information in transition countries lies in its use of international assessment as a model for domestic level assessment. Because national assessments often take place *after* international policy action, the existence of international assessment processes and products offer easily identified models for national assessors. Thus, even in the realm of environmental assessment, transition states occupy a reactive, rather than proactive, role. Technical information contained in the international assessments, and the assumptions on which they are based, serve as bases for transition state officials (and consultants) to assess if, how, and at what cost they can meet the goals and commitments contained within existing international agreements.

What Assessment on the Periphery Does Not Do

In addition to specifying what environmental assessment does in peripheral countries, it is important to specify a few things that international assessment does *not* do. By the 1980s, neither scientific researchers nor policymakers in CEE countries needed multilateral assessment processes to convince them that their countries had serious air pollution and associated acidification problems. Formal assessments did not "reveal" air pollution problems to CEE scientists or public officials. CEE officials engaged in a litany of policies demonstrating that they were aware of demonstrable air pollution damages and hazards. Among these are reforestation campaigns using hardier species, awarding workers additional hazard pay to work in heavily polluted areas, and removing children from ecologically damaged areas (see McCormick 1997).

National CEE officials did not perceive environmental policy development and implementation to be in their interest, nor did they have institutions that afforded them opportunities to reflect on these interests or the assumptions on which they were based. CEE debates, among policymakers and assessment participants, were not about the extent of the air pollution-related environmental problems, nor were they over whether acidification and related air pollution problems were costly. It was not a lack of environmental understanding, awareness, or scientific expertise that impeded environmental policy action. Rather, impediments came from factors associated with countries' ideological positions and their prioritization of industrial production and economic growth over environmental and human health concerns. An illustrative and extreme example can be found in the 1986 Chernobyl accident. Soviet nuclear power experts and bureaucrats had been aware for years that many design characteristics of Chernobyl-style nuclear power plants were dangerous and that safety procedures were lax. However, they were willing to accept human health risks, later sacrificing many lives (Shevardnadze 1991). The lack of democratic institutions and open media kept publics and would-be reformers from inciting debates that might have turned public demand into better regulatory policies. Few would say Soviet "interests" were well served by these choices.

Although LRTAP's extensive assessment activities have focused primarily on acidification over the last 20 years, CEE environmental policymakers do not appear to have framed LRTAP issues in acidification-related terms. Nor have policy issues in CEE countries been framed in terms of ecosystem protection, on which the critical loads concept is based. In general, environmental officials and consultants tend to frame LRTAP-related issues in terms of their general air quality concerns and, eventually, in terms of gaining EU membership. CEE officials' domestic air quality concerns involve wide-ranging challenges to human health and environmental protection. Acidification issues are generally framed by policymakers as one aspect of their overall air quality concerns, unlike in countries such as Germany, Norway, and Sweden, where acidification reduction is often framed as a policy end in itself. In fact, CEE environmental policymakers are generally quick to say that the reduction in acidifying emissions is a kind of added benefit of environmental policies that they need in order to pursue other goals, such as human health protection, EU membership, and aesthetic improvements.

Environmental assessments generally do not drive or enhance public debate or attract media attention in transition countries. As such, neither the results nor the timing of such assessments drive policy debates either, except in the event that they contain cost estimates for additional regulations that attract policymaker attention. Neither CEE environmental officials nor participating experts express any expectation that international environmental assessment influences policy *in the absence of other significant political interests in favor of environmental policy action*. Virtually all CEE officials interviewed for this research agreed with this statement. Furthermore, existing analysis does not attribute CEE policy development to assessment processes, but to larger political contexts and interests. Thus, international environmental assessments must support or be connected to nonenvironmental policy goals or interests (like EU membership) if they are to shape policy. Nor do assessments generally result in changes in domestic research budgets, as they sometimes do in some Western states. Not a single person interviewed in the course of this research cited increased research budgets in a CEE country in response to assessment processes or products. Once again, research budget increases, when they occur, appear to rely on justifications based on larger political commitments, such as an existing, legally binding international commitment or the goal of EU membership.[5]

In general, available evidence suggests that international assessment processes have little discernible direct impact on national policy in the countries on the European periphery. Where some influence by assessment processes can be found, it remains indirect. CEE officials tend to use the assessment processes as a way to keep track of what the big players are up to and where they are taking international discussion, but they do not use it in detail until international policy has been agreed upon. Secondary or indirect impacts detectable in this research include small increases over time in interest by scholarly communities (academics and research institutes) and some increases in interest and concern among individuals within environmental ministries.

In sum, there has been significant variance between big players and peripheral ones with respect to numerous factors associated with international LRTAP-related assessment. First, a large difference remains in the relative contribution to construction of multilateral scientific and technical consensus between the five big players and Europe's Eastern countries, particularly in the form of differential levels of effects research. The five large players drive the agendas for scientific research, assessment, and policy debate, leaving peripheral countries in a persistently reactive position. Second, the products of international assessment processes are used at different points in the policymaking process of formulating foreign and domestic policy positions. While leading states may use assessment to formulate international positions, peripheral states use it largely to assess their own ability to accomplish what the big five are pushing. Third, CEE countries lack domestic institutions linking assessment processes to policymaking common in West European states. There simply are few institutionalized formal and informal connections between policymaking (and policymakers) and scientific and technical research and assessment. Fourth, the relative salience of LRTAP issues as compared to "larger" political and economic ones differs in East and West. Transboundary pollution issues are simply regarded as less important (compared with other environmental, political, and economic issues) in the CEE region, than they are in many West European countries. Fifth, relatively high levels of participation at the international level, by themselves, do not appear to greatly influence state-level policymaking. For example, Czech and Polish participation levels (see Tables 2-3 to 2-6) are quite similar. Nevertheless, by mid-2004, the Czech Republic had become party to all eight LRTAP protocols, while Poland had ratified only the EMEP protocol.

Conclusions for Assessment Design

National participation patterns within LRTAP activities have been, and remain, highly asymmetrical. Europe's peripheral countries drive neither the political/negotiating agendas nor the scientific and technical assessment agendas within LRTAP. These agendas are related, and CEE officials and scientific experts remain largely peripheral to both. In nearly all ways and in all fora, transition country participation is identifiably lower than that of the big player countries and some other West European states. In addition, participation varies greatly across CEE itself. Poland and the Czech Republic, like Russia, participate in LRTAP assessment processes to a much larger extent than do countries such as Romania and Albania. Some transition countries, parties and nonparties to the LRTAP convention, are virtually absent from the assessment processes.

International activities require the commitment of resources, and they have frequently relied on supportive organizational, often national, sponsors (Murphy 1997). The big player countries, through their hosting of coordinating and synthesizing centers and their support of scientific and technical research, have sponsored the lion's share of LRTAP-related scientific and technological development and assessment. Such sponsorship, however, frequently results in tensions over asymmetrical influence within international organizations and programs.

The research presented in this chapter reinforces the notion that assessments are usefully viewed as communicative processes (GEA 1997). A major component—and perhaps the major message—of what is being communicated via international LRTAP assessment is the direction the big players in European environmental politics and policy are headed in the foreseeable future. LRTAP's big players are among the most influential countries in EU environmental policy development. When they reach consensus on what types of assessments they need and what kinds of policy goals such assessment will serve, officials on the periphery understand this to mean that European policy is moving in certain directions, and that states on the periphery had best pay attention. EU expansion has enhanced the importance of this communication process.

At the very least, this research suggests that analysts of international cooperation must pay much closer attention to cross-national differences in participation in scientific and technical bodies and to the international and cross-national differences in the use of various types of "environmental assessment" processes and products. As the LRTAP case demonstrates, countries that remain largely peripheral to the international environmental assessment and policymaking activities may be important sources of the environmental problems, even as their national policies remain largely unaffected by the ongoing international cooperation.

In conclusion, this examination of LRTAP's origins, assessment processes, and politics suggests four interrelated lessons for assessment design:

1. Early patterns of participation within assessment processes and initial framing of scientific, technical, and research questions are closely interrelated and often persistent over time. Early, heavily engaged participants embed their interests and priorities within assessment practices and organizations. This may discourage other potentially important participants from engaging in the assessment process. Therefore, if assessment processes are usefully viewed as communication processes, participants must carefully reflect on the many messages they may be communicating—explicitly and implicitly. When assessment processes are driven by the participation of wealthier nations and/or better-capitalized scientific and technical disciplines, it becomes clear that political and economic power may be speaking through scientific and technical assessment.

2. Linking assessment topics to "larger" political issues or state goals offers opportunities to enhance the salience of assessment processes and statements for policymakers (VanDeveer forthcoming). For example, European international cooperation around acidification and other air pollution issues has been consistently shaped by geopolitical forces such as Cold War rivalries, détente, the collapse of communist regimes, the growing environmental policy influence of the European Union, and the attempts of many former communist states to gain entry into the European Union (Carmin and VanDeveer 2005; Selin and VanDeveer 2003). Linking assessment issues to larger political, economic, and social interests may enhance the salience and policy impact of assessment findings, but it may also be a vehicle for the channeling of political and economic power through scientific and technical assessment.

3. The model of "Speaking truth to power" has clear limits. While linking assessment processes to politically powerful interests and/or organizations may enhance salience, this link does not guarantee more effective environmental policy. The LRTAP experience suggests that engaging politically important interests may improve the prospects for constructing multilateral policies within international treaties. However, this engagement may not lead to national ratification and implementation of these policies, as the low levels of CEE ratification of LRTAP protocols demonstrates.

4. Effective capacity-building efforts and attempts to broaden participation in assessment processes must include more than merely paying the costs of individuals to attend meetings (Farrell et al. 2001). The capacity to meaningfully participate in assessment processes and to make use of the information and knowledge produced within assessments calls for multiple dimensions of human resource use and organizational and institutional effectiveness (Grindle 1997). LRTAP assessment experience demonstrates that engaged assessment participation requires a minimal level of material resources and multiple forms of scientific, technical, political, and administrative expertise, as well as data gathering and monitoring capabilities, support staff, and connections to professional networks. The organizational and institutional capacities to link assessment to policymaking also vary greatly across political jurisdictions. Here too, capacity building may be needed to address gaps or institutional failures. Those who generally drive the research and policymaking agendas often fail to reflect on the different interests and concerns of other actors in multilateral assessment and policymaking efforts.

Notes

1. For one example of these important negotiating coalitions, see Patt (1999) on the adoption of the "critical loads" concept for use within LRTAP negotiations.

2. For a detailed discussion of the RAINS model and its development, see Alcamo et al. (1990).

3. IIASA/RAINS alumni involved in the development of the critical loads concept and modeling include J.P. Hettelingh, P. Kauppi, K. Kauppi, J. Kamari, M. Posch, and many young researchers and modelers who spent time working on the TAP project on contract or as a part of IIASA's annual "Young Scientists Summer Program." For a list of the many individuals involved in the development of the RAINS model, see Alcamo et al. (1990).

4. Miller (1998) addresses the complexities of international capacity building for science and technology. It focuses on programs aimed at building capacity in developing countries for the use of scientific knowledge around global climate change issues.

5. On the many linkages between LRTAP and EU air pollution assessment and policy-making activities, see Selin and VanDeveer (2003).

References

Agren, Christer. 1998. Personal communication with the author, April 13–14.

Alcamo, J., R. Shaw, and L. Hordijk (eds.). 1990. *The RAINS Model of Acidification: Science and Strategies for Europe.* Dortrecht, Netherlands: Kluwar Academic Publishers.

Backstrand, Karin. 2001. What Nature Can Withstand: Science, Politics and Discourses in Transboundary Air Pollution Diplomacy. Doctoral Dissertation. Lund Political Studies 116, Department of Political Science. Lund Sweden: Lund University.

Backstrand, Karin, and Henrik Selin. 2000. Sweden—A Pioneer of Acidification Abatement. In *International Environmental Agreements and Domestic Politics: The Case of Acid Rain,* edited by Arild Underdal and Kenneth Hanf. Aldershot, UK: Ashgate Publishing.

Bolin, Bert, et. al. 1972. *Sweden's Case Study for the United Nations Conference on the Human Environment: Air Pollution across National Boundaries.* Stockholm, Sweden: Norstadt and Sons.

Botcheva, Liliana. 1998. *Information, Credibility, and Cooperation: The Use of Economic Assessment in the Approximation of EU Environmental Legislation in Eastern Europe.* ENRP discussion paper E-98-13. Cambridge, MA: Kennedy School of Government, Harvard University.

Carmin, JoAnn, and Stacy D. VanDeveer (eds.). 2005. *EU Enlargement and the Environment: Institutional Change and Environmental Policy in Central and Eastern Europe.* London, United Kingdom: Routledge.

Castells, Nuria, and Silvio Funtowicz. 1997. Use of Scientific Inputs for Environmental Policymaking: The RAINS Model and the Sulfur Protocols. *International Journal of Environment and Pollution* 7(4): 512–525.

Chossudovsky, Evgeny. n.d. (1988/89?). *East–West Diplomacy for Environment in the United Nations: The High Level Meeting within the Framework of the ECE on the Protection of the Environment.* New York: United Nations Institute for Training and Research.

Churchill, R., G. Kutting, and L. M. Warren. 1995. The 1994 UNECE Sulphur Protocol. *Journal of Environmental Law* 7(2): 169–199.

Connolly, Barbara Mary. 1997. Organizational Choices for International Cooperation: East–West European Cooperation on Regional Environmental Problems. Ph.D. dissertation. Berkeley, CA: University of California Press.

Cowling, Ellis B. 1982. Acid Precipitation in Historical Perspective. *Environmental Science and Technology* 16(2): 110A–123A.

Darst, Robert G. 2001. *Smokestack Diplomacy: Cooperation and Conflict in East–West Environmental Politics.* Cambridge, MA: MIT Press.

Farrell, Alex, Stacy D. VanDeveer, and Jill Jaeger. 2001. Environmental Assessments: Four Under-Appreciated Elements of Design. *Global Environmental Change* (11): 311–333.

GEA (Global Environmental Assessment) Project. 1997. *A Critical Evaluation of Global Environmental Assessments: The Climate Experience.* Calverton, MD: Center for the Application of Research for the Environment.

Grindle, Merilee (ed.). 1997. *Getting Good Government: Capacity Building in the Public Sectors of Developing Countries.* Cambridge, MA: Harvard Institute for International Development.

Hordijk, Leen. 1998. Personal correspondence with the author, various dates February, April, May.

Jasanoff, Sheila, and Marybeth Long Martello (eds.). 2004. *Earthly Politics: Local and Global in Environmental Governance.* Cambridge, MA: MIT Press.

Jasanoff, Sheila, and Brian Wynne. 1998. Science and Decisionmaking. In *Human Choices and Climate Change: Volume 1: The Societal Framework,* edited by Steve Rayner and Elizabeth L. Malone. Columbus, OH: Battelle Press, 1–88.

Levy, Marc. 1993. European Acid Rain: The Power of Toteboard Diplomacy. In *Institutions for the Earth,* edited by Peter M. Haas, Robert O. Keohane, and Marc A. Levy. Cambridge, MA: MIT Press, 75–133.

McCormick, John. 1985. *Acid Earth.* London, United Kingdom: Earthscan.

———. 1989. *Acid Earth.* Second Edition. London, United Kingdom: Earthscan.

———. 1997. *Acid Earth.* Third Edition. London, United Kingdom: Earthscan.

———. 1998. Acid Pollution: The International Community's Continuing Struggle. *Environment* 40(3): 17–20, 41–45.

Mihu, Dumitu. 1998. Personal interview, Bucharest, Romania, March 3.

Miller, Clark. 1998. Extending Assessment Communities to Developing Countries. Belfer Center for Science and International Affairs discussion paper E-98-15. Cambridge, MA: Environment and Natural Resources Program, Kennedy School of Government, Harvard University.

Munton, Don. 1998. Dispelling the Myths of the Acid Rain Story. *Environment* 40(6): 4–7, 27–34.

Munton, Don, Marvin Soroos, Elena Nikitina, and Marc Levy. 1999. Acid Rain in Europe and North America. In *The Effectiveness of International Environmental Regimes,* edited by Oran Young. Cambridge, MA: MIT Press.

Murphy, Craig N. 1997. *Saving the Seas: Values, Science and International Governance,* edited by L. Anathea Brooks and Stacy D. VanDeveer. College Park, MD: Maryland Sea Grant, 255–282.

Oden, Svante. 1967. *Dagens Nyheter,* October 24.

———. 1968. The Acidification of Air and Precipitation and its Consequences in the Natural Environment. *Ecology Community Bulletin* (Swedish National Science Research Council).

OECD (Organisation for Economic Co-operation and Development). 1968. Air Management Research Group: Decisions and Conclusions of the First Session. OECD report DAS/CSI/A.68.96. Paris, France: Organisation for Economic Co-operation and Development.

———. 1977. *The OECD Programme on Long-Range Transport of Air Pollutants.* Paris, France: Organisation for Economic Co-operation and Development.

———. 1979. *The OECD Programme on Long-Range Transport of Air Pollutants: Measurement and Findings.* Second Edition. Paris, France: Organisation for Economic Co-operation and Development.

Patt, Anthony. 1999. Separating Analysis from Politics: Acid Rain in Europe. *Policy Studies Review* 16 (3/4): 104–137.

Schneider, T. 1986. *Acidification and Its Policy Implications.* Proceedings from an International Conference, May 5–9, 1986, Amsterdam. Amsterdam, Netherlands: Elsevier.

———. 1992. *Acidification Research: Evaluation and Policy Applications.* Proceedings of an International Conference, October 14–18, 1991, Maastricht. Amsterdam, Netherlands: Elsevier.

Selin, Henrik. 2000. Towards International Chemical Safety: Taking Action on Persistent Organic Pollutants (POPs). Ph.D. dissertation, No. 211. Linköping University, Linköping Studies in Arts and Sciences.

Selin, Henrik, and Stacy D. VanDeveer. 2003. Mapping Institutional Linkages in European Air Pollution Politics. *Global Environmental Politics* 3(3): 14–46.

Shevardnadze, Eduard. 1991. *The Future Belongs to Freedom.* New York: Free Press.

Social Learning Group. 2001. *Learning to Manage Global Environmental Risks.* Cambridge, MA: MIT Press.

Soroos, Marvin S. 1997. *The Endangered Atmosphere: Preserving a Global Commons.* Columbia, SC: University of South Carolina Press.

Tuinstra, Willemijn, Leen Hordijk, and Markus Amman. 1999. Using Computer Models in International Negotiations: Acidification in Europe. *Environment* 41 (9): 32–42.

UNECE (United Nations Economic Commission for Europe). 1995. *Strategies and Policies for Air Pollution Abatement.* New York and Geneva: United Nations.

VanDeveer, Stacy D. 1998 European Politics with a Scientific Face: Transition Countries, International Environmental Assessment and Long-Range Transboundary Air Pollution. ENRP discussion paper E-98-9. Cambridge, MA: Kennedy School of Government, Harvard University.

———. 2000. Protecting Europe's Seas: Lessons from the Last 25 Years. *Environment* 42 (6): 10–26.

———. 2004. Ordering Environments: Organizing Knowledge and Regions in European International Environmental Cooperation. In *Earthly Politics: Local and Global in Environmental Governance,* edited by Sheila Jasanoff and Marybeth Long Martello. Cambridge, MA: MIT Press, 309–334.

———. Forthcoming. Assessment Information in European Politics: East and West. In *Global Environmental Assessments: Information and Influence,* edited by Ronald B. Mitchell, William C. Clark, David W. Cash, and Nancy Dickson. Cambridge, MA: MIT Press.

Victor, David G., Kal Raustiala, Eugene B. Skolnikoff, (eds.). 1998. *The Implementation and Effectiveness of International Environmental Commitments: Theory and Practice.* Cambridge, MA: MIT Press.

Wetstone, Gregory, and Armin Rosencranz. 1983. *Acid Rain in Europe and North America: National Responses to an International Problem.* Washington, DC: Environmental Law Institute.

Wettestad, Jørgen. 1996. Acid Lessons? Assessing and Explaining the LRTAP Implementation and Effectiveness. Working paper WP-96-18, March. Laxenburg, Austria: International Institute for Applied Systems Analysis.

———. 1997. Acid Lessons? LRTAP Implementation and Effectiveness. *Global Environmental Change* 7(3): 235–249.

————. 2000. From Common Cuts to Critical Loads: The ECE Convention on Long-Range Transboundary Air Pollution (LRTAP). In *Science and Politics in International Environmental Regimes,* by Steinar Andresen, Tora Skodvin, Arild Underdal, and Jørgen Wettestad. Manchester, United Kingdom: Manchester University Press.

Yoder, Jennifer. 1999. *From East Germans to Germans? The New Post-Communist Elites.* Durham, NC: Duke University Press.

Annex A. List of Interviewees

NAME (listed alphabetically), institutional affiliation, interview date(s) and location.

SERENA ADLER, Director, Directorate for European Integration and International Relations, Ministry of Waters, Forests and Environmental Protection (Romania), March 5, 1998, Bucharest, Romania

CHRISTER AGREN, Director, Swedish NGO Secretariat on Acid Rain, April 13 and 14, 1998, Cambridge, MA

MARKUS AMANN, Project Leader, Transboundary Air Pollution Project, IIASA, various dates during January 1998, Laxenburg, Austria

EWA ANZORGE, Head, Department of European Integration and International Cooperation, Ministry of Environmental Protection, Natural Resources and Forestry (Poland), February 26, 1998, Warsaw, Poland

TIBOR ASBOTH, Hungarian Academy of Sciences, January 28, 1998, Laxenburg, Austria

JOHN BEALE, Head, U.S. Delegation to LRTAP WGS, U.S. Environmental Protection Agency, various dates during February 9–13, 1998, Geneva, Switzerland

LARS BJORKBOM, Chairman, LRTAP WGS, Swedish Environmental Protection Agency, February 12 and 13, 1998, Laxenburg Austria

PETER BORRELL, Scientific Secretary, EUROTRAC, January 23, 1998, Laxenburg, Austria

LASZLO BOZO, Meteorological Service (Hungary), January 30, 1998, Budapest, Hungary

RADOVAN CHRAST, Secretary of the Working Group on Effects, LRTAP Secretariat, UNECE, February 12, 1998, Geneva, Switzerland

ANTON ELIASSEN, Director, Norwegian Meteorological Institute, and Director, EMEP Meteorological Synthesis Center (EMEP-West), January 20, 1998, Laxenburg, Austria

TIBOR FARAGO, Ministry for Environment and Regional Policy (Hungary), January 29, 1998, Budapest, Hungary

JANOS GACS, Acting Project Leader, Economic Transition and Integration Project, IIASA, various dates during January 1998, Laxenburg, Austria

RAMON GUARDANS, Spanish Delegate to LRTAP WGS, DIAE/CIEMAT, February 12, 1998, Geneva, Switzerland

LEEN HORDIJK, Director, Wageningen Institute for Environment and Climate Research (former head of IIASA's Transboundary Air Pollution Project and former chair of the LRTAP Task Force on Integrated Assessment Modeling), February 4–6, 1998, Laxenburg, Austria, and various dates during April–May 1998, Cambridge, MA

BOLESLAW JANKOWSKI, Vice President, EnergSys (environmental consulting company), February 25, 1998, Warsaw, Poland

EUGENIUSZ JEDRYSIK, Environmental Policy Department, Ministry of Environmental Protection, Natural Resources and Forestry (Poland), February 26, 1998, Warsaw, Poland

TERRY KEATING, U.S. Environmental Protection Agency, various dates during 1998–2000

GER KLASSEN, Economic Analysis and Environmental Forward Studies, European Commission, DG XI-Environment, Nuclear Safety and Civil Protection, January 21, 1998, Laxenburg, Austria

MARIA KLOKOCKA, Environmental Policy Department, Ministry of Environmental Protection, Natural Resources and Forestry (Poland), February 26, 1998, Warsaw, Poland

ENDRE KOVACS, Deputy Head, Department of Integrated Pollution Control, Ministry for Environment and Regional Policy (Hungary), January 29, 1998, Budapest, Hungary

MILAN LAPIN, Chair, Department of Meteorology and Climatology, Comenius University, February 3, 1998, Bratislava, Slovak Republic

EIJA LUMME, Secretary of the Steering Body to EMEP, LRTAP Secretariat, UNECE, February 12, 1998, Geneva, Switzerland

MARTIN LUTZ, Expert-Urban Environment, European Commission, DG XI-Environment, Nuclear Safety and Civil Protection, January 21, 1998, Laxenburg, Austria

OLGA MAJERCAKOVA, Department of Hydrology, Slovak Hydrometeorological Institute, February 3, 1998, Bratislava, Slovak Republic

KATARINA MARECKOVA, Slovak Hydrometeorological Institute, February 3, 1998, Bratislava, Slovak Republic

DUMITU MIHU, Romanian Delegate to LRTAP WGS and Technical Assistant, National Center for Sustainable Development (Romania), March 3, 1998, Bucharest, Romania

JEFFREY MILLER, Executive Director, Lead Industries Association (United States), February 9 and 10, Geneva, Switzerland

IVAN MOJIK, Director, Department of Air Protection, Ministry of the Environment of the Slovak Republic, February 3, 1998, Bratislava, Slovak Republic

BRIAN MUEHLING, Member of U.S. Delegation to LRTAP WGS, U.S. Environmental Protection Agency, various dates during February 9–13, 1998, Geneva, Switzerland

TEODOR OGNEAN, Expert, Ministry of Waters, Forests and Environmental Protection (Romania), March 5, 1998, Bucharest, Romania

KRYSTYNA PANEK, Deputy Director, Department of European Integration and International Cooperation, Ministry of Environmental Protection, Natural Resources and Forestry (Poland), February 26, 1998, Warsaw, Poland

RYSZARD PURSKI, Chief Specialist, Ministry of Environmental Protection, Natural Resources and Forestry (Poland), February 26, 1998, Warsaw, Poland

HENRIK SELIN, Ph.D. Candidate and Observer of LRTAP WGS, Linkoping University, Sweden, various dates during February 11–13, 1998, Geneva, Switzerland

JANOS SUDAR, Head, Section for European Integration, Ministry for Environment and Regional Policy, January 29, 1998, Budapest, Hungary

TAJTHY TIHAMER, Energy and Environmental Consultant, January 29, 1998, Budapest, Hungary

ROBERT TOTH, Meteorologist and Hungarian Delegate to the LRTAP WGS, Department for Integrated Environmental Protection, (Hungarian) Ministry for Environment and Regional Policy, February 11 and 12, 1998, Geneva, Switzerland

Ewa Wesolowska, Department of European Integration and International Cooperation, Ministry of Environmental Protection, Natural Resources and Forestry (Poland), February 26, 1998, Warsaw, Poland

Henning Wuester, Secretary to the Working Group on Strategies, LRTAP Secretariat, UNECE, January 26, 1998, Laxenburg, Austria, and various dates during February 9–13, 1998, Geneva, Switzerland

Dusan Zavodsky, Associate Professor, Slovak Hydrometeorological Institute, February 3, 1998, Bratislava, Slovak Republic

Nandor Zoltai, Head, Department for European Integration and International Relations, January 29, 1998, Budapest, Romania

Ivan Zuzula, Deputy Director, Slovak Hydrometeorological Institute, February 3, 1998, Bratislava, Slovak Republic

CHAPTER 3

Dissent and Trust in Multilateral Assessments
Comparing LRTAP and OTAG

Alexander E. Farrell and Terry J. Keating

*T*HIS CHAPTER DISCUSSES and compares two different assessment processes concerned with regional, or subcontinental scale, flows of air pollutants: the Ozone Transport Assessment Group (OTAG) and the assessment process associated with the Convention on Long-Range Transboundary Air Pollution (LRTAP). The OTAG process occurred in the mid-1990s, while the LRTAP convention assessments discussed here occurred in the 1980s and 1990s (although LRTAP continues to support assessments today, for convenience we will use the past tense to refer to both). Both of these assessments involved multiple independent political entities (OTAG centered around the U.S. states and federal agencies, while LRTAP focused on the European nations) and therefore can be called multilateral. Although there are major differences between the two cases, there are also significant similarities between them.

Because they involved multiple political entities, OTAG and LRTAP provide an opportunity to examine the treatment of two key design elements of environmental assessments: methods for handling dissent and methods for creating trust. Dissent and trust are important considerations in the design of multilateral assessment processes because, by definition, participants are independent and usually cannot be coerced (or coerced very forcefully) into agreement. Thus, multilateral assessments generally operate on a principal of consensus, which means general, if not unanimous, agreement. Participants who do not agree can be called dissenters. In multiparty assessments, majority positions often emerge over time. Before the final agreement is reached, consensus positions are negotiated to find a version that can be accepted by all. Sometimes this is not possible, and dissenting opinion may still exist against the final consensus position. Assessment processes need methods for dealing with such dissent as consensus positions emerge and are finalized.

Trust is similarly an essential part of any process that involves individual participants, and is frequently mentioned by assessment participants as an important issue. However, trust is not so easily defined or described. Roughly speaking, the greater the trust between two people, the more willing one is to believe the other without requiring proof or verification. Less trust implies that the claims made by one are not believed as truthful until independent evidence or testimony supports it.

A key feature of both the LRTAP and OTAG assessment processes is the use of computer simulations of air pollution chemistry and transport, which are used to make predictions about the effect of control strategies on future air quality. The act of modeling requires the collection and integration of information of various types, including emission profiles, meteorology, ambient air quality, control technology, and economic development. Because of these requirements, modeling exercises are often the places in which dissent emerges, where unacknowledged assumptions often become most clear, and where technical limitations can most strongly influence the policy debate (Keating and Farrell 1999). Therefore, our analysis focuses on the activities of the main decisionmaking body and the principal modeling group for each assessment. In this chapter, first OTAG is discussed, then LRTAP, and conclusions are drawn at the end. Readers unfamiliar with LRTAP may find it useful to refer to Box 1-2 in Chapter 1 and the history of LRTAP described in Chapter 2.

The OTAG Process

In the 1970 amendments to the Clean Air Act (CAA), the U.S. Congress established a national air quality management framework based on conjoint federalism that remains largely in place today (Portney and Stavins 2000, 77–124). Under this framework, Congress assigned responsibility for establishing National Ambient Air Quality Standards (NAAQSs) to the newly created federal Environmental Protection Agency (EPA). Responsibility for determining how to achieve those goals was left to the states, which were required to prepare State Implementation Plans (SIPs) that define many of the controls that would enable each state to attain the NAAQSs. A key part of SIPs were (and remain) efforts to reduce the amount of nitrogen oxides (NO_x), a key precursor to ozone formation, emitted by vehicles, industry, and electric power generators.

Twenty years after passage of the 1970 CAA amendments, 112 million people (45 percent of the U.S. population) still lived in areas that did not attain the ozone NAAQSs. Many of them lived in the Northeast, an area stretching roughly from Washington, D.C., to Boston. This area of chronic high ozone concentrations had caused the northeastern states to adopt stringent control programs for many sources, but the states began in the early 1990s to be concerned about air pollution coming in from "upwind" midwestern, southeastern, and central states, which had fewer problems with ozone themselves and more lax ozone control regulations (Farrell et al. 1999). The upwind states were naturally skeptical of claims that they should control more for the benefit of "downwind" states in the Northeast. The locally oriented framework of the CAA exacerbated these differences.

In 1990, Congress passed a new set of CAA amendments that required states with serious ozone problems to submit new SIPs by November 1994. Of the 17 states that were required to do so, only 1 did. The primary reason was that most of these states had come to believe that a considerable amount of NO_X in their states came from sources, primarily electric power generators, located upwind (generally, to the west and south) of them. These states felt it was either technically impossible, politically infeasible, or unfair for them to control their own NO_X emissions further while upwind states did little to reduce their NO_X emissions. Scientists had long recognized that such transport of ozone and ozone precursors occurred, but they had not quantified the degree very well and, of course, had only a limited amount to say about the politics or fairness of upwind versus downwind NO_X controls.

The failure of the SIP process could have led to EPA's imposition of federal implementation plans and financial sanctions or a string of lawsuits from environmental advocates. However, these methods were not pursued for a number of reasons, including concerns about political reprisals, a belief that state environmental agencies were the appropriate level of government to address this problem, and an interest in replacing the highly combative method of making policy with a more cooperative approach (DiIulio and Kettl 1995; John 1994; Lester 1994). No less important was the stunning Republican victory in the 1994 elections, led by Newt Gingrich in the House of Representatives, which was very interested in reversing or dramatically changing what they saw as unnecessary or overly burdensome on industry (Gillespie and Schellhas 1994). Thus, instead of lawsuits, this crisis led to a "gentlemen's agreement" between key states, industrial groups, and environmentalists to hold off on lawsuits. Instead, they agreed to engage in a "consultative process . . . to reach consensus on the additional regional, local, and national emission reductions that are needed for the remaining rate-of-progress requirements and attainment [of the ozone NAAQSs]" (Nichols 1995).

Organization

To implement the new agreement, Mary Gade, then director of the Illinois EPA and vice-chair of the newly formed Environmental Council of the States, invited environmental commissioners from the eastern United States to join her in an assessment effort (Gade 1995). OTAG was launched at a meeting held on May 18, 1995, and the assessment process quickly began to take on a character and direction of its own. By August 1995, there were more than 300 participants, and by 1997 about 1,000 people were engaged in the process. OTAG eventually cost more than $20 million, and became the largest photochemical modeling effort undertaken up to that time.

Because OTAG was begun with relatively little planning, the original organization was quite simple and contained experts from different organizations who previously had little chance to interact. The structure shown in Figure 3-1 was not planned out in advance, rather it evolved to address the various needs and issues identified by the participants. It grew "top-down," as areas of controversy were identified for analysis. As described below, this process was the key method for addressing dissent in the OTAG process.

Membership in OTAG's Policy Group, the official decisionmaking body, was limited to the heads of the state environmental agencies (political appointees)

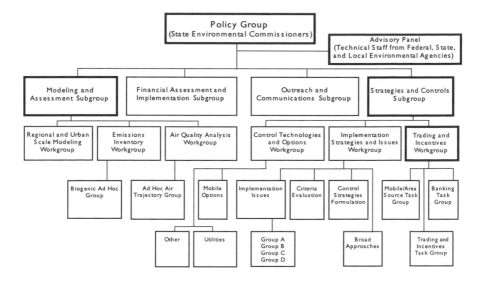

Figure 3-1. *Organization of OTAG (original components in heavy boxes)*

and the heads of the air quality departments in each state (professional staff), while the subsidiary bodies were open to all. The leadership of key subgroups was balanced between leaders from upwind and downwind states, while industry led some of the workgroups that quickly evolved to address specific issues. At least one EPA representative participated in each body, and where national policy was involved, an EPA staffer typically served as a co-chair.

For most of the OTAG process, decisionmaking was by consensus, although the leadership sometimes exerted a bit of executive privilege to end debate and make temporary decisions to enable the group to move on. This worked satisfactorily because OTAG participants understood that they could come back to work out compromises on contentious issues in smaller groups or via conference call before any final decision was made, a feature of assessments Noelle Eckley Selin has labeled "dependable dynamism" (Eckley 2002; Selin, Chapter 4). One major benefit of such dependable dynamism is that dissent need not stop an assessment process, because it could allow for the assessment process to change (and usually enlarge) to address the question at issue while still proceeding with other work. In the case of OTAG, this effect worked well partly because all participants felt that it would be against their interests to be blamed for the failure of OTAG, and partly because they knew that OTAG was not where the final decisions were going to be made. Most OTAG participants felt it was quite likely that the final decisions about controlling regional ozone would be made at the federal level and almost certainly would be reviewed by the federal courts. Thus, OTAG participants were willing to compromise but were always aware that they would see each other in court eventually.

Consensus became the basis for the final recommendations of OTAG, although for most of the process, some sort of voting procedure was generally assumed (Keating and Farrell 1999, *81–87*). The voting procedure was debated in May 1997, and the Policy Group agreed by a simple majority vote to use consensus decisionmaking (Keating and Farrell 1999, *85, 189*). As a voluntary process, perhaps this may seem inevitable in hindsight, but it was something of a change in U.S. environmental policy, where voting (in Congress and in administrative and legal decisionmaking) predominates.

These choices in organization and decisionmaking allowed for a sense of participation and transparency because any interested organization could join at the technical level, and the means by which information flowed and most decisions were made was clear, if complicated. As discussed in the Information section below, these choices were crucial in establishing trust during the OTAG process.

Analysis

Much of the central focus of the OTAG process was on the activities within the Regional and Urban Scale Modeling (RUSM) Workgroup, which was chaired by Mike Koerber (Technical Director of a consortium of midwestern states, including Illinois) and Joe Tikvart (U.S. EPA 1998; Keating and Farrell 1999, *40–52*). RUSM was charged with developing computer model simulations of the transport of ozone and its precursors across the eastern United States and to use these simulations to evaluate the air quality benefits of control strategies identified by the Strategies and Control Subgroup. Four modeling centers conducted the analysis: EPA, the Midwest Modeling Center (MMC), the Southeast Modeling Center (SMC), and the Northeast Modeling and Analysis Center (NEMAC). These modeling centers drew on resources and personnel from various organizations and relied upon periodic meetings, conference calls, and electronic communications to coordinate their internal efforts.

The MMC, the first modeling center, was established at the Lake Michigan Air Directors Consortium (LADCO), where Koerber worked and which was the logical place to focus the modeling effort, given its existing capabilities and interest in performing the regional analysis. It was also a convenient choice from a political perspective: LADCO represents both upwind and downwind states in the Midwest, where controls were likely to be recommended. Because air quality modeling is expensive and time consuming, other groups were encouraged to step forward to create additional modeling centers to help spread the work around and potentially to increase the number of simulations that could be accomplished.

However, as the modeling began to get underway in November 1995 at EPA and the MMC, dissent emerged. OTAG participants from the Northeast and Southeast began to be concerned that their regional interests would not be represented in the modeling being performed by EPA and the MMC. As a result of these concerns, two coalitions—one composed of northeastern states led by New York and another composed of southeastern states and utilities led by Georgia and North Carolina—announced their intentions to establish two additional modeling centers. Each of these two new centers brought new ideas to the process, including the use of analytical approaches and industry funding, neither of which were initially welcomed by the other OTAG participants.

In forming the NEMAC, the northeastern states proposed to perform supplemental analyses using new analytical approaches in addition to those agreed to in the OTAG modeling protocol. This idea was initially met with resistance from the workgroup chairs, who wanted to maintain consistency between the centers with respect to models, inputs, and assumptions. There was a general fear that if the centers were allowed to pursue their own analyses that the cooperative process could devolve into a battle between regional centers each using different models that emphasized their own regions' interests and making intercomparison difficult (Koerber and Tikvart 1997). In the end, each of the different centers did produce some of their own analyses independent of the other centers. However, these analyses were seen to provide an even wider diversity of analyses from different perspectives, which enriched the overall technical discussion. A key reason for this enrichment is that the different analyses and their results were introduced into the quality control process discussed below.

The southeastern states joined in a partnership with several southern electric utility companies to fund the SMC, which later received EPA funding as well. The use of private sector funds was controversial. Some OTAG participants feared that industry would be able to manipulate the predictions of the models or simply delay the modeling process. In some instances, industrial participants in the SMC did advocate for delay to resolve some technical uncertainties. However, the industrial participants also frequently stepped forward with the money to pay for consultants to study and resolve these uncertainties, which might otherwise have gone unsettled. Although some participants consistently viewed the work of the SMC as suspect and tainted by industry money, for the most part, it was perceived as credible.

The RUSM Workgroup itself provided the overall direction of the modeling effort, primarily in frequent conference calls led by Koerber. Due to the large size of the group (at times, more than 100 people) and the tight schedule required by the overall deadline for the process, Koerber often listened to discussion and debate for a fixed period of time and then made a decision based on his own judgment and his perception of the group consensus. Several key decisions were made in this way, including choices about which model to use, how large a domain to analyze, what the grid resolution should be, which episodes to choose, and so forth. While his decisions were not always popular, his leadership was respected. Although air quality models dominated the OTAG process, some other methods were used as well. One group took a new approach focusing on statistical analysis and visualization of air quality and meteorological observations, which provided insights complementary to those generated by the air quality models to the analysis of air quality data, which was considered illuminating (Guinnup and Collom 1997). It was a significant departure from the modeling-based approach used in OTAG, which had emerged from the SIP planning process, and instead used an observational approach to gain insight into the regional patterns and flows of photochemical pollution. Another approach was undertaken by a group at EPA, which used an optimization model to examine the costs of NO_X control in the OTAG region under different policy scenarios (ICF-Kaiser 1996). Without the existence and impetus of the OTAG process, it is likely these new techniques would not have been developed until later, or possibly not at all.

However, not all alternative analyses were helpful. The best example was probably the Midwest Ozone Group (MOG), a set of midwestern industrial companies, which sponsored numerous analyses that used a different model than that used by the participants in the OTAG process. MOG argued that individual states or subregions had little impact on problem nonattainment areas. This argument was characterized as a "race to insignificance" in which some states and industries argued that modeling results demonstrated that their incremental contribution to ozone transport and nonattainment was negligible, while other states and industries argued that the same results demonstrated the opposite conclusion (Colburn 1997). Importantly, MOG did *not* present its methods or results to the RUSM Workgroup for review and approval, and therefore was outside of the quality control mechanisms used in the OTAG process. Thus, it should not be surprising that MOG's analysis was widely decried by other OTAG participants, who had previously accused MOG of misrepresenting the OTAG process (Hawkins 1996). Interestingly, MOG's "race to insignificance" argument and the analyses that supported it received little attention in the final recommendations of the Policy Group.

Information

OTAG participants took advantage of a variety of media to exchange information, ranging from an exhausting schedule of face-to-face meetings to frequent conference calls, periodic newsletters, and constant electronic communications. Especially at the workgroup level, these electronic communication techniques helped to increase productivity and the speed of reaching consensus decisions. The ability to use these tools became a necessary prerequisite for effective participation and was at least as important as being able to travel to face-to-face meetings. Electronic communication also increased access to information and thus transparency.

One of the key features of OTAG was that information "filtered up" to the Policy Group in parallel formal and informal pathways. A formal pathway might lead, for instance, from the Biogenics Ad Hoc Group, to the Emissions Inventory Workgroup, to the Modeling and Assessment Subgroup, and finally, to the Policy Group. At each step along this pathway, a technical expert (or experts) would present information (usually the results of data collection or modeling performed by one of the minigroups or modeling centers), receive comments, perform further analysis if necessary, and then be authorized to present the results to the next higher group. The informal pathways operated privately to deliver more quickly the same information to the same members of the next higher group. This allowed for like-minded participants to review and debate the information and to prepare for the formal public meeting, when even more review and debate would take place. The larger organizations participating in OTAG (e.g., EPA and the electric power industry) had representatives on most of the 23 working groups and would convene privately to discuss what was going on across OTAG at any given point. This process of filtering up through repeated public discussions of particular issues, shadowed by private analyses and communications, served as a quality control method. It also allowed participants to gain trust in the OTAG process, because it could be verified. The main result was that by the time information was presented to the Policy Group, it had been thoroughly discussed and vetted so all participants generally regarded the information as both legitimate and credible (and because most of the analysis had been driven by questions and dissent from higher levels in the organization, the information was generally salient as well). That is, by the time information was presented to the Policy Group, members (especially the political leaders) had already been briefed by their own technical expert(s) and knew what to expect and what the political implications of the technical information were.

Thus, there was little new information presented at Policy Group meetings, making them a bit like play-acting, but the fact that these meetings were public and widely watched (by stakeholder groups and the trade press) meant that they were a key means of certifying that the information was legitimate and credible. For participants, the process was transparent because they could check for themselves where information had come from and how decisions were made. Importantly, every organization did *not* need to participate in each debate, as long as like-minded (and truly trusted) colleagues were able to participate and represent their viewpoint. Thus, as long as one or two downwind states participated in a minigroup, there was no need for all of them to send representatives. This lowered the cost of participating in OTAG, especially for smaller states.

An important exception to the filtering up process proves the rule. The growing prominence of OTAG led to pressure for a two-day "stakeholder presentations" meeting of the Policy Group that occurred near the end of the OTAG process. The leadership felt it needed to hold this meeting to give participants (and latecomers) a chance to present their cases, using whatever information they chose. Some of those who were able to make presentations appreciated the opportunity to express their perspective. Other OTAG participants reported that the introduction of new information from outside the OTAG process at this point tended to undermine, not improve, stakeholder arguments because it was impossible to verify the new information as credible. Thus, the stakeholder presentations had little impact on the subsequent debate on policy recommendations.

Outcomes

Just as OTAG was largely born of a political crisis, politics also engendered its closure. When the 1996 elections largely repudiated the anti-regulatory agenda of the 104th Congress, pressure to conclude the OTAG process began to grow. The Policy Group's negotiations culminated in a meeting on June 2 and 3, 1997, in which the key recommendation on NO_X controls for electric power plants was worked out. The debate over this recommendation was stretched out over two days and eventually was resolved by a decision to recommend a range of controls, from the status quo (minor reductions for some plants) to 85 percent emission reductions, the most stringent proposal. In addition, the report recognized that interstate ozone transport could be a significant problem that deserved further analysis.

This outcome resembled the sort of "least common denominator" outcome common in multilateral environmental agreements between nations, in which participants commit to nothing beyond their pre-existing plans (e.g., do nothing or control emissions significantly). And as with such international agreements, the leaders of OTAG had managed a significant accomplishment simply by obtaining an agreement on a contentious environmental issue and establishing a potential framework in which to further develop it. Indeed, given that OTAG was largely a technical analysis of an issue that had important questions of fairness embedded in it (i.e., appropriate levels of control that should be imposed on local sources and regional sources), it is hard to imagine a more conclusive outcome emerging. Internationally, such initial agreements (like the original LRTAP convention discussed next) can lead to more study and more binding agreements on environmental protection. This was not to be, however, for OTAG. As the conditions that led to a cooperative multilateral approach faded, regulatory and legalistic approaches based on federal law re-emerged. Immediately after OTAG had concluded, the states began filing lawsuits against each other and EPA proposed a new rule aligned with the most stringent recommendations.

The extensive technical analysis conducted during the OTAG process was more important than the somewhat vague policy recommendations. The analysis improved and broadened existing knowledge about ozone formation and transport by producing consistent, high-quality emissions inventories for many states, improved air quality data analysis techniques, and extensive regional-scale photochemical modeling. Ozone transport, per se, was no longer questioned by policymakers—the debate shifted to what to do about it. While OTAG did not really change what was known about ozone transport, OTAG did change who knew it. Moreover, OTAG-based analysis turned up later in the policy process—it was the technical basis for EPA's subsequent NO_X SIP Call, which was designed to achieve by 2004 emission reductions on the order of those sought by EPA and the northeastern states in the OTAG process (U.S. EPA 1998).

The technical efforts during the OTAG process created new analytical capacity at the state level that has continued to contribute to improving the understanding of regional air quality and developing management strategies. During the OTAG process, a number of states increased their technical staff and resources. Furthermore, OTAG gave air quality professionals from different states and EPA a chance to work together, many for the first time, creating a lasting professional network. Many states have continued to cooperate formally and informally in developing technical analyses of regional ozone, fine particles, and haze.

LRTAP Assessments

The LRTAP convention, a classic example of an international environmental agreement that includes a multilateral assessment process, emerged from concerns in Europe about international flows of air pollution. Initially in the late 1960s, these concerns focused on sulfur dioxide (SO_2) and acid rain and led the Organisation for Economic Co-operation and Development (OECD) to organize a special study on the topic (OECD 1968). Within this study, a group from the Norwegian Institute for Air Research (NILU), under the direction of Norwegian meteorologist Anton Eliassen, played a central role in coordinating a monitoring network across northwestern Europe and developing air pollution models. Since then, Eliassen and his colleagues, and the air quality models they have developed, have strongly shaped air pollution assessment in Europe (Eliassen and Bartnicki 1998; Hov 1998).

The OECD research helped lead to the 1979 Convention on Long-Range Transboundary Air Pollution, which identified the general problem of international air pollution and committed the signatories to a "permanent negotiating process" designed to develop follow-on protocols that would address specific problems associated with the international flow of pollution from one country to another. Most Western European nations were parties, as were the Soviet Union and most Central and Eastern European nations. Many other forces were at work, of course, including Cold War politics and European integration, as discussed by VanDeveer in Chapter 2 and others (Boehmer-Christiansen and Skea 1991; Levy 1994). These studies detail the succession of protocols implemented under the overall LRTAP convention that began with an agreement to study the issue, and then to simple, uniform cuts in emissions (e.g., 30 percent reductions for all signatories). These agreements were also least common denominator agreements that, for the most part, embodied commitments that were already part of the domestic policies of the larger, more powerful signatory countries. In some cases, smaller countries agreed to reduce emissions, largely because they viewed participation in LRTAP protocols as part of a larger set of international activities that were very valuable (see Chapter 2).

Over time, LRTAP protocols have become increasingly complex. Rather than requiring uniform emission cuts, second generation LRTAP protocols required emission cuts based on achieving specific environmental goals as determined through a complex modeling approach. However, even these later protocols still generally did not push the major participating countries beyond their domestic agendas. In order to provide for a comparison with the OTAG case and a contrast to prior studies of LRTAP, this section will focus on the modeling conducted as part of the LRTAP process, and it will focus on modeling assessments of acid rain and photochemical smog.

Organization

During the 1980s and 1990s, scientists and government officials of signatory nations accomplished virtually all of the substantive work of LRTAP through a system of smaller coordinating bodies, as shown in Figure 2-1 (Chapter 2). The policymaking group of LRTAP was the Executive Body (EB), which was made up of government officials from each party to LRTAP (i.e., from each country that has ratified LRTAP), which met once a year. The EB provided general guidance and usually operated on a consensus basis. Directly below the EB were the Steering Body of the Cooperative Program for Monitoring and Evaluation of the Long-Range Transmission of Air Pollutants in Europe (EMEP), the Working Group on Effects (WGE), the Working Group on Strategies (WGS), and the Working Group on Abatement Techniques (WGAT; recently, the WGS has become the Working Group on Strategies *and Review,* and the WGAT has been eliminated).

The working groups and EMEP Steering Body were open to representatives from all countries that had signed the convention and had two principal tasks: to manage collective research projects and to negotiate and draft proposed text for LRTAP protocols. The former task was conducted by the WGE, the EMEP Steering Body, and the WGAT, the latter by the WGS, which was the formal negotiating body for LRTAP. Many people on the research–oriented work groups also served on the WGS, and those who did made an effort to separate their roles as science manager in the former and as negotiator in the latter.

The LRTAP approach differed slightly from that used in OTAG in that the WGS had no direct counterpart, but in both processes, the scientific managers had a leadership role in developing policy proposals. And in both processes, private discussions between scientists and political leaders served as quality control and as a means of developing trust in the process. Proposed protocols were adopted at meetings of the environment ministers of signatories to the convention and then ratified by each nation's parliamentary body. The ministerial meetings thus served a purpose similar to the Policy Group in OTAG.

Below the working groups were various task forces, coordination centers, and international cooperative programs, which were also open to any willing participant. These groups were responsible for conducting specific analytical or research tasks on behalf of all the parties and for synthesizing research from different countries in various areas.

The LRTAP atmospheric modeling research, which is the science at the heart of disputes about transboundary flows of pollution, was managed by EMEP. This research task began with developing standardized procedures for monitoring meteorological and environmental conditions within participating nations. The monitoring results were collected by the Chemical Coordination Center (CCC) hosted by NILU. The CCC shared this information with two Meteorological Synthesizing Centers (MSCs), one in Oslo and the other in Moscow. Since the mid-1990s, MSC-West in Oslo has focused its work on regional pollution, and MSC-East in Moscow has focused on heavy metals and persistent organic pollutants. The EMEP synthesizing centers were funded by contributions from parties to the EMEP Protocol and by the national governments that hosted the centers.

The MSCs were responsible for developing atmospheric chemistry models to generate source–receptor relationships for all of Europe. These source–receptor relationships can be arranged into a transfer matrix (sometimes called a "blame matrix" because it assigns the blame for pollution to a particular location; Tuovinen et al. 1994). These matrices were a central part of the discourse within LRTAP because they quantified the international flow of pollution.

The modeling used in LRTAP goes beyond that used in OTAG in significant ways: Not only were atmospheric models used, but the results of these models (the transfer matrices) were combined with information about acceptable pollution levels (called critical loads) developed by the WGE and with data on control costs that were developed by the WGAT (since the elimination of the WGAT, this task was taken over by the International Institute for Applied Systems Analysis [IIASA] in Laxenburg, Austria, in its role as the Center for Integrated Assessment Modeling). The Task Force on Integrated Assessment Modeling (TFIAM) coordinated this modeling effort. By combining all these elements, the TFIAM provided LRTAP negotiators with information on the costs of the various methods for achieving desired environmental goals related to critical loads.

While the TFIAM used a variety of analytical tools, most of its work focused on the Regional Air Pollution Information and Simulation (RAINS) model, which is a constrained optimization that minimizes the costs of reducing emissions for various scenarios using the EMEP source–receptor relationships (Alcamo et al. 1990). Such modeling approaches are also used in the United States—for instance EPA's Integrated Planning Model, used in the OTAG process, follows this approach, although EPA's model only computes emissions, not effects. As discussed below, the RAINS model played a central role in European negotiations (Hordijk 1991).

Analysis and Outcomes

Because the LRTAP process has lasted for 25 years, it has had time to produce a series of analyses and outcomes, which are described below chronologically.

For LRTAP, the scientific issues in question in the 1970s (leading up to the convention) involved the flow of SO_2 and NO_X emissions into areas with soils and surface waters that were susceptible to acidification, which occur mainly in Scandinavia. The main sources of these emissions were thought to be the United Kingdom and Germany, both of which used significant amounts of coal. These two countries denied that the problem was their fault and resisted pressure for controls of their emissions, which were seen as costly. However, they were willing to join the LRTAP convention, in part because it did not require emission controls and in part because they could always choose not to join subsequent protocols. However, in the early 1980s, things changed dramatically. The discovery of significant damage to the Black Forest (called "Waldsterbern" or "forest death") quickly changed the German position to one of strong support for environmental protection, while in England the discovery and development of the North Sea gas fields gave Prime Minister Thatcher an opportunity to reduce the country's usage of coal and thus break the political power of the miners union.

The analysis conducted at this time was fairly straightforward. A key issue was how to gather the data needed to produce the transfer matrices. A significant question of trust arose—the Soviet Union was unwilling to give spatially detailed emission data, fearing this might provide Western nations with militarily useful information. The solution to this problem was to have the Soviets only supply data on the flow of pollutants across their borders. This was considered acceptable because this limitation did not really affect the main analytical question, which was about the relationship between sources in the United Kingdom and Germany and receptor sites in Scandinavia.

Following the initial assessments, most of which relied on air quality sampling and modeling, the first generation LRTAP protocols to limit emissions were adopted. These agreements were relatively simple. For instance, the 1985 Sulfur Protocol lowered transboundary fluxes of SO_2 by 30 percent, while the 1988 NO_X Protocol halted increases in NO_X emissions. These uniform reductions were not thought to be sufficient to achieve all European environmental objectives, and they would have had quite uneven impacts on different parts of the continent. Rather, although it may not have been clear at the time, these reduction commitments were least common denominator agreements in that the emission reductions agreed to had essentially already been decided upon for domestic reasons—environmental reasons for Scandinavia and Germany and political reasons for the United Kingdom—in the countries that ratified the protocols (Wettestad 1997). Critics would also remark that these agreements had "no scientific basis," and in a sense they were right. The limits in these protocols were arbitrary, based as much on what was politically acceptable as on what was ecologically necessary. The negotiators of the first generation protocols acknowledged this deficiency and made commitments to return to the negotiating table to develop "an effect-oriented scientific basis" for deciding on future emission limits.

The main response was to develop an "effects-based" approach that led to the development of the scientific and assessment organizational structure described above. An effects-oriented approach was designed to identify the least-cost mix of emission controls necessary to achieve environmental goals specified with respect to critical loads—levels of pollutant exposure below which no significant adverse effects occur (Metcalfe et al. 1997, Posch et al. 1997). This approach is embodied in the RAINS model, which performs a constrained optimization based on atmospheric source–receptor relationships defined by EMEP, critical loads defined by WGE, and information about future emissions and the costs and efficiencies of available control technology.

The development of the RAINS model began in 1984 as the major focus of the Transboundary Air Pollution (TAP) project at IIASA. At the time, IIASA was one of the few institutions in the world in which scientists from both sides of the Cold War worked cooperatively, and the governance structure of IIASA gave the major countries significant abilities to observe and guide the institution's activities. These features were important in building trust in the RAINS model.

The IIASA team worked hard to ensure the RAINS model was considered credible and legitimate, and did so, in part, by using methods that tended to increase trust in the process and could handle dissent. From the mid-1980s on, IIASA researchers and model developers held demonstrations and workshops involving policymakers and non-IIASA researchers and modelers all over Europe. The RAINS developers at IIASA made sure the model worked on older, less powerful computers that were available in Eastern Europe and the Soviet Union (and later in the independent states of the former Soviet Union). IIASA's association with the RAINS model added extra legitimacy to the work that other "national" models (such as those from the Netherlands and the United Kingdom) could not match. IIASA's multinational character seems to have allayed some officials' fears regarding the "national" character of other research (see Chapter 2). This process also created trust in the same way that it was created in OTAG, through a process of verification of group activities by national experts.

From the late 1980s onward, the output of the RAINS model was used by LRTAP negotiators to explore different goals and policy objectives. The TAP (and later the Task Force on Integrated Assessment Modeling [TFIAM]) team brought

the model to negotiating meetings and responded to questions from the negotiators directly. Most of the questions revolved around the expected costs of control
for the achievement of different environmental goals. Originally, the objective of
the Scandinavians and other pro-environment countries was protection of all sensitive ecosystems, that is, reducing emissions below critical loads. However, the
costs of meeting the absolute critical loads were considered too high, so negotiators explored relative goals, such as decreasing the gap between current conditions
and achievement of critical loads by 50 percent (that is, they examined scenarios
in which only half the surface area exceeding critical loads in the current conditions would be brought within this limit). These "gap closure" scenarios proved to
be politically viable and served as the starting point for the negotiation of national
emission ceilings that were eventually embodied in the second-generation
LRTAP protocols, such as the 1994 Second Sulfur Protocol (Alcamo et al. 1990;
Hordijk 1991).

Following the Second Sulfur Protocol, the LRTAP negotiators and assessment
group began to look toward a second NO_X Protocol as well. There was a growing consensus led by the scientific community that a coordinated approach was
needed to manage the multiple impacts associated with atmospheric fixed nitrogen, including acidifying deposition, anthropogenic eutrophication, and photochemical oxidants (Grennfelt et al. 1994). Within EMEP, models that had been
used to predict source–receptor relationships for sulfur oxides and NO_X were
modified to predict source–receptor relationships for ozone concentrations and
ammonia deposition. Thus, by the mid-1990s, there was strong interest in integrating ozone effects into the RAINS model (which had previously focused
exclusively on sulfur) to enable its use for both LRTAP negotiations and the
European Union's emerging air quality management framework. To incorporate
ozone into the RAINS model, IIASA developed a simplified version of the
EMEP ozone model, which was too complex to include in the RAINS optimization framework directly. This "model of the model" uses a statistical fit of
the outputs of the EMEP ozone model to relate emissions in each country to
long-term average ozone exposures (Heyes et al. 1995, 1997). The resulting version of the RAINS model was used to support the negotiations of the 1999
Multipollutant/Multi-Effects Protocol, which can be thought of as a third-
generation model in that it accounted for effects from multiple pollutants on
multiple environmental goals.

At the same time in the mid- and late 1990s, the European Union was rapidly
assuming a more important role in environmental policymaking. The RAINS
model and other parts of the LRTAP environmental assessment apparatus were
used to develop EU policies. Thus, because of its application in the negotiation
of the Second Sulfur Protocol, the RAINS model significantly influenced European thinking about transboundary air pollution, primarily because it provided a
framework for exploring costs and benefits of different international policies to
individual countries.

However, dissent emerged from time to time regarding the dominance of the
RAINS model. The assessment activities associated with LRTAP were not really
centralized—they relied on and coordinated national efforts, but they could not
specify what individual countries could or could not do with their resources.

There was, however, an implied centralization in that the participating nations were close neighbors and interacted in many different formats; noncooperative behavior was generally frowned upon but could not be prohibited. In addition, it is common in research to develop alternative hypotheses and test them with alternative models. Thus during the late 1980s and early 1990s, two national groups developed their own optimization models: London's Imperial College developed the Abatement Strategies and Assessment Model (ApSimon et al. 1994), and the Stockholm Environmental Institute developed the Coordinated Abatement Strategy Model (Stockholm Environment Institute 1991). Thus, the response to dissent in this case was mainly to conduct more research. All three models were used to inform the negotiations leading to the Second Sulfur Protocol, but the RAINS model emerged as a dominant tool, and IIASA emerged as a focal point for integrating information on air pollution sources, effects, and costs of control across Europe. This dominance was due, in part, to the fact that the other models demonstrated no appreciable advantage over RAINS and, in part, to the continued commitment of various participants in supporting the integrated, multinational efforts at IIASA.

Thus, in LRTAP we observe an interactive process of environmental assessment and policy decisions extending over a relatively long time, which has created a series of ever more stringent emission control agreements that have been supported and shaped by ever more sophisticated analysis. Several reasons have been given for this trend: Some assert that the high quality of the assessment has been most important (Tuinstra et al. 1999). Others suggest that the process itself has been most important, especially in that LRTAP agreements tended to lock in emission reductions that were going to occur anyway for domestic reasons. Still others maintain that the ability to defer difficult decisions until a later date (if need be) appears more important (Eckley 2002; Wettestad 1997).

Our own analysis of the longevity of LRTAP emphasizes flexibility in matching the commitments in the increasingly complex effects-based protocols to the results of the RAINS model. This flexibility can be seen in accommodations in the commitments that range from adjusting the emission ceilings upward for some economically disadvantaged countries in Southern or Eastern Europe, to writing separate but parallel sets of requirements for the United States, Canada, and Russia. The most recent relevant protocol, the 1999 Multipollutant/Multi-Effects Protocol contains three distinct sets of commitments: one for Russia, one for North America, and one for the rest of Europe. This approach permits each nation to maintain control of its own policy agenda while still being a participant in good standing in the international agreement. However, it took more than a decade of development and use of the LRTAP assessment tools and process to reach this point.

One last avenue for dissent opened up only *after* the agreement was made: a nation now may choose not to sign or ratify a protocol. This may be the most radical means of dissent, and Table 2-2 (Chapter 2) shows how often it is employed. For example, Ireland, Poland, and the United Kingdom did not ratify some of the early protocols, the Russian Federation has not ratified several of the later protocols, and the United States did not ratify either of the sulfur protocols. While this course of action ensures nations are in charge of their own destinies, this sort of dissent can have its own costs. For instance, the failure of the Russian Federation to ratify a protocol that contained special accommodations for them has led other LRTAP participants to discount Russian views in subsequent negotiations.

Information Flows

Because of LRTAP's status as a "permanent negotiating process," and because the principal negotiators in the WGS are often one and the same as the assessment team leaders, information from LRTAP assessments has been included automatically into policy decisions. This has led to the development of long-term relationships within LRTAP, and all assessments are conducted with the knowledge that there will be subsequent interactions among the participants. In general, national participation has occurred in the subsidiary bodies of LRTAP, which has created a process in which national-level data and analysis can be presented and reviewed by technical experts before being passed on to the WGS. Thus, most dissent is worked out in these expert groups, sometimes resulting in adjustments to the analysis or interpretation, so that what is passed on already benefits from consensus.

The EB, on the other hand, is made up of a completely different set of people who do not have as frequent interaction nor the same personal relationships. The meetings of the EB are highly planned and structured events. Generally all the substantive text has already been agreed to, all of the analysis and negotiation have already taken place, and little (if any) new information is presented. Nonetheless, the EB meetings are important in that these are the places in which national governments make formal agreements on environment with one another. In many ways, the LRTAP EB meetings are very much like the OTAG Policy Group meetings.

In the 1990s, the LRTAP and EU processes operated largely in parallel, and since 2000 the European Union has taken an even larger role. Early on, the European Commission contracted with IIASA to run RAINS to support analysis of possible EU Directives, indirectly subsidizing the LRTAP analyses for a number of years. Indeed, the reports for the ozone assessments performed for the European Union were sometimes identical to those conducted under the LRTAP convention, albeit with a different name on the cover. In some cases, differences existed in assessments because the European Union sometimes specified different scenarios, and because the nations in LRTAP differed from those in the European Union. The vast majority of assumptions and analytic techniques, however, remained constant. Thus, the EU Directives contain differentiated emission targets for each country, just as the second-generation LRTAP protocols do.

Conclusions

Both the OTAG and LRTAP assessment processes examined here created a shift in their respective issue domains, and both developed effective and, we argue, similar methods for building trust and handling dissent.

Over the course of the OTAG process, the technical modeling and analysis results shifted the theme of the policy debate at the state level from whether or not long-range ozone transport occurs to how it should be controlled. Through participation in the OTAG process, states developed improved technical capabilities and built a professional network that will improve the ability to address regional air pollution problems in the future. In the LRTAP case, the assessment process may not have led to major new commitments by large, powerful countries, but the interaction of environmental politics at the international and national scales seems to have accelerated the process of reducing emissions. The assessment activities undertaken under LRTAP helped significantly by reducing concerns about the economic implications of unilateral emission cuts, by providing important information about environmental conditions and technological options that might not have been available otherwise, and by verifying (mostly through EMEP) that international commitments were being carried through. Two changes in the European issue domain that were especially important were the recognition by countries outside Scandinavia (especially Germany) that they had domestic interests in controlling transboundary pollution flows, and the recognition that multiple pollutants were important on the regional scale.

These shifts were the result, in large part, of credible and convincing analysis conducted as part of the participatory assessment process. The credibility of the analysis was derived in large part from the trust that participants developed for the individuals who conducted the analysis, the methods used in the analysis, and the institutions or overall decisionmaking process in which the analysis was conducted and presented.

Trust in the individuals who conducted the analysis developed through repeated interactions throughout the overall assessment process. Through many presentations and discussions of the methodologies and results of the technical analyses, participants in the process gained familiarity with the leaders of the principal modeling groups. By witnessing how the analysts responded to individuals' concerns, the participants were able to judge the analysts' expertise and biases. Over time, the leaders of the principal modeling groups developed reputations among the participants for being fair and objective in their roles of representing the group consensus. This trust only went so far, however, because it was granted only in the context of an assessment process that allowed for constant evaluation and verification.

Trust in the methods used in the analysis also developed with familiarity. In both the LRTAP and OTAG contexts, the models that were used were quite complex, involved many assumptions, and had important limitations. Through participation in the planning of the analysis, participants were able to understand why certain assumptions were chosen and why better information or techniques were not available or practical. Once some of the analysis was conducted, participants in the process began to see how the results could be used to inform policy decisions. Once the utility of the analysis was demonstrated, participants wanted more analysis to address additional questions. Through repeated interactions with the analysts, participants learned about the limitations of the methods, and more generally, what was analytically useful and what was not. The ability of the available data and methods to address certain questions analytically but not others likely shaped the policy options that were ultimately considered.

Above all else, however, trust in the institutions and decisionmaking processes within which the assessments were conducted was the most important feature of both assessments. The keys to developing this trust were structures that allowed individual or like-minded participants to review and discuss data, methods, and results privately to verify their accuracy; structures that took dissent into account by further research or modifications; and structures that left the ultimate decisions about what course of action to take up to each nation or state. In addition, trust in these assessment processes seemed to develop from a belief that the decisionmaking procedures followed in the assessment process provided adequate time for participation, and it came from an understanding that each of the assessment processes was not the final or only venue for seeking an alternative policy outcome. In the case of OTAG, participants knew that any recommendations developed by the assessment process would be fed into a separate federal rulemaking process, which would be subject to further public comment and debate and, ultimately, challenge in the courts. In the case of LRTAP, participants understood that the parties to the convention were engaged in a continual series of negotiations, and that issues that were not addressed in a given agreement could be revisited in later negotiations.

While the development of multiple layers and types of trust contributed to the success of the two assessment processes considered here, the assessment processes were not without dissent. In both contexts, however, the participatory nature of the institutions or processes within which the assessments were conducted was important to the expression and resolution of dissent. In both cases, interested individuals or groups were able to voice dissent and suggest alternative methods as part of the routine operation of the assessment process. Regardless of whether this changed the analysis or decision, being able to dissent was important to the process. As long as participants were given the opportunity to present their own points of view, they were often willing to accept the majority opinion or approach. However, where dissent pointed to changes in methods or institutions that would improve the credibility or usefulness of the assessment, the changes were often made. Thus, the institutions and methods employed in both contexts evolved over time through the constructive expression of dissent.

In both contexts, dissent was sometimes expressed by organizations outside the process. While internal expressions of dissent sometimes led to constructive change, the participants in both assessment processes mostly ignored external expressions of dissent. This disregard for outside criticism stemmed from participants' perspective that individuals who had not participated in the assessment process but offered critiques of the analysis conducted in the assessment were unlikely to have the same understanding of the context in which the participants made analytical choices. If they were not aware of the choices available to the participants, they were likely to come to different conclusions. Furthermore, individuals who had not participated in the process may not have had the level of familiarity with the models, their limitations, or their utility within the context of the specific decisionmaking process. Therefore, it is not surprising that nonparticipants may have believed that the analysis produced in an assessment was less credible or useful than did those who participated in the process. It is interesting to note that such external dissent is more likely to be expressed in newspapers and published reports, whereas internal dissent is expressed in group discussions and presentations that are generally poorly documented for later analysis.

It is through participation in the assessment process that trust is developed and that dissent is expressed and resolved. It is through the development of trust and the resolution of dissent that the analysis conducted in an assessment comes to be seen as credible and legitimate, which helps create shifts in issue domains. Therefore, engaging the appropriate experts, stakeholders, and decisionmakers, and designing institutions for constructive interaction are keys to the design of successful assessment processes.

References

Alcamo, J., R. Shaw, and L. Hordijk. 1990. *The Rains Model of Acidification: Science and Strategies in Europe*. Dordrecht, Netherlands: Kluwer Academic Publishers.

ApSimon, H.M., R.F. Warren, and J.J.N. Wilson. 1994. The Abatement Strategies Assessment Model—ASAM: Applications to Reductions of Sulphur Dioxide Emissions across Europe. *Atmospheric Environment* 28(4): 649–663.

Boehmer-Christiansen, S., and J. Skea. 1991. *Acid Politics: Environmental and Energy Policies in Britain and Germany*. New York: Belhaven Press.

Colburn, K.A. 1997. Memorandum on OTAG recommendations from K.A. Colburn, director, New Hampshire Department of Environmental Protection. June 30.

DiIulio, J., and D. Kettl. 1995. *Fine Print: The Contract with America, Devolution, and the Administrative Realities of American Federalism*. Washington, DC: Brookings Institution.

Eckley, N. 2002. Dependable Dynamism: Lessons for Designing Scientific Assessment Processes in Consensus Negotiations. *Global Environmental Change—Human and Policy Dimensions* 12(1): 15–23.

Eliassen, A., and J. Bartnicki. 1998. Personal communication between A. Eliassen and J. Bartnicki, Deputy Director General and Senior Scientist, Norwegian Meteorological Institute, Oslo, and T.J. Keating. March 11.

Farrell, A.E., R. Carter, and R. Raufer. 1999. The NO$_x$ Budget: Costs, Emissions, and Implementation Issues. *Resource and Energy Economics* 21(2): 103–124.

Gade, M. 1995. Correspondence to ECOS members proposing OTAG from M. Gade, Illinois Environmental Protection Agency, Springfield, IL. April 13.

Gillespie, E., and B. Schellhas (eds.). 1994. *Contract with America: The Bold Plan by Representative Newt Gingrich, Representative Dick Armey and the House Republicans to Change the Nation*. New York: Times Books.

Grennfelt, P., O. Hov, and R. Derwent. 1994. 2nd Generation Abatement Strategies For NO$_x$, NH3, SO2 and VOCS. *Ambio* 23(7): 425–433.

Guinnup, D., and B. Collom. 1997. Telling the OTAG Story with Data: Executive Summary of the OTAG Air Quality Analysis Workgroup, (on file with the author). Produced for the OTAG Air Quality Analysis Workgroup by Washington University Center for Air Pollution Impact and Trend Analysis, St. Louis.

Hawkins, D. 1996. Correspondence to Merrylin Zaw-Mon from D. Hawkins, Natural Resources Defense Council, Washington, DC. Sept. 23.

Heyes, C., W. Schopp, M. Amman, and S. Unger. 1995. *Towards a Simplified Model to Describe Ozone Formation in Europe*. Laxenburg, Austria: International Institute of Applied Systems Analysis.

Heyes, C., W. Schoepp, M. Amann, I. Bertok, J. Cofala, F. Gyarfas, Z. Klimont, M. Makowski, and S. Shibayev. 1997. *A Model for Optimizing Strategies for Controlling Ground-Level Ozone in*

Europe (interim report). Laxenburg, Austria: International Institute for Applied Systems Analysis.

Hordijk, L. 1991. Use of the RAINS Model in Acid Rain Negotiations in Europe. *Environmental Science and Technology* 25: 596–603.

Hov, O. 1998. Personal communication with T.J. Keating. March 12.

ICF-Kaiser. 1996. OTAG Trading Analysis with EPA/IPM: Policy Case Paper, U.S. Environmental Protection Agency. Special report. Fairfax, VA: ICF-Kaiser.

John, D. 1994. *Civic Environmentalism: Alternatives to Regulation in States and Communities.* Washington, DC: Congressional Quarterly Press.

Keating, T.J., and A. Farrell. 1999. Transboundary Environmental Assessment: Lessons from the Ozone Transport Assessment Group. Knoxville, TN: National Center for Environmental Decision-Making Research, 210.

Koerber, M., and J. Tikvart. 1997. Memorandum: Maverick Modeling. Memorandum to Regional and Urban Scale Modeling Workgroup and Implementation Strategies and Issues Workgroup, from M. Koerber and J. Tikvart, Environmental Council of the States, Washington, DC. February 4.

Lester, J.P. 1994. A New Federalism? Environmental Policy in the States. In *Environmental Policy in the 1990s: Toward a New Agenda,* edited by N.J. Vig and M.E. Kraft. Washington, DC: Congressional Quarterly Press, 51–68.

Levy, M.A. 1994. European Acid Rain: The Power of Tote-Board Diplomacy. In *Institutions for the Earth: Sources of Effective International Environmental Protection,* edited by P.M. Haas, R.O. Keohane, and M.A. Levy. Cambridge, MA: MIT Press, 75–132.

Metcalfe, S.E., J.D. Whyatt, R.G. Derwent. 1997. Multi-Pollutant Modelling and the Critical Loads Approach for Nitrogen. *Atmospheric Environment* 32(3): 401–408.

Nichols, M. 1995. Memorandum: Ozone Attainment Demonstrations. Memorandum to Regional Administrator, U.S. Environmental Protection Agency Regions I–IX, from M. Nichols, Office of Air and Radiation, U.S. Environmental Protection Agency, Washington, DC. March 2.

OECD (Organisation for Economic Co-operation and Development). 1968. Air Management Research Group: Decisions and Conclusions of the 1st Session, Paris, 2–4 October 1968. Paris, OECD Directorate for Scientific Affairs.

OTAG (Ozone Transport Assessment Group). 1998. OTAG Technical Supporting Document. Washington, DC: U.S. Environmental Protection Agency.

Portney, P.R., and R.N. Stavins (eds.). 2000. *Public Policies for Environmental Protection.* Washington, DC: Resources for the Future.

Posch, M., J.P. Hettelingh, P. DeSmet, and R. Downing (eds.). 1997. *Calculation and Mapping of Critical Thresholds in Europe: Status Report 1997.* Bilthoven, Netherlands: National Institute for Public Health and the Environment, LRTAP Coordination Center for Effects.

Stockholm Environment Institute. 1991. *An Outline of the Stockholm Environment Institute's Coordinated Abatement Strategy Model.* York, United Kingdom: University of York.

Tuinstra, W., L. Hordijk, and M. Amman. 1999. Using Computer Models in International Negotiations: Acidification in Europe. *Environment* 41(9): 32–42.

Tuovinen, J.P., K. Barrett, and H. Styve. 1994. *Transboundary Acidifying Pollution in Europe: Calculated Fields and Budgets 1985–93.* Oslo, Norway: Norwegian Meteorological Institute.

U.S. EPA (Environmental Protection Agency). 1998. Finding of Significant Contribution and Rulemaking for Certain States in the OTAG Region for Purposes of Reducing Regional Transport of Ozone. Rule 63: 57356.

Wettestad, J. 1997. Acid Lessons? LRTAP Implementation and Effectiveness. *Global Environmental Change* 7(3): 235–249.

CHAPTER 4

Applying Assessment Lessons to New Challenges

From Sulfur to POPs

Noelle Eckley Selin

ASSESSMENT DESIGNERS SEEKING to make assessments more effective often look to what has worked before for ideas and models. Previous experience in conducting scientific assessments of environmental issues is a source of many lessons—both positive and negative—about how to construct and manage these processes. For example, participants in international scientific assessments often view the Montreal Protocol on Substances that Deplete the Ozone Layer (1987) as having a particularly successful scientific assessment process, and therefore it is often the source of assessment-related lessons. However, in the midst of designing assessments, those drawing lessons often are not able to analyze or judge whether lessons from assessments conducted in other situations on other issues are truly applicable to the context and issues at hand. It may be possible that case-specific factors dominate, such as exactly who led the assessment or the specific scientific issues in question. Therefore, these designers risk drawing the wrong lessons—by incorrectly assuming that what was effective elsewhere will be effective in their situation.

Clearly, a process that adopted *in toto* all of the institutions and processes of scientific advice in another issue area, without regard to the unique context of the issue at hand, would compromise its ability to incorporate expert advice effectively. In contrast, it is equally improbable that each issue must develop such processes in isolation, without the benefit of previous experiences, and disregard as issue-specific all qualities that might have influenced effectiveness in other forums. Identifying those lessons that are robust across issue areas—those qualities of assessment processes that contribute to credibility, salience, and legitimacy in a variety of different settings—is a central challenge of learning from experience and of learning how to design effective assessments in the future.

Little research is available to help assessors and policymakers analyze the applicability of such lessons. Because the participants, interests, and issues vary so dramatically over different environmental issues, it is difficult to find examples of processes that are comparable across different issues, particularly in an international context. Even if assessors and policymakers initially borrowed elements from other issue areas, the dynamics of international negotiations virtually assure that this process would be changed significantly by the time it is implemented.

In a recent example, however, international negotiations have addressed quite different environmental issues in a framework that was developed around very different sorts of issues. This is the case of two recent protocols to the Convention on Long-Range Transboundary Air Pollution (LRTAP): the 1998 Aarhus Protocols on Persistent Organic Pollutants (POPs) and Heavy Metals.[1] Both of these protocols addressed environmental issues quite different from the transboundary acidification issues that the LRTAP convention had addressed in the past. Most participants consider the earlier acidification-related assessments and protocols to have been effective in various ways, leading to the question of whether these processes and institutions were similarly effective when applied to an issue very different from that for which they were designed. If so, what sorts of assessment design features tend to contribute to effectiveness across these very different issues? This chapter compares the scientific assessment processes that informed three of the protocols to the LRTAP convention: the 1985 and 1994 protocols on sulfur emissions, and the 1998 protocol on POPs, asking whether the assessment processes were effective and what lessons can be drawn from this experience about the effectiveness of assessment across different issues (see Box 1-2 in Chapter 1 on the structure of LRTAP; for further background on LRTAP, see Chapters 2 and 3). It will do so by investigating whether they succeeded in being credible, salient, and legitimate to all participants, examining specifically the design choices surrounding participation and the interface between scientists and negotiators.

Data for this comparison are drawn from personal interviews with negotiators and scientists involved in the assessment and negotiating processes for the LRTAP protocols considered. In addition, primary documentation including assessment reports, meeting documents, and written communications among parties was analyzed.

Can Assessment Designers Learn from LRTAP?

Whether those designing assessment processes might draw positive lessons from the LRTAP experience depends upon whether these assessment processes were actually effective—were they credible, salient, and legitimate to policymakers? Even a brief survey of LRTAP indicates that both policymakers and analysts see the convention as a particularly successful endeavor in providing scientific advice to policymaking.

LRTAP is often cited as a model of effective science–policy collaboration. In his description of the convention, Levy (1995) writes, "The LRTAP process integrated knowledge-building processes artfully with the task of negotiating international regulations." Wettestad (1997) writes that LRTAP's "diverse scientific–political complex, with the well-functioning [European Monitoring and Evaluation Programme (EMEP)] monitoring system as a solid core, provided information crucial to the progress of the process, especially in the early 1980s." Those who have participated in the process of LRTAP's assessment and negotiations cite science (that is, scientific data, information, and assessments) as a strong basis for their work, and they deem LRTAP's assessment one of the more effective they have been involved in. One negotiator said in an interview about the LRTAP process: "Over the 20 years that the convention has existed, it has built up quite a network and support system to develop good scientific work."

In negotiations over the 1985 sulfur protocol, for example, scientific assessment facilitated establishing a credible and legitimate scientific basis for negotiating an international agreement on transboundary pollution. The 1985 sulfur protocol committed its parties to a 30 percent reduction in sulfur emissions (from 1980 levels) by 1993; this policy decision was informed by an assessment process that not only established acidification as a problem, but also used modeling to establish conclusive links between sources of pollution and locations of deposition. Several delegates cited a role for science in getting the issue on the agenda, and identifying the sources of pollution. "When this convention came about, certainly it was scientific findings that were in the bottom," said one negotiator. "It was a very long fight to get acceptance that emissions . . . had a spread that went hundreds and perhaps even thousands of miles, and in the European context at least that meant it was transboundary." Another negotiator noted "the science came in acidification to prove that it was in a sense the United Kingdom and Poland that were sending their dirt to Sweden." While some countries questioned the credibility of the science (and this stalled the negotiations), assessment clearly played a role in setting a basis for further action, and it was effective enough for many parties to act upon.

Negotiations over the 1994 sulfur protocol were based on the concept of "critical loads" and an effects-based approach to management—that is, emissions reductions are based on cuts relative to thresholds for the effects of environmental pollutants (McCormick 1998). The critical loads approach represented an effort to incorporate assessment of the impacts of pollutants into decisionmaking on protocols. This was new: the 1985 protocol focused almost entirely on the concept of limiting emissions without dealing in a detailed way with the issue of impacts. The idea of critical loads grew out of scientific work by affected countries, particularly Sweden. During negotiations, delegates made use of assessment tools such as integrated assessment models, the preeminent one being the Regional Acidification Information and Simulation (RAINS) model developed at the International Institute for Applied Systems Analysis (IIASA; Hordijk 1991). The critical loads were set at a level that would protect the most sensitive 5 percent of ecosystems; delegates negotiated the degree to which parties would reduce emissions relative to the critical loads—these were termed "critical targets." The resulting protocol aimed to achieve a "60 percent gap closure" between sulfur deposition and critical loads in Europe (Thommessen 1997). In this way, a political compromise—the idea of the critical target and the 60 percent figure—was negotiated relative to a product of scientific assessment (the critical load). Negotiators from at least most European parties accepted the critical loads concept and saw the results of the integrated assessment models as particularly credible. The models, as well, were extremely salient—parties were able to evaluate scenarios that calculated the costs of remedial action on sulfur and that measured the effects of such action on levels of sulfur relative to critical loads. One of the factors that contributed to the RAINS model's salience was its interactive nature: negotiators used these models directly to predict the outcomes of various levels of proposed regulatory action.

The negotiation over a protocol on persistent organic pollutants, as well as one on heavy metals, represented a departure for LRTAP from traditional air pollution issues (Selin and Eckley 2003). POPs are compounds whose properties of high toxicity, persistence in the environment, and bioaccumulation in living organisms make them a risk to the environment and human health far from the locations of their use and emission (Eckley 2001). These protocols addressed a problem of very different qualities from sulfur and nitrogen, and even from volatile organics. In contrast to sulfur, which can be traced using relatively straightforward atmospheric models from emission to deposition, POPs can revolatilize from environmental reservoirs. Therefore, definitive links between the sources of POPs and the receptors suffering effects are very difficult to establish. In addition, where the effects of acidification are relatively observable in the environment, POPs exert often-unseen toxic effects on the environment and human health. And where previous LRTAP protocols addressed unwanted byproducts of industrial processes, several POPs are commercially produced chemicals—such as pesticides—that were designed and produced for the very properties that make them of environmental concern.

During negotiations over the 1998 POPs protocol, negotiators' views of the effectiveness of assessment were similarly positive, although the type of scientific assessment that informed the POPs negotiations was quite distinct from the role of assessment in the sulfur protocols. Whereas the sulfur protocols were built from modeling, and eventually a detailed integrated assessment, scientific assessment of POPs took the form of "state-of-knowledge" reports (i.e., assessments describing the latest science on the issue) and technical data on substance properties. Another significant difference is that in the case of POPs, assessments were essentially national in character, rather than multinational. This issue is discussed in depth later in the paper.

The scientists and policymakers associated with LRTAP began to consider POPs in 1989, when the convention's Executive Body charged a group of experts to prepare a discussion paper on POPs. A task force on POPs, set up by the Executive Body, met four times between 1991 and 1994, and it produced a substantiation report—an assessment report outlining the state of scientific knowledge on POPs emissions, transport, impacts, and abatement techniques. The main differences between the process of scientific assessment in the POPs case and in the sulfur case were the emphasis in the POPs case on basic science, and the procedure by which scientific assessment work occurred in the context of the convention before negotiations on a protocol.

Similar to the first sulfur protocol, scientific assessment of POPs did help to put the issue onto the international agenda. A North American negotiator said of the protocol "in a very real sense, science was the basis and proved to us the need for us to go out and have a protocol for this." One of the leaders of the early scientific work spoke of its credibility: "Most of the conclusions we made . . . were accepted by a broader public, basically because it was founded on the scientific background." Assessment of POPs was salient, especially in the establishment of criteria for selecting which substances would be addressed by the protocol, as well as in deciding upon a procedure for adding substances to the protocol via amendments in the future. Substances were evaluated and ranked using numerical criteria, which took into account their potential for long-range transport, bioaccumulation, and toxicity (Rodan et al. 1999). Although some of the data that went into substance evaluation were questioned on the basis of their credibility (as will be explored further below), on the whole, the concept of numerical criteria—which emerged from a scientific assessment process and was based on scientific data—was both salient and legitimate to most parties. The parties agreed that the criteria were a useful and proper way to evaluate these substances, although they differed on the specific cutoff points and the extent they viewed these criteria as determinative.

Even a cursory examination of these three protocols seems to indicate that there is indeed something to learn from LRTAP's assessment process, and that the process seems to work across the very different issues of sulfur and POPs. These ideas form the basis for a more detailed examination of the assessment processes involved. What elements of LRTAP's assessment processes contributed to negotiators finding these very different assessments credible, salient, and/or legitimate?

Lessons for Design

The source of lessons for assessment designers from the LRTAP protocols' assessment processes comes in identifying the design elements that contributed to these assessments' credibility, salience, and legitimacy across different issue areas. Two categories of design choices seem to have been especially influential across the LRTAP sulfur and POPs protocols in promoting assessment effectiveness: participation and the structure of the science–policy interface. Participation was important, but not simply by ensuring that it was "international"—in fact, results of this comparison show that fully institutionalized international assessments are not always necessary for effective assessment. The structure of the science–policy interface was unique and very influential in the structure of the communication process and in the dynamism or adaptability of the negotiations with respect to changing science.

Participation

Interviews with participants in the LRTAP process strongly indicate that international participation—the representation of all parties in relevant assessments—was seen as a key influence on the credibility and legitimacy of those scientific assessments during negotiations. In general, participants espoused the view that the more international the process of scientific assessment, the more likely it was to be accepted. One participant stated this view in the case of evaluating substances in the POPs protocol, positing that a single country performing a scientific assessment runs the risk of having the assessment's conclusions challenged based on the country's political positions, and thus appears less credible. The argument can be made that EMEP's (and therefore LRTAP's) characteristic as an international institution and not a domestic program helped its conclusions to be widely accepted, and therefore the formal internationalization of science had a significant influence on the policy process.

However, the POPs experience shows that citing "international science" as the key influence on the LRTAP assessments' effectiveness is likely too simplistic an explanation. The POPs protocol assessment experience shows that completely international assessment processes are not a necessary prerequisite for an assessment to be credible and legitimate. The issue of participation is significantly more complex than an initial interpretation of LRTAP would lead one to believe. Particularly, the initial scientific work in the POPs protocol negotiations was almost exclusively national (much of the initial scientific work was performed by Canada). Although the task force was international, a small subset of LRTAP countries participated on a voluntary basis, and the process was pushed largely by the influence of Sweden and Canada—who had particular concerns about POPs chemicals due to their presence in arctic ecosystems.

In interviews, one delegate was explicit about the differences between the scientific process for POPs work and earlier protocols: "In acidification we have institutionalized the science. We have EMEP, we have the effects community doing their programs, and they make their findings public, and they have reports, they have meetings, etc. In developing the POPs protocol, you have lead countries doing certain aspects of the work. . . . And this is different science than we are at the moment using in acidification." The results of lead country assessment in the POPs protocol assessment process were brought in to the negotiations through international task forces and working groups; however, the assessments themselves often remained unchanged. However, the science put forth by the task force was seen as credible and legitimate even by parties who did not participate in the assessment activity.

A closer look at the sulfur protocols, particularly the first sulfur protocol, shows that the use of national assessments in this way is not unprecedented in the LRTAP convention. Indeed, although well-institutionalized assessments were established early on in the sulfur assessment process, many assessments done on acidification issues were national (Cowling 1982). Many of these early national assessments contributed to the early conceptualizations of acidification as a problem. However, the POPs protocol assessment process shares similar characteristics with the processes in the sulfur protocols, at least the trappings of internationalization. For example, national assessments entered the process largely under international cover (at least those assessments that sought to form a basis for further decisionmaking). While "national" science portrayed as such is suspect on the basis of its legitimacy as well as its credibility, sufficient vetting of national assessments through international institutions can provide the necessary legitimacy and credibility upon which further decisionmaking can be based.

Science–Policy Interface

A peculiar characteristic of the interactions between scientists and policymakers in the LRTAP process is the closeness (in some cases, overlap) between the two categories. This unique construction of the science–policy interface, common among the sulfur and POPs protocols, seemed to contribute to effectiveness across these different issues.

It is often difficult to tell in the LRTAP assessment process who is participating as a negotiator and who is considered a scientific expert. Often, participants play more than one role. One participant identified himself in an interview as a "scientific advisor/delegate." Negotiators have also been looked to as scientific interpreters in negotiations—one participant was asked to answer several scientific questions during the final stages of the POPs negotiation. A comment by a LRTAP negotiator spoke to the way in which science is viewed: he noted that from a policy perspective, science is not so definitive that "policy people ought not to mess" with it. In his view, therefore, policymakers need not sit back and allow scientific information to dictate conclusions; policymakers can and should shape the direction of assessment activity.

Two elements of the structure of the science–policy interface in LRTAP assessments seem to have had particularly significant influences on the assessments' effectiveness across the sulfur and POPs cases: the iterative communication processes between scientists and negotiators, and the dynamism or adaptability inherent in the structure of the negotiations and LRTAP's scientific assessment processes.

Iterative Communication One of the hallmarks of the LRTAP assessment process is its iterative communication process. A senior negotiator said that the LRTAP convention offers a chance for policy advisers to participate in development of science both in preparatory work and in negotiations, and also for scientists to gain understanding of the political process. He characterized this interactivity as the *modus vivendi* of the convention. The repeated communication and interaction between technical experts and policymakers influenced effectiveness through increasing the assessment's credibility, legitimacy, and salience. By providing intermediate corrections and reality checks to ongoing assessment processes, communication between assessors and policymakers ensured that the assessment continued to be salient and legitimate to its targeted audience. By giving policymakers the opportunity to learn about assessment step-by-step, repeated communication increased the assessment's credibility to them. Iterative communication also helped to ensure that assessors and policymakers shared realistic expectations about what the assessment could provide, reinforcing the resulting assessment's credibility and salience to its users.

Negotiators had much to say about the communication process by which policymakers learned the technical details of particular issues. The early assessment work of the POPs protocol was marked by repeated communication between the LRTAP Working Group on Effects (WGE) and the assessment body of the working group's task force. A senior member of the WGE said of this process in an interview that the WGE was used to slow down the process of deciding what to do about POPs. Repeated communication between the task force and WGE was used to help parties understand the need for the LRTAP convention—widely considered an acidification convention—to address the POPs issue. The senior WGE participant credited this communication process, particularly the time element, as an important factor in influencing the credibility and authority of the science behind the POPs negotiations; through this process, communication influenced the broad feeling of consensus that emerged. The repetition of the WGE activities, therefore, served to legitimize the scientific assessment work that was occurring on a lead-country basis within the task force, as well as increase its salience to the tasks at hand.

To look particularly at the influence of this communication process, one might ask what would have happened had the repeated communication between the WGE and the task force not been able to occur. There would have been little opportunity for an LRTAP institution to evaluate the work of the task force before it issued its final report. Given the skepticism of some involved in the LRTAP context regarding what they felt was an acidification convention addressing issues outside its purview, the WGE would not have been able to delay action and get used to the idea. Therefore, the assessment might have suffered from a lack of legitimacy. Alternatively, if the issue did somehow succeed in getting past these obstacles and onto the LRTAP decisionmaking agenda, weaknesses in an assessment report's policy salience might have opened it up to criticism from policymakers. However, the communication between the task force and the WGE in early phases gave those preparing the report a systematic way to find out what policymakers needed more information on, and the report was then less susceptible to such criticism later.

Similar processes facilitated effective outcomes in the cases of the sulfur protocols. In the case of the early LRTAP scientific consensus, early assessments such as the Organization for Economic Cooperation and Development's work of the 1970s were scientific enterprises conducted in a political forum. The degree of consensus was therefore reinforced by political activities in communication with scientific ones. Levy (1995) cites LRTAP's "collaborative science" as important for "advancing the state of consensual knowledge" of acidification. He notes that the WGE and EMEP oversaw collaborative research. In the case of the second sulfur protocol, Alcamo, Shaw, and Hordijk (1990) note that a primary guideline for development of the RAINS model was its formulation as a collaborative effort by analysts, experts, and potential users. A scientific participant confirmed that during the ongoing process of sulfur protocol negotiations, more policymakers came to understand the details of the model.

In all of these cases, communication was most effective when it allowed policymakers and scientists repeatedly to shape the construction of assessments in ways that made them more legitimate, credible, and salient. Repeated policy guidance about what scientific information was needed and could be provided has promoted effectiveness in the case of LRTAP protocols. Such iterative communication ensured that the results of the assessment process were salient to policy by influencing the ways in which scientific information was conceptualized. Through these influences, communication was able to facilitate the agreements on science that set a basis for further work.

Ensuring Realistic Expectations of Science Another way in which iterative communication facilitated the effectiveness of LRTAP's assessment processes was in giving negotiators realistic expectations about the information and recommendations that could be provided by scientists. LRTAP assessment processes have promoted effectiveness by not asking too much of science; that is, negotiators did not demand more of science than science could provide. Ensuring that policymakers have realistic expectations about what assessment can provide can increase the resulting assessment's credibility and salience to them.

Demanding too much of science—or, not recognizing the limitations of current scientific knowledge—is a well-recognized pitfall of assessment. In many cases, if policymakers had pushed the science to make conclusions prematurely, the effectiveness of the assessment process would have suffered. In the LRTAP protocols specifically, had policymakers pushed science further than scientific principles would allow, the credibility of science in the process would have suffered, and the assessments would have been less effective. Policymakers, indeed, were conscious of the limitations of science: they did not look to science to answer questions that would have entered areas of extreme uncertainty. For example, had policymakers insisted on an effects-based analysis of sulfur at the time of the 1985 sulfur protocol, the science added to the process would have been extremely uncertain: the state of scientific knowledge would not have been seen as credible by the policymakers. Participants' consciousness of the limitations of science was created and reinforced by LRTAP's iterative communication process between scientists and decisionmakers.

In the POPs case, there were two areas in which policymakers stopped short of asking science to address questions that it could not answer well—in conducting substance assessment and selection, and in determining whether there would be critical loads for POPs. In selecting substances, policymakers did not push science to make assessments of substances for which sufficient data were not available. One delegate noted that the scientific assessment process simply left out substances where data were lacking. This prevented uncertainties about marginal substances from becoming an unwarranted focus of negotiators' concern. During the POPs negotiations, policymakers and scientists also discussed whether critical loads were a viable concept to use in addressing POP substances. Policymakers were realistic in their acknowledgment that the state of the science would not support a critical loads analysis of POPs—and indeed that the POPs problem may always be incompatible with a critical loads-type approach.[2]

An example of how communication facilitated the setting of realistic expectations for critical loads analysis can be seen in the communications between the WGE, the Working Group on Strategies (WGS), and the assessment activity on critical loads for POPs. The WGE and the WGS encouraged the exploration of methodologies for effects-based approaches for POPs. A scientific workshop held in November 1997 on critical limits for heavy metals and POPs, jointly sponsored by the Netherlands and Germany, explored this issue; it recommended that further negotiations draw a clear distinction between risk assessment procedures and critical loads approaches that included mapping. The risk assessment procedures were recommended for POPs. In light of that scientific assessment, there was a level of agreement among negotiators not to push science toward a critical loads approach, and there was a realization of the limitations of such analysis for these substances. Had negotiators pushed scientists too hard on answering questions based on an inappropriate analytical framework, the credibility of scientific research in those areas would have suffered. The ability of policymakers to communicate and establish reasonable expectations of science knowledge, therefore, contributed to the effectiveness of assessment activity, through ensuring its credibility and salience.

Dynamism in Assessments and Policy Processes In analysis of the factors that contribute to the effectiveness of LRTAP's assessment processes, the dynamism seen in the assessment and policy (or negotiating) processes stands out. LRTAP's assessment and policy processes have responded quite actively to changes in science and to modifying science-based conclusions at later times.

Assessment processes can be designed as more or less adaptable or dynamic. For example, they can range from virtually inadaptable or static (in which all decisions are final) to completely adaptable (in which any and all decisions could be changed at any time). Either of these extremes would likely signal an ineffective process: in the former case, changes in information or preferences cannot be taken into account; and in the latter case, policies have no measure of predictability, making rational efforts for compliance nearly impossible.

The LRTAP convention, however, is an instrument that seems to strike an effective balance between shortsighted rigidity and unpredictable fluidity. The convention itself is an instrument that has changed repeatedly over its 20-year history by the addition of successive, substantive protocols. In its assessment processes, new scientific information can be incorporated into policy negotiations.

All three protocols examined in this chapter shared a sense of dynamism in their assessment processes. Countries pushing for action to reduce sulfur emissions viewed the 1985 sulfur protocol as a first step, to be followed by further reductions within the framework of the convention. The second sulfur protocol occurred within a tradition of iterative protocol negotiations; it was a revisitation of the sulfur issue with additional science and regulatory measures. In the context of the LRTAP convention, delegates often refer to the first sulfur protocol (as well as many other first protocols on particular substances) as a "first-generation" protocol. The second sulfur protocol is referred to as a "second-generation" protocol. The structure of the convention itself, the repetitive addressing of particular issues within it, and even the terminology of generations, shows that adaptability, or dynamism, is an integral part of the LRTAP convention. LRTAP's POPs protocol was envisioned and designed as a dynamic instrument, and it embodied many of the convention's ideals of adaptiveness. Although the protocol regulates a list of 16 POPs,[3] LRTAP's Executive Body has set out criteria and a procedure for adding new substances to the agreement (UNECE 1998). The decision to incorporate this dynamism into the POPs protocol was made early in the process, and it means that further substances can be regulated in the future without the negotiation of an entirely new protocol.

The particular type of adaptation that contributed to effectiveness in the LRTAP assessment processes was a quality of "dependable dynamism"—the ability of the process to postpone decisions that relied on controversial science, with confidence that they would indeed eventually be addressed (Eckley 2002). Such difficult decisions could be informed later by further research and additional scientific information. This sort of process allowed parties to facilitate decisionmaking and consensus-building in two ways. First, the fact that parties believed a decision taken was not necessarily the final airing of a particular issue meant they were more willing to base current actions on information they considered less credible, with the knowledge that future action would be based on additional information. Second, dependable dynamism allowed parties to make compromises along a temporal dimension—that is, to agree to keep options open for future revisions, in return for negotiating tradeoffs. These processes facilitated decisionmaking in all three LRTAP protocols examined in this chapter.

In the first sulfur protocol, negotiations were largely based on political considerations rather than on scientific analysis. Negotiations aimed to set a sulfur reduction target to which all parties would agree. The decision to choose a 30 percent reduction was not based on science—it was based on political negotiations, and the negotiations stalled when some parties used the lack of science as justification for arguing against further action. Björkbom (1997) notes that the United States and the United Kingdom cited inconclusive scientific evidence as reason for not joining the so-called "30-percent club" of countries that had agreed to the emissions cut. However, the assumption that the sulfur issue would be revisited ultimately allowed parties to view the 30 percent reduction goal as a first step. One delegate observed that the countries opposed to regulation "were citing an unsatisfactory scientific base for such a thing, and of course they were right, but the function was to simply start the ball along." The science, in the delegate's opinion, did not have to be completely determinative because the process could, and would be, revisited later. Therefore, dynamism contributed to the effectiveness of the assessment process by lowering the threshold of scientific credibility seen as necessary for further action.

In the case of the POPs protocol, a similar mechanism was at work when delegates were making final decisions about which substances would be included in the final protocol list and which would be left out. The initial list of substances for negotiation was developed through a screening process conducted during the early assessment work under the LRTAP protocol. The initial list of substances selected for inclusion, identified by the screening process set up by the initial assessment task force and the subsequent Preparatory Working Group, incorporated a clear methodology evaluating long-range atmospheric potential, environmental persistence, bioaccumulation potential, and toxicity for more than 100 substances. The Preparatory Working Group presented 14 substances to negotiators for inclusion on the initial list, and the group recommended further evaluation and sought policy guidance for 6 others. The establishment of selection criteria for persistence, bioaccumulation, toxicity, and long-range transport put forth a clear framework by which substances were ranked and ordered. Although this ranking was certainly not purely scientific, parties generally accepted the concept that a substance's inclusion should be based at least partly on scientific properties and not solely on political tradeoffs.

The most illuminating example of the effect of a confidently dynamic assessment process on influencing the progress of the negotiations occurred in the case of pentachlorophenol, one of the 14 substances recommended for inclusion by the Preparatory Working Group. Every delegate interviewed about the role that scientific assessment played in the progress of POPs protocol negotiations mentioned the debates over pentachlorophenol as an example in which science influenced policy. Pentachlorophenol, a wood preservative, is widely regulated in Europe. Several European countries considered the preservative to be a substance that should definitely have been included in the protocol. Late in the negotiations of the protocol, however, the United States, prompted by new data from industry, reexamined the initial data that had prompted pentachlorophenol's inclusion.[4] The United States asserted that the new data indicated that pentachlorophenol did not satisfy the criteria for inclusion. At a meeting of the WGS in September 1997, technical experts from the U.S. Environmental Protection Agency and the Chemical Manufacturers Association presented the new industry data on pentachlorophenol and their interpretation that these data did not support its inclusion in the protocol. The industry science presented by the United States was greeted with skepticism by Europeans. A senior member of the U.S. delegation observed that when the United States raised questions about pentachlorophenol, basically all European countries "laughed at us," because the substance was so heavily regulated in Europe. Asked how credible he found the information presented by the United States, a senior delegate from Sweden said: "We had some problems with that. I mean, our sincere opinion was that this substance should be in the protocol." He noted, however, that the United States had indicated that it would not sign the protocol if pentachlorophenol were included; the inclusion of the United States in the agreement was seen as critical by certain key players in the negotiations, including Sweden. U.S. participation was viewed as crucial in part because the LRTAP POPs protocol was seen as setting a precedent for negotiation of a global POPs agreement, which began in 1998 and was completed in 2000.[5]

At the end of the debate, pentachlorophenol was left off the initial list of substances regulated under the POPs protocol. A key factor in negotiations that led to this decision was that the POPs protocol was envisioned and designed as a dynamic instrument in the context of the LRTAP convention. It was understood among negotiators that a substance left off the protocol in its first incarnation could later be added via a negotiated procedure relying on both scientific criteria and political decisionmaking. Because the protocol was dependably dynamic, a decision not to include a substance on the initial list did not mean that the substance would never be addressed. For this reason, countries favoring pentachlorophenol's inclusion could agree to put off a decision on the substance until further scientific work could be performed and more evidence could be employed, without agreeing that pentachlorophenol would never be regulated internationally. In fact, the issue of pentachlorophenol was addressed in the section of the protocol dealing with research, development, and monitoring. Parties agreed to encourage research on "levels of persistent organic pollutants generated during the life cycle of timber treated with pentachlorophenol." The inclusion of pentachlorophenol in the research section of the protocol was an additional way that science helped to encourage countries' agreement.

In the case of the POPs protocol, the option to put off a decision lowered the barrier to countries accepting U.S. data. One delegate mentioned that "we decided rather early that we tried to . . . find a solution, tried to find a compromise and a protocol, and then anticipate that there would be a second step." A

Swedish delegate made it clear that the adaptability of the protocol factored into his country's decision to forego action on certain controversial substances. After mentioning his problems with the science and with the political ramifications of the United States not signing the protocol with pentachlorophenol included, he added:

> At the same time, we said, let's go and look upon this once more after the protocol. . . . not only pentachlorophenol, but also other substances such as chlorinated paraffins that we wanted in. Further substances. So now the next step is once more to go back to science and really look if there are loopholes that we have to some way or other try to close. If necessary, bring in further research to assess what's been going on.

The option to put off a policy decision on certain substances facilitated the use of U.S. science as sufficiently credible for moving the policy process along, and it thereby contributed to scientific effectiveness in the POPs case. If delegates had not had this option, the science put forth by the United States would likely have been questioned more rigorously by countries favoring regulation, because not addressing pentachlorophenol would have been a final decision; this would have stalled the negotiating process further.

Offering a Temporal Dimension for Compromises Dynamism in LRTAP assessment processes offered a way for compromises to be made along another dimension—that is, it added a temporal dimension for compromise on both science and policy options. In the second sulfur protocol, the use of the concept of critical loads—and the corresponding concept of gap closure—represented a way in which this sort of dynamism facilitated science-based decisionmaking and reinforced the adaptive nature of the agreement. Although the protocol only mandated a 60 percent gap closure, the critical load remained as a goal to be achieved eventually. By compromising on gap closure, parties were able to agree to some reductions, with the idea that the critical load was still the goal. Parties were therefore able to compromise not only on what reductions were to be undertaken at present, but also what reductions might be undertaken in the future. The critical load itself, in addition to the history of repeated protocol negotiations, gave negotiators confidence in the goal of further reduction.

The facilitation of compromise over time was particularly evident in the POPs negotiations. Many issues of contention in the POPs negotiating process, particularly between North Americans and Europeans, were based on underlying differences about the nature of precaution and unacceptable risk. Where, for example, the United States regulates substances based on calculated evidence of risk determination, Europeans tend to take a more precautionary approach (Eckley and Selin 2004). A U.S. negotiator expressed the difference between U.S. and European regulatory policy as a difference between regulations based on hazard and on risk. He defined hazard as the quality set inherent to a particular substance without regard to dosage or exposure pathways, and he defined risk as a step beyond hazard in which the substance poses actual harm to the environment or human health. These different policy approaches set up a dynamic in which one set of parties believed that science was sufficiently credible for including a certain set of substances, compared with another set of parties that believed science was too uncertain. A compromise between the two sides would most likely result in the exclusion of some substances from regulation. However, the dynamism built into the LRTAP process allowed this compromise to benefit from another dimension—countries could agree to revisit those substances at a later time, a decision that satisfied both sides. Had negotiators been forced to compromise during a process in which decisions could not be revisited, the negotiations would have faced the pitfall of possibly irresolvable debates about whether the science on these additional substances was credible enough to support inclusion on the protocol. This lack of dynamism would have stalled negotiations by promoting debates on the uncertain substances, while other, more certain substances would have remained unaddressed.

How, then, was LRTAP able to encourage dependable dynamism? One method seems to be its history of repeatedly addressing issues. In the first sulfur case, delegates had already reconvened to negotiate a substantive protocol under an existing framework agreement, which gave delegates confidence in the institutional longevity of LRTAP as a convention. The POPs protocol was negotiated against a background of a long institutional history, which included the first second-generation protocol and a second one (the multi-pollutant protocol) already in the pipeline. This history gave delegates the confidence that an issue postponed would not be permanently shelved or forgotten. A related influence has been the setup of LRTAP institutionally as a convention-protocol framework, which has lowered the presumed barriers to collective action in the future.

Conclusions

For policymakers who hope to draw lessons from previous policy experiences, determining the conditions under which a program might work effectively is crucial. The examination of scientific assessment processes in LRTAP protocols can offer just such an observation. Similar processes of institutional setup and the science–policy interface, particularly communication processes and institutional dynamism, contributed to effective outcomes in all three protocols examined in this chapter. LRTAP's assessment processes were indeed effective at addressing environmental problems very different from those for which they were designed.

In most negotiations over transborder environmental issues, policymakers seek to draw lessons both across different issue areas and across different contexts. Although the conclusions from this chapter cannot offer a concrete basis for both of these types of generalizations, it can suggest what questions policymakers might ask about assessment process design, and it can propose certain design elements to which policymakers should pay special attention. In particular, the results of this analysis suggest that policymakers should look carefully at elements of process in designing assessment. They might examine the degree to which international participation is necessary for the task at hand, pay particular attention to the structures and institutional arrangements for communication, and consider the degree of dynamism built into the assessment process. These specific factors contribute to effectiveness across very different environmental issues. The

LRTAP experience suggests three key hypotheses about the correspondence between elements of assessment design and effectiveness in general, which might serve as the basis for such lessons:

- **Fully institutionalized, international assessments are not a necessary prerequisite for a credible and legitimate assessment process.** A more international process does not always lead to a more effective outcome. The different degrees of international participation in the POPs and sulfur assessment processes resulted in similarly effective outcomes; it is possible to ensure the credibility and legitimacy of a process to a variety of participants, regardless of whether they were involved in producing the scientific data. This finding suggests that for some purposes, policymakers need not construct fully international institutions for science—assessments that seek to form a basis for future policy progress (e.g., assessments of the "state-of-the-world" sort) may be sufficiently legitimate and credible if they use national science with international cover or if they are vetted by fully representative institutions.

- **Iterative communication, in which scientists and policymakers discuss assessment activity, can increase the effectiveness of scientific advice processes.** The finding that iterative communication processes influenced effectiveness in LRTAP protocol assessment processes is significant, because it challenges a common view that science and/or scientists should always be disinterested, should not attempt to directly influence policy, and should be discouraged from direct contact with policymakers. One example of an assessment process in which scientists and scientific activities are by design distinctly separate from negotiations is the Intergovernmental Panel on Climate Change process. In the case of the three LRTAP protocols examined, exactly the opposite setup contributed to an effective outcome. Iterative communication allowed policymakers to provide intermediate corrections and reality-checks to ongoing assessment processes, and it therefore ensured that assessments continued to be salient and legitimate to their target audience. By giving policymakers the opportunity to learn about assessment step-by-step, repeated communication increased credibility of the assessment to them. Iterative communication also helped to guarantee that assessors and policymakers shared realistic expectations about what assessment could provide, ensuring that the resulting assessment would be both credible and salient to its users.

- **A more dynamic process of assessment and policy negotiation, in which every policy decision is not necessarily taken as final, can contribute to the effectiveness of assessment processes and science-based decisionmaking.** Dynamism—here defined as the ease with which policymakers may put off particular policy decisions for addressing in the future—emerged as a common element influencing effectiveness across the three LRTAP assessment processes examined. The results of this analysis suggest that when policymakers are able *with confidence* to assure those seeking further actions that issues will be addressed later, the policy process can move forward by lowering the threshold of credibility necessary for decisionmaking, and by offering a new temporal dimension of compromise on science–policy decisions. The existence of a history of repeatedly addressing issues—as seen in LRTAP's convention–protocol framework—contributes to delegates' confidence that issues put off will indeed be addressed later.

To policymakers designing assessment processes, the results of this comparison between LRTAP sulfur and POPs assessment processes can offer some suggestions for drawing particular lessons. Policymakers who are interested in improving the effectiveness of scientific advice processes, could look to some of the LRTAP design elements presented above to inform lesson-drawing. Designers might want to set up processes that encourage communication between scientists and policymakers, and they may want to design institutional frameworks that facilitate this communication. They might also consider setting up an assessment process such that science-based decisions can be confidently postponed for later iterations of assessment or negotiations. If assessment designers are to draw better lessons in the future from previous experiences, looking at LRTAP, and the hypotheses above, could be one place to begin.

Notes

1. The Aarhus protocol entered into force on October 23, 2003.

2. Because POPs are bioaccumulative, very low levels in the environment can accumulate to very significant levels in organisms. Also, POPs can revolatilize from environmental sinks and subsequently pose potential problems in other locations.

3. The 16 substances subject to the LRTAP POPs protocol are aldrin, chlordane, chlordecone, dichlorodiphenyltrichloroethane (DDT), dieldrin, endrin, heptachlor, hexabromobiphenyl, hexachlorobenzene (HCB), hexachlorocyclohexane (HCH), mirex, polychlorinated biphenyls (PCBs), toxaphene, polycyclic aromatic hydrocarbons (PAHs), dioxins, and furans.

4. Interestingly, the data in question had originally been provided by the U.S. Environmental Protection Agency.

5. The Stockholm Convention on POPs entered into force on May 17, 2004.

References

Alcamo, J., R. Shaw, and L. Hordijk (eds.). 1990. *The Rains Model of Acidification: Science and Strategies in Europe.* Dordrecht, Netherlands: Kluwer Academic Publishers.

Björkbom, L. 1997. Protection of Natural Ecosystems from Transboundary Air Pollution in Europe. Paper presented at the SCOPE UK meeting: Effective Use of the Sciences in Sustainable Land Management at the Royal Society. February 21, 1997.

Cowling, E.B. 1982. Acid Precipitation in Historical Perspective. *Environmental Science and Technology* 16: 110A.

Eckley, N. 2001. Traveling Toxics: The Science, Policy, and Management of Persistent Organic Pollutants. *Environment* 43(7): 24–36.

———. 2002. Dependable Dynamism: Lessons for Designing Scientific Assessment Processes in Consensus Negotiations. *Global Environmental Change* 12: 15–23.

Eckley, N., and H. Selin. 2004. All Talk, Little Action: Precaution and its Effects on European Chemicals Regulation. *Journal of European Public Policy* 11: 78–105.

Hordijk, L. 1991. Use of the RAINS Model in Acid Rain Negotiations in Europe. *Environmental Science and Technology* 25(4): 596–603.

Levy, M. 1995. International Cooperation to Combat Acid Rain. In *Green Globe Yearbook 1995,* edited by H.O. Bergesen. Oxford, United Kingdom: Oxford University Press, 59–68.

McCormick, J. 1998. Acid Pollution: The International Community's Continuing Struggle. *Environment* 40(3): 17.

Rodan, B.D., D.W. Pennington, N. Eckley, and R.S. Boethling. 1999. Screening for Persistent Organic Pollutants: Techniques to Provide a Scientific Basis for POPs Criteria in International Negotiations. *Environmental Science and Technology* 33(20): 3482–3488.

Selin, H. and N. Eckley. 2003. Science, Politics, and Persistent Organic Pollutants: Scientific Assessments and their Role in International Environmental Negotiations. *International Environmental Agreements: Politics, Law and Economics* 3(1): 17–42.

Thommessen, O.B. 1997. Convention on Long-Range Transboundary Air Pollution (LRTAP). In *Green Globe Yearbook 1997,* edited by O.B. Thommessen. Oxford, United Kingdom: Oxford University Press, 84–88.

UNECE (United Nations Economic Commission for Europe). 1998. Executive Body Decision 1998/2 on Information to be Submitted and the Procedure for Adding Substances to Annexes I, II or III to the Protocol on Persistent Organic Pollutants. Aarhus, Denmark: United Nations Economic Commission for Europe.

Wettestad, J. 1997. Acid Lessons? LRTAP Implementation and Effectiveness. *Global Environmental Change* 7(3): 235–249.

CHAPTER 5

Making Climate Change Impacts Meaningful

Framing, Methods, and Process in Coastal Zone and Agriculture Assessments

Marybeth Long Martello and Alastair Iles

S CIENTIFIC ASSESSMENTS ARE integral in identifying and characterizing the social and biophysical manifestations of environmental change. These analyses link European acid rain with dying forests, the ozone hole with dramatic increases in skin cancer, and desertification with African famine. Nowhere have such assessments played a more prominent role than in the area of climate change. Climate impact assessments are social processes aimed at making climate change phenomena tangible and tractable for policymakers, researchers, and the public. Participants in these processes collect, synthesize, interpret, and communicate knowledge about the likely ramifications of climate change. Impact assessments analyze how changes, for example, in temperature and elevated carbon dioxide concentrations are likely to affect matters of more direct concern to human well-being such as weather patterns, crop yields, sea level, disease vectors, and water availability.

"Climate change impact" has long been an important organizing concept in climate change research and assessment at national and international levels. In 1975, a U.S. Department of Transportation study projected changes in agriculture and natural resources associated with stratospheric flight (CIAP 1975). Four years later, the National Research Council's Climate Research Board produced a study examining the effects of climate change on temperature, precipitation, and soil moisture, while the National Defense University launched a series of reports on climate change and world food security. During the 1980s, impact studies began to proliferate (e.g., Chen 1981; NRC 1983; Schneider and Chen 1980; U.S. EPA 1983) and to examine not simply how climate change might affect crop yields or ocean levels, but how people interact with social and ecological phenomena through such responses as migration and water management (e.g., Kates et al. 1985; Parry et al. 1988). Farmers, for example, may respond to climate change through decisions about crop selection, crop location, and irrigation practices. Similarly, the threat of sea-level rise depends on how people, businesses, and governments develop coastal regions and on what defenses (e.g., building infrastructure) they undertake. More recently, the Intergovernmental Panel on Climate Change (IPCC) devoted one of its three working groups to analysis of climate change impacts (see Box 1-3 in Chapter 1 on the IPCC assessment process).

As a long-standing focus of scientific inquiry, "impact" appears, at first impression, to designate a taken-for-granted, stable concept—a class of identifiable socioecological phenomena, amenable to particular forms of measurement and analysis. However, a closer look at American and international impact assessments across time and issue areas reveals that scientists, assessors, and policymakers have defined and evaluated impacts in widely varying ways since the early 1970s, when climate change effects first began to garner interest (Long and Iles 1997). Early assessments tended to characterize impacts in terms of linear, cause-and-effect relationships. Later models developed by Kates et al. (1985), Parry et al. (1988), and the IPCC (1992, 1995) assumed that impacts involved complex, interactions between social and natural systems. These approaches captured synergistic, multicausal, and nonlinear climate–society relationships and accounted for previously overlooked social, economic, and nonclimatic factors affecting human behavior. The organization of impact analyses, especially prior to the mid-1990s, tended to mirror familiar market and ecosystem sector classifications (e.g., agriculture, water resources, forests, and coastal zones). In 1978, for example, the National Academy of Sciences' International Workshop on Climate Change concluded that climate impact studies should "determine the effects of climate on the various sectors of the natural world and on human economic and social life, such as hydrology, agriculture, and energy." By the 1990s, the IPCC was investigating more than 25 impact sectors (IPCC 1995). Impact research in each of these sectors has involved different forms of scientific and nonscientific expertise, meanings of "impact," and methodological tools (Long and Iles 1997).

In this chapter, we seek to understand some of the changes and variations associated with the impact concept and their implications for science and policy, as reflected in American and international assessment processes between the early 1970s and the mid-1990s. Based on a comparison of impact assessments for agriculture and coastal zones, we identify possible linkages between assessment framings,[1] methodologies, and processes, and the salience of climate impact categories. Salience is evidenced, in part, by the prominence of knowledge about impacts in debates and decisions about climate change among scientists, government officials, natural resource managers, economic producers, and the public in local, national, and international settings, and by political activism around particular impact concepts. Agriculture and sea-level rise represent two of the most prolific and most established impact assessment fields. Yet each has enjoyed different degrees of salience in climate change science and policymaking. Climate change impacts associated with sea-level rise have inspired potentially affected groups like the Alliance of Small Island States (AOSIS) to organize and act in international and domestic political arenas.[2] Agricultural impacts, in general, have not attracted the same level of political activism.

One might argue that differences in the salience of agricultural and sea-level-rise impacts simply reflect the different magnitude of projected impacts and varied levels of uncertainty in assessment-generated knowledge.[3] One might expect, for example, that severe effects, known with confidence, are more likely to serve as the basis for political mobilization and to feature in policy discourse than are impacts that are less serious and more uncertain. Agricultural impacts such as fluctuating rainfall or crop yields are more varied and uncertain over the longer term than are the more dramatic and arguably more knowable effects of sea-level rise such as inundation of islands. We suggest, however, that to focus solely on uncertainty and the nature of projections is to miss at least two important factors. First, assessments are social processes that entail varied problem definitions, analytical practices, and forms of participation. Second, these differences shape and are shaped by a given assessment's claims about the effects of climate change, the certainty with which these effects are estimated, and the nature of public interest and public involvement in the assessment process. While prominent assessments in the area of sea-level rise framed impacts in terms of "vulnerability," agricultural impact assessments tended to emphasize adaptation to climate change. We argue that these different ways of framing impacts had important implications for who took part in assessment processes, the methodologies they employed, the knowledge they generated, and the salience of this knowledge.

The following sections present case studies of agriculture and sea-level rise assessments. In each case, we examine impact framings and assessment methods and processes prior to the mid-1990s and their connections to the salience of impact categories for decisionmakers and members of the public.

Agricultural Impacts

More than 20 years ago, the economist Thomas Schelling remarked that agricultural effects of climate change constitute "the only readily identified potential impact of significant magnitude on future living standards" (Schelling 1983, 474). Many decisionmakers had worried about similar food security problems associated with crises in the 1970s (see e.g., Kates et al. 1985; NDU 1978, 1980, 1983; Rockefeller Foundation 1976; Rosenberg 1993). Many agreed with Schelling, and agriculture became, perhaps, the most researched area of climate change impacts. Despite this long history of analysis, agricultural impacts did not feature prominently in climate policy debates or in agricultural policy and decisionmaking internationally and within the United States through the 1990s (Smit et al. 1996). Nor were these impacts integral to political mobilization among relevant publics such as farmers and natural resource managers. We suggest that this apparent disjunction between agricultural impacts and potentially at-risk populations is related, in part, to the tendency of assessors to imagine climate change impacts at regional and global scales, to rely on computer models and econometric analyses that are generally not amenable to lay participation, and to assume that adaptive capacities are not problematic and that the redistribution of resources globally will compensate for loss of agricultural productivity in certain areas. (As discussed below, this situation is beginning to change with initiatives such as the U.S. National Assessment of Climate Change Impacts [NAST 2000].)

Interest in agricultural impacts follows a history of government programs, research institutes, and scientific networks devoted to understanding climate-agriculture–food security interactions. In response to the Dust Bowl, the long period of drought and land degradation in the 1930s, the U.S. federal government began to build an extensive network of agricultural research stations and academic programs whose work included analysis of weather and climate-related issues (Danbom 1988). International bodies such as the Food and Agricultural Organization similarly engaged in climate-focused projects. Starting in the late 1970s, following the initial agricultural impact assessments, scientific researchers applied computer models to evaluate climate change effects, assuming (through their experience with weather modeling) that this was the best way to forecast long-range change.

Computer modeling was centrally important to agricultural impact assessments through the 1990s. Increasing reliance on general circulation models of climate (GCMs), interdisciplinary assessment processes (especially in the United States and Europe), and the linking of crop and economic models co-evolved with the identification and framing of impacts. With computer models, researchers could expand the types and complexity of the impacts they investigated by linking numerous elements such as climate projections, crop and economic dynamics, technological changes, and even farm-level and government decisionmaking (Easterling 1996). Given the coarse geographic resolution of most GCMs, however, studies tended to support analysis at regional (e.g., across hundreds of square miles) and global scales, with limited attention to local conditions.

The centrality of computer modeling in agricultural impact assessment meant that assessors in this area were usually specialized experts from fields such as agrometeorology, plant physiology, agronomy, economics, and climatology. Because most of these experts were from developed countries, the majority of agricultural impact studies focused on these countries.[4] Additionally, computer models depended on particular types of agricultural, geographical, and climate data that, prior to the late 1990s, tended to be more abundant in developed countries. Farmers and local experts in developing countries had a wealth of observations and knowledge to potentially contribute, yet their contributions were not in forms readily useable in computer modeling. Despite limited cross-fertilization across national boundaries, modeling did allow for rich cross-disciplinary interactions among scientists from many different fields. Integrated agricultural impact studies, for example, combined GCM outputs, plant physiology models, and global trade models (Rosenzweig and Parry 1994) and brought together climatologists, plant experts, and economists. Yet few agricultural assessments prior to the early 1990s involved the participation of nonscientists such as farmers, government officials, and natural resource managers.

In turn, agricultural impact assessors predominantly used adaptation framings to structure and report on their research. Adaptation refers to the degree to which people can adjust to projected climate changes through alterations in social practices, decisionmaking processes, and institutional, economic, and physical infrastructure system structures (IPCC 1995). Adaptation frames have important implications for science and policy, in that they emphasize the human reactions to climate change impacts. Using these frames, researchers typically ask: What adaptations are needed to respond to (and, indeed, shape the very nature of) projected impacts, what societal resources are required, and how will adaptations affect productive activity? Over the last 20 years, agricultural assessment processes have continually refined methods for modeling adaptation to climate change (Parry et al. 1988; Rosenberg et al. 1993; Rosenzweig and Parry 1994; Smith and Tirpak 1989; Smithers and Smit 1997).[5]

Past studies of adaptation tended to assume that humans could engage in a variety of technological, social, behavioral, and physiological adjustments. These studies generally did not examine the resources and infrastructure required by such adjustments (Rosenzweig and Hillel 1998). Early studies treated adaptation as exogenous to climate change effects. Adaptations were thought to be something people did in response to climate change such as migrating, abandoning farms, or altering farm practices (NDU 1978; Rockefeller Foundation 1976; see also Easterling 1996). According to a 1983 U.S. Environmental Protection Agency (EPA) assessment, there was a possibility that "some negative effects will be mitigated, depending on the success and speed of efforts to adapt economic activity to altered climatic conditions." Framing adaptation as a response to climate change, rather than as an effect of climate change was also evident in the early work of the IPCC, when analysis of adaptation options became the responsibility of Working Group III (responses) instead of Working Group II (impacts).

In the early 1990s, however, Working Group II decided to examine impacts and adaptation together, reflecting a new view of the ways in which adaptation could shape the very nature of climate change impacts. This new view emerged from studies of the late 1980s, which began to integrate impacts and adaptation by incorporating assumptions about adaptation into impact modeling. Some early assessments identified changes in planted crop varieties as a climate change impact. Yet the crops that farmers plant do not spontaneously transform without changes in farmer decisions and behaviors or perhaps changes in elements of the broader agricultural system such as in market demand or government support. Such interactions were increasingly incorporated into agricultural models. At the International Institute for Applied Systems Analysis (IIASA), Parry et al. (1988) led the first major international project to systematically model short-term adjustments to climate change in 11 regional case studies worldwide.[6] The study addressed adaptation in two primary ways: as farm-level adaptation and as governmental responses from the local to international level. The IIASA assessors modeled adaptation by, in part, incorporating empirical evidence of local perceptions of climate change and social behavior. Inputs into biophysical models of first-order impacts (e.g., crop growth and animal productivity) were derived from climate scenarios based on GCM outputs and the observational record. Economic model results also estimated effects on farm profits, regional employment, and regional gross domestic product as a result of climate change.

In the late 1980s, the U.S. EPA commissioned a large study of potential climate change impacts on four regions in the United States (the southern Great Plains, the Southeast, the Great Lakes, and California). Researchers used three different GCMs to generate equilibrium climate change scenarios to serve as inputs to crop simulation models. Simple agronomic adjustments (e.g., earlier planting, increased irrigation, changes in crops) were considered for some regions. These studies suggested that, in some cases, adaptation could prevent losses in agricultural productivity (largely specified in terms of crop yields and livestock growth) otherwise caused by climate change. But they still treated adjustments in ad hoc ways, with only limited attention to the conditions under which adaptive behaviors might occur (Easterling 1996). The studies also viewed agriculture as largely an isolated sector, apart from broader social and economic systems.

In 1993, the U.S. Department of Energy sponsored a regional study of climate change impacts on natural resources in a four-state area: Missouri-Iowa-Nebraska-Kansas (MINK). The MINK study was the first to model the combined effects of increased carbon dioxide concentrations and adaptations. However, the MINK analysis examined only a small set of global and regional economic interactions (Easterling 1996). The study evaluated first-, second-, and third-order effects of climate change for current and future (year 2030) scenarios by examining climate change effects on baseline resource productivity. The study also examined ways in which primary enterprises (e.g., farmers and water resource districts) might react to these first-order effects and linkages between primary enterprises and the broader economy (Rosenberg 1993). Assessors used a crop model to simulate effects of climate change on crop productivity. Likely adaptive measures and technological changes were identified through literature review and sensitivity analysis and were then represented in the crop model (Rosenberg 1993).

In the mid-1990s, the IPCC (1995, 1997) recommended ways to improve adaptation analysis, largely through integrated modeling and comparisons among alternative research approaches. According to the IPCC, agricultural impact assessments could benefit from incorporating a much broader set of technological changes including those associated with seasonal changes and sowing dates; different crop varieties or species; new crop varieties; water supply and irrigation systems; management; tillage; and improved short-term climate prediction. The IPCC also noted the importance of addressing interactions overlooked in many studies, namely those between adaptive capabilities, technical advances, institutional finance, and information exchange at local and regional scales. To accommodate these factors, the IPCC (1995) suggested that adaptation options and economic responses be directly integrated in models, rather than treated as separate from the models. The IPCC recommended that researchers account for adjustment costs and examine dynamics involving farmer perceptions and expectations, as well as the effects of education about climate change and farm investments.[7] The IPCC remained focused on improving computer models with data that scientists had chosen and collected.

While researchers pursued extensive refinements in adaptation analysis, several factors limited the political salience of agricultural impact assessments in the United States and internationally. In particular, some agricultural assessments assumed technology transfer would occur readily. The embedding of adaptation into impact models sometimes obscured the challenges people might experience in particular regions. Some assessments reflected an "affluent society" perspective in overlooking the costs of adaptation (Harvey 1993). These studies often suggested that new techniques, tools, and crop varieties would develop and diffuse in response to the anticipation of climate change. However, the much greater availability of resources, capacity, and technology in industrial countries meant that the specific adaptations that assessors recommended would be much more likely to facilitate adaptive responses to climate change in industrial countries than in developing countries.

Many assessments did not address the social practices and political and economic conditions that make technology transferable and useful across different settings (Jasanoff 1994). For example, deploying a computer system to track and evaluate climatic variability requires a computer-friendly culture, people with training in computer use, electricity, and continued maintenance. The availability of such resources differs across settings. Hence, the successful transplanting of such technology requires more than moving equipment or know-how from one geographic location to another (for example, as evidenced by the Green Revolution). Yet assessors often assumed adaptations to be universally applicable (e.g., Johnston and Clark 1982), supported by farm equipment, techniques, and resources available and usable anywhere in the world.

Similarly, some agricultural impact studies portrayed inevitably shifting agricultural zones as simultaneous with climate change (e.g., Carter et al. 1991; Rosenzweig 1985). By suggesting that climate change alone gives rise to such large-scale transformations, however, blurs the categories of "impact" and "adaptation." Such blurring obscures human–environment interactions, such as decisions about what crops to plant, where to plant them, and how to manage them. It implies that large, widespread classes of farmers across broad geographical regions will make similar decisions regarding farming practices and crop patterns (see Smit et al. 1996). Assessors may overlook the losses that farmers are willing to incur before deciding to abandon or move their farms, the costs of relocating, investing in new technologies, establishing new farming businesses, and, perhaps, the costs and hardships consumers experience as production diminishes in existing zones. Without an account of the costs and changes that make up a shifting agricultural zone, the shift itself appears automatic and unproblematic. The distributive effects of climate change are ignored.

The dominance of adaptation frames in agricultural impact assessments helped shape the current broad consensus among modelers (but not necessarily everyday decisionmakers) that, in the face of climate change at doubled carbon dioxide equilibrium conditions, global agriculture could be maintained relative to baseline production over the next century. Studies projecting little change in global food production might have supported complacency among policymakers and populations who could eventually experience the deleterious effects of climate change–agriculture interactions (e.g., IPCC 1990, 1995; NRC 1983). Studies at subglobal levels, however, suggested much greater variability in the nature and direction of changes and in the spatial distribution of these changes. Agricultural impacts were expected to be both favorable and adverse, such that some regions would benefit overall while other regions would be disadvantaged. The IPCC report in 1995 noted that, while global production averages would not decline significantly, regional effects of climate change on agriculture could be quite negative.

Given modeling capabilities and narrow impact framings (e.g., in terms of crop yield), the implications of climate change for agriculture at sub-regional levels remain very difficult to project. For example, the U.S. National Assessment of 2000 characterized regional (namely, "Midwest" and "West") trends in crop outputs, precipitation, storm frequency, and ecosystem shifts. The study focused primarily on crop yields, yet it "did not fully consider all of the consequences of possible changes in pests, diseases, insects, and extreme events resulting from climate change" (NAST 2000). Nor did it take account of how its projection of greatly increased pesticide use with climate change might affect ecosystems and, ultimately, crops. These aspects are important in relation to the everyday experiences of many farmers, especially in developing countries.

Despite the abundance of agricultural impact assessments between the 1970s and mid-1990s, politicians, policymakers, and farmers did not organize to any large degree around climate change issues. There were almost no agriculture-based lobbies pushing for policymaking on climate change grounds in most developed countries. One rare exception is the American Farmer Bureau, but this group opposed climate change policies on the grounds that they would make farming more expensive, for instance by increasing fuel costs. Agricultural groups tended to respond to concrete problems in the short term, such as trade, genetically modified crops, fuel inputs, and water supplies, even though climate change could profoundly affect many of these issues.

This disconnect between available knowledge about agricultural impacts and potentially vulnerable communities is likely related to the framing of agricultural impacts in terms of adaptations, the computer-dominated methodologies used for evaluating them, and assessment communities primarily composed of modelers (albeit from diverse disciplines). Most assessments concentrated on changes that would happen decades away such as temperature change and altered crop patterns, rather than on problems that farmers could readily identify with in the present, such as drought and pests.[8] Similarly, although many of the most hard-hit agricultural areas are expected to be in developing countries, few assessment processes took place in these areas or addressed local issues of particular concern to their inhabitants. Conversely, most studies in the past focused on global and regional levels and assumed that large-scale distributive adjustments would take place when needed. Embedding adaptation into impacts further obscured the costs of climate change that may otherwise motivate everyday decisionmakers to engage in debates about climate change and to take action at subregional levels. A lack of engagement between researchers with potentially at-risk rural communities reinforced established assessment practices, making it less likely that future assessments would connect with, and derive input, from these people.

Coastal Zone Impacts

The coastal zone, like agriculture, has been a high-profile topic in impact assessment communities for more than 25 years. Early on, impact assessments presented sea-level rise as the most clearly foreseeable direct effect of climatic change. Unlike agricultural impacts, coastal zone impacts featured prominently in climate change policy debates in both domestic and international arenas during the 1990s. Moreover, these impacts were associated with the emergence of an active coalition of at-risk countries, AOSIS. We argue that the salience of coastal zone impacts is intertwined with the framing, methodology, and participation characterizing coastal zone assessments. We do not mean to suggest that impact assessments gave rise to AOSIS. Instead, we argue that political interest in sea-level rise and the characterization of coastal zone impacts in assessment processes arose jointly. In contrast to studies in the agricultural sector, coastal zone assessments (especially in the early 1990s) emphasized the vulnerability of at-risk communities. In doing so, they examined issues considered relevant by these communities (especially in small island states), employed a wide range of analytical techniques at local and regional scales, and involved more diverse participants in terms of expertise and background.

Because sea-level rise impact assessments grew out of geology and glaciology studies of Antarctica and other ice-covered regions, early assessors were often oceanographers, geologists, and paleoclimatologists (see also the discussion in Chapter 6). They examined possible melting of the West Antarctic Ice Sheet (Mercer 1978), thermal expansion and alpine glacier melting (Bolin 1977; Schneider and Dickinson 1974), and the effects of these processes on globally averaged sea-level rise. In the 1970s and 1980s, most assessors were from developed countries, notably the Netherlands and the United States, both of which have particularly low and long coastlines (IPCC 1990, Schneider and Chen 1980). Their analyses focused on issues of relevance to these countries, such as flooding of low-lying lands and intrusion of salt water into freshwater. Nonetheless, while computer models provided important data about potential sea-level rises, assessors relied on other types of data and analytical methods reflecting the physical and geographic conditions with which rising seas interacted.

Assessments have presented a wide range of sea-level rise estimates over the past 30 years with varied implications for response strategies. Early projections of sea-level rise ranged from several meters to tens of meters (SCEP 1970; U.S. EPA 1983). At these levels, adaptation was not considered a feasible response to sea-level rise. Between the early 1980s and mid-1990s, global projections decreased markedly and stabilized around 0.5 to 1.0 meter, because scientists concluded that neither the West Antarctic Ice Sheet nor the Greenland ice cover would melt significantly. These comparatively modest projections indicated that sea-level rise would be manageable in developed countries. Assessments, therefore, focused on technological adaptive measures. Dutch planners, for example, historically sought to reclaim low-lying land from the North Sea and to control the risks of flooding by sea surges through extensive engineering projects (Hekstra 1986). Their assumptions and analytical methods, such as determining flood contours and deploying defensive mechanisms, were readily transferred to the climate change arena. For example, Dutch and American participants in a 1983 National Academy of Science study specified three categories of response: retreat (leaving or preventing development in areas that will be inundated); accommodation (building piers or changing building codes); and protection (defending population centers and economic activities with structures such as dikes, sea walls, and tidal barriers, or via restoration or creation of dunes or wetlands; NRC 1983). These adaptive strategies continued to appear in assessments as possible responses to climate change (IPCC 1990, 1995). American impact research also emphasized the economic dimensions of sea-level rise. EPA, for example, estimated the costs of various adaptation strategies in the United States (e.g., Smith and Tirpak 1989; Titus and Narayanan 1996). These studies analyzed whether specific response measures would be cost effective when compared with the potential effects of sea-level rise and increasingly likely extreme events.

Adaptation analyses remain part of sea-level rise assessment. Other framings, however, have become increasingly prominent. Beginning in the late 1980s, the IPCC introduced new approaches to sea-level rise assessment with the aim of persuading policymakers, coastal managers, and potentially affected populations to incorporate sea-level rise issues into their decisionmaking. The IPCC set up its Coastal Zone Management (CZM) Subgroup, which asserted that information about impacts needed to be made relevant to managers and their concerns, including coastal development, pollution run-off, and water availability (IPCC 1992; WCC 1993). The new methods focused on understanding the vulnerability of populations living in coastal areas.

Vulnerability focuses attention on the characteristics of a system and the ways in which these characteristics make it possible (or not) for a system to respond to change. Vulnerability and adaptation framings differ in important ways with regard to their science and policy implications. Vulnerability refers to the extent to which climate change may harm a system, and it depends on the sensitivity and adaptability of the system (IPCC 1995). Vulnerability frames emphasize the distributive costs of climate change impacts. Who, for example, will suffer the most from impacts, and where will these impacts be experienced? Group A might be considered less vulnerable to inundation than Group B because of A's institutional flexibility and resources. Similarly, within a given human population, the degree of risk of death from heat waves depends on age, income, access to health care, and geographical characteristics (McMichael et al. 1996). Societies and localities differ widely in the resources and institutional structures available for responding to climate change (Kates et al. 1985). Agricultural studies between the early 1980s and mid-1990s tended to emphasize net global and regional changes in agricultural productivity, thereby glossing over the challenges that might beset particular countries or regions. By contrast, assessments of vulnerability enabled analysis of potentially at-risk populations, examined relevant strengths and weaknesses of social systems, and often highlighted the asymmetrical distribution of impacts locally and regionally. Vulnerability approaches also appeared to facilitate the involvement of a broader array of participants in assessment processes and greater attention to subregional and local levels.

In 1988, the CZM Subgroup began to develop what became the Common Methodology on Vulnerability Assessment and sought to apply it in many settings worldwide. These endeavors involved scientists from Japan, Australia, and the South Pacific, as well as the United States and the Netherlands. These researchers also included geographers and anthropologists. The subgroup defined vulnerability as "a nation's ability to cope with the consequences of an acceleration in sea-level rise and other coastal impacts." The common methodology was aimed, in part, at encouraging states, especially developing ones, to begin planning for future sea-level rise in the context of present needs (WCC 1993). It set out a flexible approach for identifying physical, ecological, and socioeconomic factors likely to be sensitive to sea-level rise and for evaluating a country's capacity to implement responses within a broad framework of coastal zone management.

In contrast to agricultural assessments that tended to be carried out by scientists who employ computer models, the Common Methodology on Vulnerability Assessment offered a more open-ended, assessment process that engaged local-level coastal management specialists and diverse analytical tools. The common methodology required the compilation of extensive information about specific geographic areas and the people who inhabit them. These characteristics are evidenced in experimental studies conducted during the early 1990s by the South Pacific Regional Environmental Programme (SPREP) in cooperation with the IPCC. This work included six detailed case studies of the Federated States of Micronesia, the Cook Islands, Kiribati, the Marshall Islands, Tokelau, and Western Samoa (IPCC 1995). These studies asked about the resources that could or could not be substituted, the cultural objects that might be lost, and the changes in food supplies that might result with sea-level rise. Assessors relied on ethnographic and community resource mapping tools, for example, to find out what local people thought were important for their ways of life (e.g., SPREP 1992, 1993, 1994). Many researchers came from the islands themselves. In contrast to agricultural impact studies, which are often not part of a policy planning effort, these vulnerability assessments were part of an emerging, ongoing management approach across the South Pacific, to which local people contributed.

Vulnerability framings facilitated analysis of social factors and helped lead to more politically meaningful climate change projections. The common methodology case studies, for example, examined how extreme events (such as cyclones and tsunamis) might affect island states. These analyses resonated with island peoples, especially those who had experienced destructive weather events, such as the cyclones that devastated several Pacific islands in the early 1990s. Thus, assessors began to frame impacts in terms of present effects that might be exacerbated by climate change in the longer term (e.g., SPREP 1993). They could, then, argue that taking measures to respond to climate change would also help manage these effects here and now, thereby improving their ability to reach and communicate with everyday decisionmakers who tend to focus on short-term goals.

New techniques for studying sea-level rise impacts also allowed for analysis of a broader range of social and ecological factors. A team of Australian, Japanese, and Pacific researchers developed the "Sustainable Capacity Index" to evaluate the vulnerability of island states in terms of a wide range of cultural, social, agricultural, and industrial effects flowing from sea-level rise (Yamada et al. 1995). The Sustainable Capacity Index expanded impact assessments beyond the traditional emphasis on physical and economic effects. Early vulnerability assessments had focused on determining which areas to defend on a cost and benefit basis, or how best to minimize the costs of planned retreat. But people living on many Pacific islands may experience social and cultural dislocation when forced to move elsewhere and abandon their traditional fishing and agricultural practices. Their dislocation may differ greatly from that of people living along coasts in industrialized countries. The index assisted assessors in recognizing and analyzing these often-overlooked considerations.

With the standardized yet flexible common methodology, small island nations could evaluate their own vulnerability and relate these to similar situations on other islands (e.g., SPREP 1994). Small island states increasingly participated in defining the risks they faced, which then supported political organization among at-risk groups worldwide. New ways of talking about sea-level rise appeared in the arguments made by small island states in the international arena. In 1994, the Conference on the Sustainable Development of Small Island Developing States declared: "Small island developing states are particularly vulnerable to natural as well as environmental disasters." The conference further determined that small islands require special assistance under the United Nations Framework Convention on Climate Change (UNFCCC) because of their "peculiar vulnerabilities and characteristics" (CSDSIDS 1994). AOSIS has also invoked the vulnerability concept to call for more stringent emission reductions during international climate negotiations. Hence, these countries are citing their vulnerability as a basis for a new moral framing in political debates.

The advent of new concepts such as vulnerability, new techniques of impact analysis, and more diverse participation in sea-level rise assessments reveals important innovations. These developments occurred alongside growth in the visibility and voices of at-risk populations and changes in the way the IPCC conceptualizes impacts associated with sea-level rise. Between the IPCC's 1990 report and its 1995 assessment, the title of the sea-level rise chapter changed from "Coastal Zones and Sea-Level Rise" to "Coastal Zones and Small Island States." The IPCC's new chapter heading signaled increased emphasis on the human dimensions of impacts and a certain legitimacy for island states as threatened communities.

Conclusions and Recommendations

One might assume that the degree of political mobilization around a particular category of climate change impact depends solely on the severity and certainty of the projected effects of climate change. We argue, however, that this type of explanation is an oversimplification. How and by whom impacts are defined are integral to determinations of severity, uncertainty, and who counts as at-risk populations. We suggest that the salience of and political mobilization around impacts shape and are shaped, in part, by impact framings, analytical methodologies, and assessment processes (especially with regard to participation) employed in impact knowledge-making. The relatively low profile of agricultural impacts and activists in climate change policy debates prior to the 1990s was related, in part, to the nature of agricultural impact science, which addressed questions that are generally not of immediate concern to farmers, and which relied on highly specialized computer modeling, embedded hidden assumptions about adaptation, and the distribution of costs and resources. A higher level of political activity around climate-induced sea-level rise accompanied impact framings that emphasized human-centered notions of vulnerability, varied and generally more accessible methods of analysis, and the engagement of local communities and resource managers in assessment processes.

If impact assessments are aimed, in part, at creating knowledge that is meaningful for and useful to decisionmakers and the public, then two measures of an effective climate change assessment are salience of the assessment and social action built upon the assessment process and its findings. It follows, then, that a comparison of agricultural and sea-level sectors prior to the mid-1990s offers three important lessons to assessors for assessment design. Based on this comparison we can speculate that assessment processes are likely to be more effective if they frame impacts in ways that are meaningful to potentially at-risk populations and relevant to their ways of life, employ varied scientific methods, engage diverse participants and their respective realms of knowledge, and take shape via a multistage process.

One important lesson concerns framing. If assessors frame impacts such that they are relevant to peoples' existing concerns and ways of life, assessors can increase the salience of impact assessments, as well as the willingness of decisionmakers to act on the findings. Prior to the mid-1990s, most agricultural impact studies focused on future projections of climate change, treated climate change in isolation from other environmental and social factors, and assumed that societies could readily adjust to climatic change. By contrast, vulnerability analyses of coastal zones entailed more holistic analyses of particular human–environment systems, paying close attention to their conditions, challenges, and capabilities in the present. These investigations identified linkages between future climate change and existing climate variability, and other environmental, social, economic, and institutional factors. They examined adaptation in the context of vulnerability, a framing device that focused attention on the socioecological systems expected to encounter adverse change.

Framing coastal zones as vulnerable was relevant to coastal zone management efforts concerned with what could be done to anticipate and respond to inundation, storm surges, and salt water intrusion. These analyses required knowledge about human–environment systems in coastal zones and sometimes involved local people from islands in assessment processes. Coastal zone assessments that attracted and facilitated this involvement provided means for weaving climate change concerns into the present everyday activities of some small island state inhabitants. These assessments also provided decisionmakers with opportunities to influence assessment processes for coastal zone issues. We suggest that this involvement of nontraditional assessment participants and their priorities enhanced the salience of coastal zone assessments in the eyes of coastal zone inhabitants. Farmers and agricultural managers, on the other hand, tended not to be included in agricultural impact studies because their observations and needs are seen as not relevant or amenable to agricultural impact methodologies. Changes in framing could improve the perceived salience and relevance of these analyses to ground-level decisionmakers, and help change assessment methods and processes as well.

Another important lesson is that reliance on more varied methodologies and forms of knowledge could usefully connect an assessment process with diverse experts, audiences, and users. This appears to be the case for coastal zone assessments. Rather than depend exclusively on computer modelers, climate scientists, and other scientific experts, vulnerability assessors in the coastal zone also engaged coastal zone managers and local residents. Coastal zone studies of vulnerability examined local-to-global dynamics of environmental and social change. They recognized that so-called "local knowledge" about the priorities, weaknesses, and capabilities of human–environmental systems was essential for understanding what climate change might mean to the present generation and how its effects might play out in the future. These activities both necessitated and supported the use of varied analytical methodologies. This diversity of methods helped to make assessment processes accessible and relevant to a wider array of assessment contributors and audiences. In general, a forecasting-by-analogy approach to understanding environmental change could be one way of diversifying conventional methods for impact analysis. This approach combines observational, historical, and ethnographic research methods in an analogy or case study approach "based on the premise that while the climate of the future may not be like the climate of the present, societal responses to climate change in the near future will most likely be like those of the recent past and the present" (Glantz 1988). This field methodology could enhance the credibility and salience of assessments in the eyes of local decisionmakers (Glantz 1988). It could also aid assessments for developing countries, which may lack the kinds of data readily integrated into computer models, yet be rich in other forms of knowledge.

The sea-level rise experience also suggests that effectiveness will increase if assessments are designed and developed over time as multiple-stage processes with opportunities for reflecting on and improving weaknesses in assessment design and practices. The IPCC and SPREP vulnerability assessments across the South Pacific during the 1990s are examples of longer-term assessment processes. These initiatives allowed for periodic reflection on and modifications to assessment processes while fostering development of a semi-standardized yet flexible impact assessment framework (such as the Common Methodology on Vulnerability Assessment in the coastal zone area) that is applicable worldwide, but can be tailored to particular settings and can be updated with experience.

By the late 1990s, many assessors were grappling with ways to try to apply the insights of coastal zone assessments to many impact sectors. Both the IPCC and the U.S. National Assessment processes featured vulnerability as an important framing device. They also adopted a regional focus for analyzing vulnerability. In 1997, for example, the IPCC examined the relative vulnerability of eight regions worldwide to agricultural impacts of climate change. The IPCC acknowledged that vulnerability varies across scales, populations, and research approaches, and the panel identified three categories of vulnerability: farmer and farm sector, regional economic, and hunger. This work continued in the IPCC's Third Assessment Report of 2001. Similarly, vulnerability was central to analysis of agriculture, water resources, health, and forests in the U.S. National Assessment, which was sponsored by the U.S. Global Change Research Program (NAST 2000). This process relied on dialogue between assessors, decisionmakers, and affected populations (of the kind found in coastal zone sector assessments).

However, just as certain forms of agricultural and sea-level rise impact analyses are converging, traditional approaches to agricultural impact assessment are perhaps becoming increasingly relevant to powerful actors in the agricultural sector in industrial nations. Large-scale analysis of agricultural impacts, while not particularly useful to the family farmer, is increasingly applicable to industrial agriculture. Thus, computer modeling of complex systems may provide valuable data about climate change to large-scale producers who themselves use models to design their operations. Yet the scale of industrial agriculture, notably in the United States, with its massive beef feedlots, large wheat and maize fields, irrigated fruit orchards, widespread use of pesticides, and rapid processing facilities tends to disconnect food science, government policy, and farming practices from the great diversity of ecological and agricultural conditions at ground level (Kimbrell 2002). Traditional impact assessments, then, may not provide an adequate picture of the underlying vulnerabilities, whereas more diverse methods may better illuminate these.

In conclusion, assessment practitioners need to be aware of how their frames, methods, and processes can shape the meaningfulness of their findings, not only for fellow scientific experts, but also for the lay actors who will ultimately choose and implement policies. Whether and how people understand and take action on climate change is likely to be influenced by the ways in which assessment processes impart meaning to impacts. This insight applies to many forms of environmental assessments, not just to climate change impact studies. There is much scope for innovative linkages between science, policy, and politics if assessments are open to the kinds of experimentation that we have observed in the coastal zone sector. Climate change impacts are continually being made meaningful through the ways in which scientists, farmers, coastal managers, island dwellers, governments, and the lay public are (or are not) brought together to share their data, analyses, and priorities. Recognizing this can lead to more effective assessment and greater collective action on one of humanity's greatest challenges.

Notes

1. Framing refers to "the perceptual lenses, worldviews, or underlying assumptions that guide communal interpretation and definition of particular issues" (Miller 2000). For example, framing climate change as a problem of carbon dioxide emissions has different science and policy implications than does framing climate change as a problem of consumption. Similarly, climate change impacts might be framed in terms of changes on a global scale such as shifts in food production and distribution, or alternatively as risks to localities in the form of weather damage to crops. Each way of framing suggests a different set of problem definitions and policy responses. Divergent constructions of impacts affect the questions that researchers and policymakers ask: Who/what is at risk? Who/what is responsible? Who/what should policies be about? Who takes an active interest and how? Not surprisingly, then, the varied ways of analyzing, defining, and communicating knowledge about climate change impacts are likely to give rise to information that is more (or less) salient, credible, and legitimate to decisionmakers in different settings.

2. AOSIS is a coalition of small island and low-lying coastal countries that functions primarily as an ad hoc lobby and negotiating voice for small island developing states; see http://www.sidsnet.org/aosis[0].

3. For example, uncertainty is contingent on how assessors frame climate impacts and focus on specific environmental and social factors when developing methodologies and models. Assessors in the agricultural impact area often emphasize changes in crop yields (rather than,

say, farm ecosystems), thus embedding specific types of uncertainty in their models. Consequently, they are less able to provide data and analysis regarding variables not addressed via their methods. Uncertainty also reflects, to some extent, the regions, time frames, social systems, productive activities, and ecological interactions that assessors have chosen to address.

4. The Intergovernmental Panel on Climate Change points out, for example, that relatively few agricultural impact studies exist for countries in sub-Saharan Africa, the Middle East, North Africa, and Latin America, which are likely to be most susceptible to climate change (IPCC 1995).

5. The focus here on adaptation is not meant to imply that concerns about and interest in vulnerability have never shaped or informed agricultural impact studies. Even the first investigations into possible effects of climate change on agriculture address both vulnerability and adaptation. During the 1970s, droughts led to widespread famine in the Sahel and East Africa. Drought and winter snow cover in the Soviet Union also resulted in crop failures, and they created a crisis in world grain trade (Rockefeller Foundation 1976). At that time, the U.S. government began to view climatic impacts on agriculture as a national security issue, where world food production was potentially vulnerable to climate variations in the context of Cold War politics. As one of the first major studies of agricultural impacts, the Rockefeller Foundation (1976) assessed how a potential temperature decrease might affect food production. The study analyzed whether predictions of climate variations could inform timely world action and whether it was possible to develop crop varieties versatile enough to survive extreme climatic conditions. Vulnerability and adaptation frames were also central in the design of the National Defense University studies of the late 1970s and early 1980s (NDU 1978, 1980, 1983). These assessments examined the world grain economy to the year 2000 and its policy implications for the United States and developing countries. Understanding the sensitivity of U.S. agricultural production to long-term climate change (both warming and cooling trends) was a priority in this research.

6. Climate scenarios derived from GCM outputs and the observational record were inputs into biophysical models of first-order impacts (e.g., crop growth and animal productivity). Economic models were used to estimate effects on farm profits, regional employment, and regional gross domestic product as a result of climate change.

7. Debates continue over how to investigate and account for adaptive responses. While impact researchers apply increasingly sophisticated models, these models are based on a limited understanding of how humans might respond to climate change (Smithers and Smit 1997). Most early studies did not account for changes in land-use management as a result of climate change. This approach has come to be known as the "dumb farmer" approach, and in response, some researchers developed "genius farmer" models, assuming that farmers can detect climate change signals and respond immediately. More recently, assessments have used more realistic models of farmer behavior based on ethnographic surveys in the field. But there is great uncertainty regarding whether farmers respond in strategic and autonomous ways, or if other dynamics are at work (Smit et al. 1996).

8. Researchers increasingly note the importance of extreme events in climate change scenarios. However, there is little information about the properties of these events and how people respond to them (Smit et al. 1996).

References

Bolin, B. 1977. The Impact of Production and Use of Energy on the Global Climate. *Annual Review of Energy* 2: 197–226.

Carter, T.R., J.R. Porter, and M.L. Parry. 1991. Climatic Warming and Crop Potential in Europe: Prospects and Uncertainties. *Global Environmental Change* 1: 291–312.

Chen, R. 1981. Interdisciplinary Research and Integration: The Case of CO_2 and Climate. *Climatic Change.* 3: 429–447.

CIAP (Climatic Impact Assessment Program). 1975. *Monograph 5: The Impacts of Climate Change on the Biosphere.* Washington, DC: U.S. Department of Transportation.

CSDSIDS (Global Conference on the Sustainable Development of Small Island Developing States). 1994. *Report of the Global Conference on the Sustainable Development of Small Island Developing States:* Bridgetown, Barbados, April 26–May 6, 1994. New York: United Nations.

Danbom, D.B. (ed.). 1988. *Publicly Sponsored Agricultural Research in the United States: Past, Present, and Future.* Berkeley, CA: University of California Press.

Easterling, W. 1996. Adapting North American Agriculture to Climate Change in Review. *Agricultural and Forest Meteorology* 80: 1–53.

Glantz, M.H. (ed.). 1988. *Societal Responses to Regional Climatic Change: Forecasting by Analogy.* Boulder, Colorado: Westview Press.

Harvey, D. 1993. Comments on "An Empirical Study of the Economic Effects of Climate Change on World Agriculture." *Climatic Change* 24: 273–275.

Hekstra, G.P. 1986. Will Climatic Change Flood the Netherlands? Effects on Agriculture, Land Use and Well-Being. *Ambio* 15: 316–326.

IPCC (Intergovernmental Panel on Climate Change). 1990. *Climate Change: The IPCC Impacts Assessment. Report from Working Group II to IPCC.* Canberra, Australia: Australian Government Publishing Service.

———.1992. *Global Climate Change and the Rising Challenge of the Sea.* Response Strategies Working Group, Coastal Zone Management Subgroup. Geneva, Switzerland: Intergovernmental Panel on Climate Change.

———. 1995. *Climate Change: Impacts, Adaptations and Mitigation of Climate Change: Contribution of Working Group II to IPCC.* Canberra, Australia: Australian Government Publishing Service.

———. 1997. *The Regional Impacts of Climate Change: An Assessment of Vulnerability.* Summary for Policymakers. Geneva, Switzerland: Intergovernmental Panel on Climate Change.

Jasanoff, S. (ed.). 1994. *Learning from Disaster: Risk Management after Bhopal.* Philadelphia: University of Pennsylvania Press.

Johnston, B., and W.C. Clark. 1982. *Redesigning Rural Development: A Strategic Perspective.* Baltimore, MD: Johns Hopkins University Press.

Kates, R. W., J.H. Ausubel, and M. Berberian (eds.). 1985. *Climate Impact Assessment: Studies of the Interaction of Climate and Society.* Chichester, United Kingdom: John Wiley and Sons.

Kimbrell, A. (ed.). 2002. *Fatal Harvest: The Tragedy of Industrial Agriculture.* Washington, DC: Island Press.

Long M., and A. Iles. 1997. Assessing Climate Change Impacts: Co-evolution of Knowledge, Communities, and Methodologies. Discussion paper E-97-09, Environmental and Natural Resources Program, Kennedy School of Government. Cambridge, MA: Harvard University.

McMichael, A.J., A. Haines, and R. Sloof (eds.). 1996. *Climate Change and Human Health.* Geneva, Switzerland: The World Health Organization, World Meteorological Organization, and United Nations Environment Programme.

Mercer, J.H. 1978. West Antarctic Ice Sheet and CO_2 Greenhouse Effect: A Threat of Disaster. *Nature* 271: 321–325.

Miller, C.A. 2000. The Dynamics of Framing Environmental Values and Policy: Four Models of Societal Processes. *Environmental Values* 9(2): 211–233.

NAST (National Assessment Synthesis Team). 2000. *Climate Change Impacts on the United States: The Potential Consequences of Climate Variability and Change.* A Report of the National Assessment Synthesis Team, U.S. Global Change Research Program. New York: Cambridge University Press.

NDU (National Defense University). 1978. *Climate Change to the Year 2000.* Washington, DC: National Defense University.

———. 1980. *Crop Yields and Climate Change to the Year 2000.* Washington, DC: National Defense University.

———. 1983. *The World Grain Economy and Climate Change to the Year 2000: Implications for Policy: Report on the Final Phase of a Climate Impact Assessment.* Washington, DC: National Defense University.

NRC (National Research Council). 1983. *Changing Climate.* Washington, DC: National Academy Press.

Parry, M.L., T.R. Carter, and N.T. Konijn (eds.). 1988. *The Impact of Climate Variations on Agriculture.* Dordrecht, Netherlands: Kluwer Academic Publishers.

Rockefeller Foundation. 1976. *Climate Change, Food Production and Interstate Conflict: A Bellagio Conference.* New York: The Rockefeller Foundation.

Rosenberg, N.J. (ed.). 1993. *Towards an Integrated Impact Assessment of Climate Change: The MINK Study.* Boston: Kluwer Academic Publishers.

Rosenberg, N.J., P.R. Crosson, K.D. Frederick, W.E. Easterling, M.S. McKenny, M.D. Bowes, R.A. Sedjo, J. Dramstadter, L.A. Katz, and K.M. Lemon. 1993. The MINK Methodology: Background and Baseline. In *Towards an Integrated Impact Assessment of Climate Change: The MINK Study,* edited by N.J. Rosenberg. Dordrecht, Netherlands: Kluwer Academic Publishers. Reprinted from *Climatic Change* 24: 1–2 (1993).

Rosenzweig, C. 1985. Potential CO_2-Induced Climate Effects on North American Wheat-Producing Regions. *Climatic Change* 7: 367–389.

Rosenzweig, C., and D. Hillel. 1998. *Climate Change and the Global Harvest: Potential Impacts of the Greenhouse Effect on Agriculture.* New York: Oxford University Press.

Rosenzweig, C., and M. Parry. 1994. Potential Impact of Climate Change on World Food Supply. *Nature* 367: 133–138.

SCEP (Study of Critical Environmental Problems). 1970. Man's Impact on the Global Environment. Cambridge, MA: MIT Press.

Schelling, T. 1983. Climate Change: Implications for Welfare and Policy. In *Changing Climate: Report of the Carbon Dioxide Assessment Committee.* Washington, DC: National Academy Press.

Schneider, S., and R.S. Chen. 1980. Carbon Dioxide Flooding: Physical Factors and Climatic Impact. *Annual Review of Energy* 5: 107–140.

Schneider, S.H., and R.E. Dickinson. 1974. *Review of Geophysics and Space Physics* 12: 447.

Smit, B., D. McNabb, and J. Smithers. 1996. Agricultural Adaptation to Climatic Variation. *Climatic Change* 33: 7–29.

Smith, J.B., and D.A. Tirpak. 1989. The Potential Effects of Global Climate Change on the United States. EPA-230-05-89-050. Washington, DC: U.S. Environmental Protection Agency.

Smithers, J., and B. Smit. 1997. Human Adaptation to Climatic Variability. *Global Environmental Change* 7: 277–290.

SPREP (South Pacific Regional Environment Programme). 1992. *The Pacific Way: Pacific Island Developing Countries' Report to the United Nations Conference on Environment and Development.* Noumea, New Caledonia: South Pacific Commission.

———. 1993. *Climate Change and Sea Level Rise in the South Pacific Region: Proceedings of the Second SPREP Meeting.* Apia, Western Samoa: South Pacific Regional Environment Programme.

———. 1994. *Phase II: Development of Methodology.* Environment Agency Japan. Overseas Environment Cooperation Centre Japan. Apia, Western Samoa: South Pacific Regional Environment Programme.

Titus, J.G., and V. Narayanan. 1996. The Risk of Sea Level Rise. *Climatic Change* 33: 151–212.

U.S. EPA (Environmental Protection Agency). 1983. *Can We Delay a Greenhouse Warming?* edited by S. Seidel and D. Keyes. Washington, DC: Strategic Studies Staff, Office of Policy Analysis, Office of Policy and Resources Management.

Watson, R.T., M.C. Zinyowera, and R.H. Moss (eds.) 1998. Summary for Policymakers: The Regional Impacts of Climate Change: An Assessment of Vulnerability. Cambridge, MA: Cambridge University Press.

WCC (World Coast Conference). 1993. *Preparing to Meet the Coastal Challenges of the 21st Century: Report of the World Coast Conference,* November 1–5. Noordwijk, Netherlands: Ministry of Transport, Public Works and Water Management.

Yamada, K., P.D. Nunn, N. Mimura, S. Machida, and M. Yamamoto. 1995. Methodology for the Assessment of Vulnerability of South Pacific Island Countries to Sea Level Rise and Climate Change. *Journal of Global Environmental Engineering* 1: 101–125.

CHAPTER 6

Dealing with Uncertainty
How Do You Assess the Impossible?

Anthony Patt

O NE WAY OF ASSESSING an uncertain future is to translate it into a probability distribution of potential outcomes. When all of the potential outcomes are expressed in terms of a single unit—such as amount of money, temperature change, or human population—one can also generate variables such as the mean and standard deviation to describe the distribution. Even when one cannot express the different outcomes in unitary terms, one can still describe the distribution in terms of its central tendency and its extremes: events that one predicts with very high and very low likelihoods, respectively.

Which variables are the most salient depends on the problem at hand. For instance, if I decide to invest in the stock market in a stable blue-chip company, my choice of company could depend on predictions of average growth or expected level of dividends. By contrast, if I want to invest in a high risk Internet startup, which I do not expect to maintain a stable stock price, I would want information such as the chances that the company will fail (making the stock worthless), the chances that the company will be acquired during the next year (making me a shareholder in a different company altogether), and the chances that it will become the next Microsoft (making me rich). Indeed, were a financial analyst to frame the Internet stock solely in terms of its central tendency, and not the chances of qualitatively different types of outcomes, she would be doing her job poorly; for naïve investors, she would be misleading them about the true nature of the stock, whereas for savvy investors, she would be telling them information lacking salience.

Some environmental problems are like the stable blue-chip company, well described by the central tendency. Many more are like the Internet stock, where the extremes dominate the decision problem—we do not fear so much what we know, but what we do not know. In these cases, a primary goal of assessment should be to help bring to light potential surprises (Clark 1989), and failure to do so can lead to permanent losses of credibility (Freudenburg 1996). As Lee (1993) states: "What does it mean to be prudent when there is uncertainty? First, recognize the possibility of surprise. Second, plan and act to detect and to correct avoidable error." Effective and prudent assessments of uncertain outcomes will draw attention to events in the future that could indicate that a particular outcome is more or less likely than previously thought; these assessments will help decisionmakers to understand both the state of knowledge and the sources of uncertainty.

However, many assessments of global environmental change have failed to shed light on surprises or to promote the understanding of uncertainty. Assessments commonly err by reporting only those scientific findings that pertain to the expected outcome, as opposed to more extreme, though less likely, events (Kasperson and Kasperson 1996; Van der Sluijs 1997). Other assessments, such as the Third Assessment Report of the Intergovernmental Panel on Climate Change (IPCC), have reported on extreme events, but only after internal and external prodding (Moss and Schneider 2000; Patt and Schrag 2003). The observation that extreme events are often missing from scientific assessments can be made precisely because these extreme events do appear in the basic scientific literature. It is not a failure of science and scientists to predict possible extreme events, but rather a failure of environmental assessments to include these predictions. Something is at work in the assessment process, as distinct from the scientific process, that leads to this omission.

This chapter attempts to explain why omission of extreme events is likely to occur, and more importantly, to make predictions about in what kind of assessments it will occur. The chapter suggests that there are different types of assessments, with different goals, and with different abilities to offer insight into why many assessments might adopt a strategy of ignoring the tails. This chapter suggests classifying assessments by their purpose: advisory, advocacy, and agreement (i.e., consensus-seeking). Advisory assessments tend to focus on extremes to the extent justified by their mandate and the state of scientific knowledge. Advocacy assessments tend to assess extremes, but in a biased fashion. Consensus-seeking assessments, both because of their purpose and because of their typical organizational structure, tend to avoid assessing extreme outcomes.

To test these hypotheses, I present a content analysis of scientific assessments of climate change, focusing on a particular event for which assessments could provide useful information: a sudden large rise in sea level due to the rapid deterioration of land-based polar ice. Ideally, a climate change policymaker would want to know, to the extent possible, how likely such an event would be—is it 1 in 10 million, 1 in 10 thousand, or 1 in 10—and when it might occur. There may be no credible answer to these questions, because there are so many different relevant factors, few of which are well understood, and none of which has occurred in recent history. Given this uncertainty, information would be salient if it described not only the likelihood of the events themselves, but also the sources of the uncertainty, and what possible events, such as the calving of large Antarctic icebergs, would be consistent with the hypothesized causes of polar meltdown.

This chapter offers several useful lessons. To the users of assessments (those who turn to assessments for the best available information on a problem) this chapter suggests the importance of looking beyond one kind of assessment for all the answers. For instance, users should be aware that agreement-seeking assessments may have a bias toward omitting coverage of extreme events, and instead look to advisory assessments for needed information. To the designers of assessments (e.g., program managers at government agencies that fund research), this chapter suggests the importance of considering the incentives that the assessors themselves will face. It may be that to obtain information on the tails of the distribution, one would have to structure the assessment, and define its intended audience, to enable assessors to see their role as advisory rather than agreement-seeking. To the assessors themselves, this chapter suggests that there may be a tradeoff between providing salient, useful information on highly uncertain topics and actually building consensus in the short-term. Assessors should reflect on the implicit reasons for undertaking the assessment, and they should understand that their reasons can lead to a bias in how they frame the problem. By considering their implicit goals and potential biases, assessors may be able to seek creative solutions to the problem of providing information that is salient, credible, and in furtherance of their objectives.

Extreme Events, Framing, and Strategy

In this section, I build a framework for understanding the difficulties of assessing extreme events. First, I discuss what is qualitatively different about extreme events, compared with their higher probability and lower consequence cousins. Next, I suggest that different assessments might be pursuing different goals and strategies. Finally, I describe a classification scheme to describe three broad types of assessments, based on their goals and strategies, and how these goals and strategies can be expected to translate into the assessments' treatment of extreme events.

Thinking about Uncertainty

People find it hard to think probabilistically. They avoid doing so even if they are able, and they engage in denial, even to their own detriment (Renn 1991; Weber 1997). The field of risk communication developed largely in response to these difficulties, initially offering the promise of helping people to make wise tradeoffs between different quantifiable risks to their health and safety, based on the measurement of past data combined with assumptions linking those data to people's behavior. Given the propensity of economists to advocate reliance on decentralized decisionmaking (Zeckhauser and Viscusi 1996), risk managers first assumed that people would use this quantitative information to make consistent choices about which risks to accept and which to shun (Leiss 1996). But a number of studies show a sharp divergence between popular opinions of risk and the opinions of so-called experts (e.g., Breyer 1993; U.S. EPA 1987; Zeckhauser and Viscusi 1996). Furthermore, empirical research in behavioral economics shows that people respond to risk and uncertainty at an emotional level (Covello 1990; Kahneman and Tversky 1979; Kammen et al. 1994; Tversky and Kahneman 1973). In their handling of uncertainty and dissent, risk communicators began to recognize the need to work hard at demonstrating to the public the salience of the risk information. Yet overselling information can lead over time to a loss of

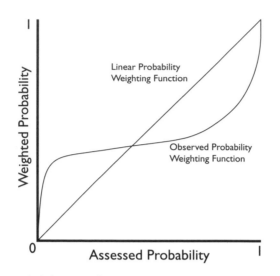

Figure 6-1. *Probability Weighting Function*

credibility and legitimacy. Risk communicators have increasingly incorporated greater public participation as a way of enhancing the salience of information while preserving their own credibility and legitimacy.

In helping people to understand the salience (or lack of salience) of risk information, assessors have run headfirst into several predictable decisionmaking biases that most people demonstrate. The first such bias is the propensity to overreact to the chance of very low probability events, and underreact to very high probability (but still uncertain) events. Behavioral economists have modeled this with a *probability weighting function,* shown in Figure 6-1 (Kahneman and Tversky 1979). It describes that people's responses to risk are not a linear function of the quantified estimates of the risk itself.

For example, imagine a coastal homeowner facing a risk of a $10,000 loss due to flooding, and consider how much money he would be willing to spend to eliminate that risk, perhaps by building a retaining wall. First, consider someone who is "risk neutral" and responds linearly to the quantified risk estimate. If the estimated probability were 1 percent, he would be willing to spend $100 to eliminate the risk. Likewise if the probability were 50 percent or 99 percent, he would be willing to spend $5,000 or $9,900, respectively. If this person were "risk averse," he would be willing to spend more than these amounts for each estimated probability; if he were "risk loving," he would spend less. Next, however, consider the homeowner that research has shown is more typical. This person might be willing to spend $1,000 to eliminate the 1 percent risk, $5,000 to eliminate the 50 percent risk, and $9,000 to eliminate the 99 percent risk. To such a person, the difference between 0 percent and 1 percent likelihood, or 99 percent and 100 percent, is much more important than the difference between 49 percent and 50 percent or 50 percent and 51 percent. Thus, he demonstrates strong risk averse behavior for low probability events, risk neutral behavior for intermediate probability events, and strong risk loving behavior for high probability events.

Events with probabilities very close to zero are the hardest of all to think clearly about, and where people's behavior shows the greatest variance. These low-probability events seem to register in people's minds as having either some positive probability, in which case people demonstrate very strong risk aversion, or else no chance at all of occurring. Into which of these two boxes the risk falls depends less on the precise probability estimate (whether it is one-in-a-million or one-in-ten-million) than on a set of attributes associated with that risk (Covello 1990). For instance, if a risk tends to cause fatalities grouped in time and space (e.g., a plane crash), is unfamiliar, or is caused by a mechanism that is not well understood, people tend to consider the risk as possible and worrisome. If, by contrast, a risk tends to cause fatalities that are scattered and nonidentifiable (e.g., automobile crashes) and is due to a familiar process that is well understood, people tend to ignore the risk. Most risks, or course, have multiple attributes, some of which are controllable and familiar and some of which are not. For these risks, it is always possible for the communicator to choose aspects of the risk that will magnify or diminish the risk in people's minds. This choice over how to frame the risk will always influence people's responses. For risks of very low probability, however, framing becomes especially important, because it can result in whether people think the risk is impossible on the one hand, or likely to occur on the other. For a single risk with an agreed-upon small probability of occurring, two different risk assessors could engender in their audience very different responses to the risk simply as a result of strategic framing decisions.

Studies of managers and administrators in the public and private sectors indicate that policymakers are not necessarily better at decisionmaking under conditions of uncertainty—including probabilities close to zero—than are lay people. As long as the risk is not one they have studied extensively, policymakers tend to fall into the same mental patterns of inconsistent risk aversion that plague others (March 1988). For example, when researchers performed experiments that involved choices over probabilistic medical diagnoses, they found that doctors not familiar with the specific diagnoses performed no more in accordance with a "rational actor" model than did members of the lay public (Tversky and Kahneman 1988). Even experts in the field can commit consistent errors. Shlyakhter et al. (1994) found that energy models were subject to consistent and predictable biases, and Gordon and Kammen (1996) found a similar result in forecasts of stock behavior—both of these involved predictions by experts in their field, although they could involve strategic behavior. Kammen et al. (1994), however, note instances in which trained experts in government, such as physicians who decide matters of public health, *are* able to perceive risks correctly, but they face a voting public that insists on another interpretation.

Assessors can help decisionmakers perceive and respond to risks more consistently. Leiss (1996) and Fischhoff (1996) suggest that helping people respond sensibly to assessments of low probability events requires involving them in the process of developing the risk estimates themselves and making them partners in the assessment through extensive public participation. Bazerman (1998) discusses techniques of "debiasing" decisionmakers once that participation is under way, causing them to "unfreeze" their decisionmaking methods.

After criticism on its early efforts to communicate extreme events, the IPCC responded by developing sets of guidelines for communicating uncertainty on the basis of advice offered by policymakers and risk communicators (Moss and Schneider 2000). In the Third Assessment Report, Working Group I of the IPCC avoided describing events by numerical probabilities, and instead used a set of words, corresponding to precise ranges of probabilities (Houghton et al. 2002). For example, the IPCC represented the probability range 0–1 percent with the words "exceptionally unlikely." Currently, for the Fourth Assessment Report, the IPCC is convening a group of experts to evaluate this practice and to suggest means of improvement.

Assessment Goals and Strategies

These practical issues of how people think about uncertainty have important implications for assessment design, as the recent IPCC work reflects. Chapter One of this book presented environmental assessments as one approach that actors in an issue domain (such as climate change) can use to change the issue domain in directions they desire. It is also possible to think of environmental assessments as games with many players, each trying to achieve various goals, well defined or not. Each player's goals depend on his or her position in the game. For instance, a representative of a small island state might want very much for the world to take actions to stop climate change, because the representative feels the potential outcome of major sea level rise is a risk not worth taking. A representative of an oil company might be less concerned with that risk and instead worry more about short-term stock price effects of aggressive climate change policies. The scientific community, often frustrated by the slow pace at which policy moves incrementally forward, may want merely to see political action of any kind, a sign that the scientists are being listened to and the issue taken seriously. Assessors, who are scientists working to transfer knowledge beyond their narrow field, likely have a set of goals aligned with the interests of the institution for which they are working.

For each of these actors, each with different goals, there are distinct strategies they could employ. Consider Figure 6-2 as representing a complex environmental issue such as global climate change, as well as two different strategies for assessment. The left-hand tail of the distribution represents outcomes that are not worrisome. By contrast, the right-hand tail represents outcomes that are very worrisome. The most likely scenario is somewhere toward the middle of the gray area in-between, where the outcomes are significant but not catastrophic. Assessment strategies can involve highlighting particular areas of the distribution. Those highlighting the central tendency create relatively few framing issues. People understand mid-range probabilities well, and therefore the major issue becomes the magnitude of the outcome predicted. Assessment strategies highlighting the tails (particularly the right-hand tail) are open to multiple interpretations. Is the significant feature of the right-hand tail that its consequences are so severe, or that its probability is so low? At what point in the tail does the

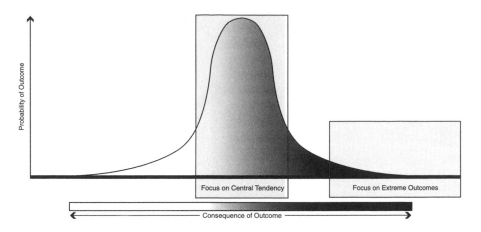

Figure 6-2. *Framing Choices for Different Risks*

probability become zero? The act of assessing the right-hand tail can raise more questions than it answers.

One can distinguish assessments by different sets of goals and intended audiences. Having done so, one can then predict respective assessment strategies. These strategies are likely to differ the most—from one assessment to the next—around those issues for which framing decisions are most likely to influence people's perceptions. For example, a relatively likely outcome of climate change is that mid-latitude temperatures will rise by a modest amount, on the order of a few degrees, and that this will bring about a set of likely effects such as ecosystem change and warmer winter weather. Different people may argue about whether such effects are good or bad, but their underlying understanding of the change in temperatures is likely to be fairly consistent. An assessment is unlikely to change people's minds about what is good and what is bad—that is more the result of their interests and goals, although it may help them to understand if something is possible, probable, or highly likely.

By contrast, consider rapid sea-level rise. In this case, people no doubt agree that the result is bad, but may hold very different beliefs about whether it is possible. People's response to the issue is largely binary—rapid sea-level rise is either possible or impossible—which means that any assessment framing strategy will likely influence their ultimate opinion about the entire climate change issue a great deal. Because there is room for strategic framing to make the most difference for these issues, one should expect assessors to pay careful attention to the framing of these issues. The actual framing issues will also likely contain greater variance.

Agreement Assessments

Some assessments are undertaken in order to reach agreement among a panel of national or global experts, especially when governments hesitate to take action because of a perceived lack of information or consensus. The purpose of these assessments is to report on the scientific consensus of a particular issue. The assessment audience is usually broadly defined, being not only policymakers but also experts in other fields, as well as the press and the general public. These types of assessments are often lengthy or semipermanent efforts, frequently made up of a panel of experts who represent a broad range of disciplines, interest groups, and stakeholders. In most cases, these experts must reach agreement on what to include in the assessment, and their work is subject to peer review. Prime examples of this type of assessment are the works of the IPCC and the U.S. National Research Council. For several powerful reasons, consensus-seeking assessments may be likely to ignore the analysis of extreme events.

Why would these assessments ignore extreme events? First, their intended audience is broad, and indeed the audience may not be informed enough to easily interpret information about extreme events. Second, discussing extreme events could run counter to these assessments' intended purpose. Consensus-seeking assessments are an attempt to put a given amount of uncertainty to rest to enable decisionmaking based on the level of knowledge that does exist. Were the assessors to focus on issues marked by high levels of uncertainty, the effect might well be for governments reluctant to take action to call for more study, rather than promote substantive policy. Were the assessors to focus on issues marked by consensus around low-probability and extreme outcomes, there would be a danger of people over- or under-reacting to the risk because of the way it is framed. Finally, these assessments are unlikely to achieve internal consensus on the proper treatment of extreme events. Consensus assessment panel members, each with their own opinion, may often find it difficult to agree on a single way to frame a particular extreme outcome. Some would insist on it being treated as a major threat. Others would insist that by discussing it as a serious potential problem, even one with low probability, they would be lending the theory an undeserved legitimacy. As experts whose reputations have been built elsewhere, members of these assessment panels have an incentive not to compromise their desired problem frame, even if it means omitting a particular issue from the assessment document.

Advisory Assessments

Some assessments give advice to a narrowly defined audience. They are usually written under contract for a group within an industry or for a department in a government agency or ministry. These assessments are undertaken in order to sort through difficult questions that relate to policy and to suggest a set of solutions consistent with the organization's goals and purposes. Many of these assessments go unobserved, because as the property of the contracting party they may not be part of the public domain. The principal–agent relationship between the advisory assessors and the group commissioning the assessment is closer than they are in the case of agreement assessments; hence, there ought to be less divergence between the definitions as to goals and problem frames. The experts preparing these assessments may represent a range of disciplines, but the purpose is to bring together minds that can sort through the issues and analyze the results for the benefit of a narrowly defined group of users. Prime examples of this type of assessment are the works of independent consulting companies and government research arms such as Directorate General Research of the European Commission. For reasons given below, advisory assessments are likely to give either detailed treatment of extreme events, or no treatment at all.

Career analysts generally prepare advisory assessments to brief decisionmakers directly. These analysts make their living putting to use their understanding of technical issues involving probability and uncertainty. Thus, they can be expected to interpret low probability extreme events relatively easily. These assessments fulfill their purpose if they are able to present the relevant information that enables the decisionmaker to make (or put off) a choice. Furthermore, they are often asked to address a specific issue or set of issues. If this issue is one for which the possibilities of extreme events is important, one would expect these assessments to give detailed treatment to the extremes. If the issue is one for which a particular extreme event is not relevant, one would not expect these assessments to cover the event at all.

Advocacy Assessments

A third type of assessment is used for advocacy. To varying extents, industry groups, nonprofit and nongovernmental organizations, and even some government agencies have interests at stake in a given issue and a particular policy outcome. While their assessments often at first glance appear to be consensus or advisory documents, in truth they represent a much narrower set of interests. Their reports may be commissioned by an organization, but if so they are meant to be read primarily outside that organization—by the general public and policymakers. For the reasons below, advocacy assessments are likely to give limited treatment to extreme events.

The audience for these assessments is not composed of experts in the field, but rather made up of people whose opinions on an issue are undecided or who are prone to persuasion. These people can be expected to poorly understand low-probability extreme events and to be heavily influenced by the framing of the issue. The purpose of these assessments is to use whatever facts necessary, framed however necessary, to further a particular interest. These assessments present evidence supporting the advocates' side of the debate as much as possible, and they discredit the evidence used by the other side of the debate—all within the constraint of maintaining credibility. Some groups attempt to magnify the likelihood in the audiences' minds of the less certain risks. They use the framing techniques discussed above to achieve this end. They do not, however, give unbridled attention to extreme events, because doing so would risk informing their audience too much, and it might overcome the audiences' very vulnerability to framing. Other groups want people to discount the possibility of low-probability extreme events, as well as the validity of their opponents' arguments. They, too, pay some attention to low-probability extreme events, using a different set of problem frames, to convince people to discount the risks and to mistrust the other side. Again, these groups do not devote considerable attention to extreme events, for doing so would risk informing their audience too much.

Case Study—Assessing the West Antarctic Ice Sheet

To examine the potential for distinguishing different types of assessments, and to test the predictions about their treatment of extreme events, I conducted a content analysis of climate change assessments written between 1981 and 1997. The analysis focuses on the extent of attention assessments paid to a particular extreme event: the deterioration of the West Antarctic Ice Sheet (WAIS), which is marked by high uncertainty. The magnitude of potential losses, however, is so large that the issue would be important for almost any probability.

In 1978, J.H. Mercer published an article in *Nature* in which he hypothesized that the WAIS could disintegrate rapidly if a number of ice shelves (the floating masses of ice at the fringe of the ice sheet) were to deteriorate (Mercer 1978). As Figure 6-3 shows, the Antarctic ice sheet falls into two general geographic regions. The large East Antarctic ice sheet covers most of the continent, and it rests on a landmass above sea level. It is a relatively stable glacial formation. The smaller WAIS covers the smaller portion of the continent. Two ice shelves, the Ross and Ronne-Filchner, potentially hold the entire WAIS together, and they prevent it from collapsing into the sea. However, these two ice shelves are themselves fragile, as the recent collapse of the Larsen ice shelf demonstrated. To survive, the ice sheets require average summer (January) temperatures to remain below 0°C. This is roughly equivalent to mean annual temperatures remaining below −5°C.

Mercer hypothesized that global warming, due to increased concentrations of carbon dioxide (CO_2) in the atmosphere, could threaten these ice shelves. He noted that predictions called for a rise in global average surface temperature of 3°C due to a doubling of greenhouse gas concentrations. However, he noted, general circulation models predicted the warming to be more pronounced at high latitudes. Thus, it is not unreasonable to predict a warming of 10–15°C in the Arctic and Antarctic regions. Such warming would put the Ross and Ronne-Filchner Ice Shelves at risk, because currently they lie at a January isotherm of approximately −10°C. If these ice shelves were to be confronted with a 0°C January isotherm, they could disintegrate and cause the rapid deglaciation of the entire West Antarctic region. Such a deglaciation would raise sea levels by as much as five meters.

As parts of Antarctica have melted, such as occurred with the rapid collapse of the Larsen Ice Shelf in the mid-1990s, the issue has received bouts of renewed attention by the news media. Mercer's scientific theory, however, began to attract a considerable amount of criticism soon after it was published. Many attacked his claim that the WAIS would disintegrate rapidly were it to lose the support of the northern ice shelves. Others suggested that the rate of additional snow accumulation on the entire Antarctic continent would exceed the rate of disintegration of the WAIS (e.g., Jacobs 1992). Finally, scientists took issue with the projection that temperatures in the West Antarctic region would rise by 10°C. In short, the WAIS issue can be characterized as one that has been increasingly seen as highly unlikely, although not altogether impossible. Its potential severity, however, makes it an issue that one would expect to appear in climate change assessments.

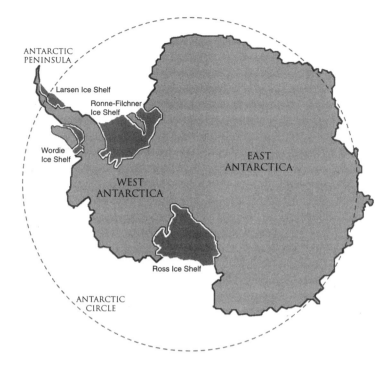

Figure 6-3. *Antarctic Ice Shelves*

Data Coding and Analysis

To see whether assessments have covered the WAIS issue, and whether their coverage correlated with their goals, I examined a set of 38 assessments of climate change published between 1981 and 1997. Consistent with the other chapters in this book, I treated as an assessment (and hence included in the data set) those social processes that sought to communicate the results of scientific knowledge to policymakers or the public. Thus, my data do not include papers appearing in journals of natural or social science that seek to expand the knowledge base or derive the biogeophysical, economic, or political implications of climate change. It does include works that seek to derive specific policy implications using interdisciplinary research and results.

The first challenge was to classify each assessment according to the three criteria of audience, purpose, and process. If an assessment was written for a limited audience, to inform them rather than persuade them, I viewed it as an advisory assessment. If the purpose of the assessment was persuasion and targeted a general audience, it was an advocacy assessment. If it was written for an audience of policymakers and the general public, to inform of the scientific consensus, and if it brought together a panel of experts in a relatively egalitarian process, I classified the assessment as an agreement assessment. The classification appears in Table 6-1.

The second challenge was to classify the degree of issue treatment—discussion of the WAIS collapse—by each assessment. I chose to distinguish three levels of treatment. I defined *no treatment* as the lack of even a mention of the possibility of the rapid deterioration of the WAIS due to climate change. I defined *limited treatment* as some discussion of the issue, without providing a detailed or balanced account of the scientific theory and the probabilities associated with the possible outcomes. I defined *detailed treatment* as a discussion of the issue from a balanced perspective, presenting the scientific theory, qualified by statements about its likelihood of occurrence (again, see Table 6-1 for assessments grouped by level of treatment).

Analyzing the data in Table 6-1, using a statistical regression model, it is possible to control for other explanatory variables, and to determine whether the effect of the assessment type agrees with the prediction from my model.

To isolate the effect of the assessment type, I controlled for two other sets of variables in the regressions. I first controlled for the year in which the assessment was published. This may be important, given the development of the theory surrounding West Antarctica and climate change. Shortly after Mercer published his paper in 1978, the issue gained widespread attention. Soon, however, other scientists began publishing findings critical of Mercer's hypothesis. I expected the assessed median probability of the ice sheet's collapse to decrease as these later findings were published, and thus to see less treatment of the issue by assessments as the years progressed from 1978 onward. I found the best fit with the data when I modeled a quadratic functional form. Hence my two time variables are *years after 1978* and *years^2 after 1978*.

I also controlled for the length of the assessment. I defined short assessments as those of 100 pages or fewer, medium assessments as those of 100 to 500 pages, and long assessments as those of more than 500 pages. Longer assessments tend to devote more attention to all of the issues involved in climate change, and thus are more likely to devote more space to issues of rapid sea-level rise. I found that the difference between short and medium assessments is significant, whereas there is no discernible difference between medium and long assessments. I therefore compare only those assessments that are short with those that are medium and long.

A major limitation of my data is that I looked only at the published reports of the assessment effort. I ignored the fact that assessment efforts often span years, and that the major impacts on policy may arise out of informal communication with policymakers during the assessment process or through media channels after the assessment is finished (an area for further research would be whether the substantive content of assessments is similar across formal and informal paths of communication). Nevertheless, these reports are useful because the published reports do constitute the official record of the assessments.

Results

Table 6-2 presents the results of the first two regression models. Model A is a multinomial logit regression, which allows one to set up the dependent variable—treatment of sea-level rise—as a choice among three possible outcomes.

Table 6-1. *Assessments and Data Coding*

	Detailed treatment	Limited treatment	No treatment
Agreement	National Research Council (1983, 1984)	IPCC (1992) Joint WMO/ICSU/UNEP Group of Experts (1981) National Research Council (1992)	Bruce et al. (1996) Chen and Parry (1987) Houghton et al. (1995) IPCC (1991, 1993) Jäger (1988) Jäger and Ferguson (1991) National Research Council (1987) UNEP (1989)
Advisory	Barnett (1982) International Institute for Applied Systems Analysis (1981) MacCracken and Luther (1985) Smith (1982) U.S. Department of Energy (1983, 1985, 1988) U.S. EPA (1983, 1988, 1995)	Smith and Tirpak (1989) U.S. Office of Technology Assessment (1991)	Environmental Energy Solutions (1996) Lashof and Tirpak (1990) Parry and Carter (1984) U.S. Office of Technology Assessment (1993) U.S. Department of Energy (1993)
Advocacy		Asian Development Bank (1995) Council on Environmental Quality (1981) Greenpeace (1997) Lyman (1990) Mintzer (1992)	George C. Marshall Institute (1996) Mintzer (1987)

Note: The table shows how each assessment in the sample was coded, both in terms of type of assessment (agreement, advisory, or advocacy) and treatment of the WAIS issue. Full citation to each assessment appears in the References section.

The left-hand column shows the coefficients associated with increased or decreased likelihood of showing limited treatment compared with no treatment. The right-hand column shows coefficients associated with detailed treatment compared with no treatment. It is difficult to interpret the magnitude of the coefficients, because their marginal impact on the likelihood of issue treatment depends on the values of all of the variables, but it is useful to focus on their significance and sign.

Nearly all coefficients are significant at the 5 percent or 10 percent level. Both advisory assessments and advocacy assessments are more likely to result in the limited treatment of the issue of rapid WAIS deterioration. The effect is larger for the advocacy assessments. Over time, more recent assessments are more likely to give the WAIS issue either detailed treatment or none at all, and the negative coefficient for the *years² after 1978* estimator indicates that this is especially true in the first few years after the scientific finding. The second column of Model A shows that advisory assessments are more likely to address the WAIS issue in greater detail, while advocacy assessments are less likely than agreement assess-

Table 6-2. *Logit Regression Results*

| | Model A: Multinomial Logit | | Model B: Logit |
	Limited treatment	Detailed treatment	Any treatment
Advisory	2.3*	5.6**	3.0**
	(1.4)	(2.3)	(1.3)
Advocacy	6.3*	-36	6.9*
	(3.4)	(1.7 x 109)	(3.5)
Years after 1978	-2.5**	-4.7**	-2.8**
	(1.2)	(1.9)	(1.2)
Years^2 after 1978	0.090*	0.18**	0.10**
	(0.047)	(0.075)	(0.047)
Short	-6.9 *	-12**	-7.8**
	(3.7)	(5.2)	(3.7)
Constant	14*	25**	17**
	(7.8)	(10)	(7.7)
Sample size	38	38	
Pseudo R^2	0.57	0.57	

Notes: Model A results are from a mutinomial logit regression, which allows for a categorical dependent variable to take on more than two outcomes. Presented are coefficient estimates on the dependent variables for two of the three possible outcomes: *limited treatment* and *detailed treatment*. In Model B, the number of coded outcomes on the dependent variable has been reduced from three to two, allowing a standard logit regression.
Asterisks denote statistically significant results: * = $p < 0.10$; ** = $p < 0.05$; standard errors are listed in parentheses.

ments to give detailed treatment. There is an unusually large standard error associated with advocacy assessments. This is due to the absence of a single advocacy assessment in which a detailed treatment of WAIS was found. Again, the time effect is the same as with the first scenario. As the *pseudo R^2* value indicates, Model A explains more than half of the variance in the dependent variables.

The standard logit regression presents results that are robust to the judgment call between limited and detailed treatment; these appear as Model B in Table 6-2. The results are similar in sign and significance to those of Model A. Both advisory and advocacy assessments are more likely than agreement assessments to treat the issue at all, and the effect is larger for advocacy assessments Again, the model explains more than half of the variance in the dependent variable. The coefficients in Model B resemble quite closely those of the left-hand column of Model A, indicating that the limited treatment component of the *any treatment* variable dominates the results of Model B.

Finally, I used a "bootstrap" technique to test the Models A and B results for robustness, because of the concern that 38 observations may be too small a sample from which to derive meaningful results. To simplify matters, I ran an ordinary least squares (OLS) regression within the bootstrap. Although OLS is not ordinarily used in the context of a binary choice dependent variable, it typically gives results that qualitatively agree, in terms of sign and significance, with the more complicated logit techniques. Because sign and significance are the features of the coefficients that are most interesting in this case, OLS provides a useful check on the logit analysis. The most important results from the bootstrap tech-

Table 6-3. *Bootstrap Regression Results*

	Coefficient	Standard error	95% lower bound	95% upper bound
Advisory	0.37	0.15	0.076	0.67
	0.37	*0.16*	*0.067*	*0.69*
Advocacy	0.59	0.18	0.21	0.97
	0.59	*0.19*	*0.19*	*0.93*
Years after 1978	-0.19	0.074	-0.34	-0.035
	-0.19	*0.073*	*-0.32*	*-0.039*
Year² after 1978	0	0.0033	-0.00020	0.013
		0.0033	*-0.00014*	*0.013*
Short	-0.53	0.16	-0.86	-0.20
	-0.53	*0.16*	*-0.85*	*-0.20*

Notes: Single least-squares regression results are presented in roman type; bootstrapped least-squares regression results are presented in italic type.
Number of repetitions in bootstrap: 1,000 sample size: 38

nique are the lower and upper bounds on the 95 percent confidence interval. These can be compared with the confidence interval bounds calculated in a single regression, using the standard error and assumptions of normality. As Table 6-3 shows, the bootstrap confidence interval lower and upper bounds lie close to those determined with the single regression. Indeed, there is no single estimator that is significant at the 5 percent level in the OLS model and not significant in the bootstrap technique. This demonstrates that the earlier results are robust to small changes in the assessments included in the data, assuming that the sample of 38 assessments itself is unbiased (Mooney and Duval 1993). The empirical results observed in the four regression models are consistent with the theoretical predictions.

Conclusions

The literature in risk communication and behavioral economics suggests that discussion of low probability events, and especially events for which the probability is unknowable, is likely to generate more controversy than it puts to rest. So how *does* one assess the likelihood of an event that is nearly impossible, or a likelihood that itself is impossible to determine? If the purpose in conducting the assessment is to generate consensus around an issue, the answer may be to not assess it at all. The case study of climate change assessments, examining their treatment of the WAIS, finds this to be a common but not universal practice. Among all types of assessments, consensus-seeking assessments avoided the issue, while advisory and advocacy assessments gave it coverage to a greater or lesser extent.

Consensus-seeking assessments may not be serving the long-term interests of the policy community, or indeed their own long-term credibility or legitimacy, when they fail to assess extreme events. In the case of climate, the policy justification for preventing or reducing climate change appears to have shifted away from anticipating a gradual warming of a few degrees spread across the globe and toward the possibility of extreme outcomes. For example, with the large El Niño events of 1991–1992 and 1997–1998 came a variety of disruptive weather events around the world, which led to the hypothesis that climate change would lead to more frequent and more intense episodes of disruptive weather. Depending on the costs of adapting to changes versus preventing them in the first place, climate change can appear to be a more or less acceptable future path for the world to embark on. By contrast, extreme events are ones for which little ex ante adaptation is possible, and hence they generate uniformly negative consequences.

It is critical that the people designing the assessment process—those undertaking the initial scoping meetings and inviting scientists to participate—be clear about what they want the assessment to achieve. If it is important to inform the assessment audience of all the issues, including those for which no credible probability estimates are available, then this goal should be clearly understood by assessment participants. If what is really desired is the building of consensus within the policymaking community, then omission of extreme events may appear to be justified. However, while an assessment omitting extreme events may be less contentious to the naive, it might also be less salient to sophisticated policymakers. Assessors who do omit discussion of extreme events should understand that this strategy runs the risk of mischaracterizing the problem as a whole, diminishing their own credibility and legitimacy in years to come. Recent efforts within the IPCC, for example, have begun to address these very issues. In preparation for the Third Assessment Report, the IPCC commissioned a background paper to examine the treatment of uncertainty (Moss and Schneider 2000), and developed separate guidelines for each of the working groups. In 2004, preparing for the Fourth Assessment Report, the IPCC convened a meeting of experts to discuss the treatment of uncertainty, and is currently at work drafting a background paper and set of guidelines to authors.

The risk-communication literature suggests that a greater emphasis on the participatory side of assessment, rather than the written report, can help to avoid this conundrum. Through involvement as partners in the processes that negotiate probability estimates, decisionmakers come to see those numbers in more moderated and less binary terms. When attention focuses more clearly on the causes and consequences of future possibilities, extreme events lose their ability to sway public opinion quite so far in either direction.

References

Asian Development Bank. 1995. *Climate Change in Asia: Executive Summary*. Manila, Philippines: Asian Development Bank.

Barnett, T. 1982. On Possible Changes in Global Sea Level and Their Potential Causes. Washington, DC: U.S. Department of Energy.

Bazerman, M. 1998. *Judgment in Managerial Decision Making* (fourth edition). New York: John Wiley and Sons.

Breyer, S. 1993. *Breaking the Vicious Circle: Toward Effective Risk Regulation*. Cambridge, MA: Harvard University Press.

Bruce, J., H. Lee, and E. Haites (eds.). 1996. *Climate Change 1995: Economic and Social Dimensions—Contribution of Working Group III to the Second Assessment Report of the Intergovernmen-*

tal Panel on Climate Change. Cambridge, MA: Cambridge University Press.

Chen, R., and M. Parry. 1987. *Climate Impacts and Public Policy.* Laxenburg, Austria: International Institute for Applied Systems Analysis.

Clark, W. 1989. Managing Planet Earth. *Scientific American* 261: 47–54.

Council on Environmental Quality. 1981. *Global Energy Futures and the Carbon Dioxide Problem.* Washington, DC: U.S. Government Printing Office.

Covello, V. 1990. Risk Comparisons and Risk Communication: Issues and Problems in Comparing Health and Environmental Risks. In *Communicating Risks to the Public: International Perspectives,* edited by R. Kasperson and P. Stallen. Dordrecht, Netherlands: Kluwer Academic Publishers, 79–124.

Environmental Energy Solutions. 1996. *Climate Change and the Insurance Industry.* Hartford, CT: Environmentally Effective Solutions.

Fischhoff, B. 1996. Public Values in Risk Research. *Annals of the American Academy of Political and Social Science* 545: 75–84.

Freudenburg, W. 1996. Risky Thinking: Irrational Fears about Risk and Society. *Annals of the American Academy of Political and Social Science* 545: 44–53.

George C. Marshall Institute. 1996. *Are Human Activities Causing Global Warming?* Washington, DC: George C. Marshall Institute.

Gordon, D., and D. Kammen. 1996. Uncertainty and Overconfidence in Time Series Forecasts: Applications to the Standard & Poor's 500 Stock Index. *Applied Financial Economics* 6: 189–198.

Greenpeace. 1997. *Polar Meltdown.* London, United Kingdom: Greenpeace.

Houghton, J., L. Meira Filho, J. Bruce, H. Lee, B. Callander, E. Haites, N. Harris, and K. Maskell. 1995. *Climate Change 1994: Radiative Forcing of Climate Change and an Evaluation of the IPCC IS92 Emission Scenarios.* Cambridge, MA: Cambridge University Press.

Houghton, J.T., Y. Ding, D.J. Griggs, M. Noguer, P.J. van der Linden, and D. Xiaosu (eds.). 2002. *Climate Change 2001: The Scientific Basis.* Cambridge, MA: Cambridge University Press.

IPCC (Intergovernmental Panel on Climate Change). 1991. *Climate Change: The IPCC Response Strategies.* Washington, DC: Island Press.

―――. 1992. *Climate Change: The IPCC 1990 and 1992 Assessments.* Cambridge, MA: Cambridge University Press.

―――. 1993. *Climate Change 1992.* Cambridge, MA: Cambridge University Press.

IIASA (International Institute for Applied Systems Analysis). 1981. *Life on a Warmer Earth: Possible Climatic Consequences of Man-Made Global Warming.* Laxenburg, Austria: International Institute for Applied Systems Analysis.

Jacobs, S. 1992. Is the Antarctic Ice Sheet Growing? *Nature* 360: 29.

Jäger, J. 1988. Developing Policies for Responding to Climatic Change. Geneva, Switzerland: World Climate Programme of the World Meteorological Organization.

Jäger, J., and H. Ferguson (eds.). 1991. *Climate Change: Science, Impacts, and Policy.* Cambridge, MA: Cambridge University Press.

Joint WMO/ICSU/UNEP Group of Experts. 1981. *On the Assessment of the Role of Carbon Dioxide on Climate Variations and their Impact.* Geneva, Switzerland: World Climate Programme of the World Meteorological Organization.

Kahneman, D., and A. Tversky. 1979. Prospect Theory: An Analysis of Decision under Risk. *Econometrica* 47: 263–291.

Kammen, D., A. Shlyakhter, and R. Wilson. 1994. What Is the Risk of the Impossible? *Journal of the Franklin Institute* 331A: 97–116.

Kasperson, R., and J. Kasperson. 1996. The Social Amplification and Attenuation of Risk. *Annals of the American Academy of Political and Social Science* 545: 95–105.

Lashof, D., and D. Tirpak (eds.). 1990. *Policy Options for Stabilizing Global Climate: Report to Congress.* Washington, DC: U.S. Environmental Protection Agency.

Lee, K. 1993. *Compass and Gyroscope: Integrating Science and Politics for the Environment.* Washington, DC: Island Press.

Leiss, W. 1996. Three Phases in the Evolution of Risk Communication Practice. *Annals of the American Academy of Political and Social Science* 545: 85–94.

Lyman, F. 1990. *The Greenhouse Trap: What We're Doing to the Atmosphere and How We Can Slow Global Warming.* Boston: Beacon Press.

MacCracken, M., and F. Luther. 1985. *Detecting the Climatic Effects of Increasing Carbon Dioxide.* Washington, DC: U.S. Department of Energy.

March, James G. 1988. Bounded Rationality, Ambiguity, and the Engineering of Choice. In *Decision Making: Descriptive, Normative, and Prescriptive Interactions,* edited by D. Bell, H. Raiffa, and A. Tversky, Cambridge, MA: Cambridge University Press.

Mercer, J.H. 1978. West Antarctic Ice Sheet and CO_2 Greenhouse Effect: A Threat of Disaster, *Nature* 271: 321–25.

Mintzer, I. 1987. *A Matter of Degrees.* Washington, DC: World Resources Institute.

———. (ed.). 1992. Confronting Climate Change: Risks, Implications and Responses. Cambridge, MA: Cambridge University Press.

Mooney, C., and R. Duval. 1993. *Bootstrapping: A Non-Parametric Approach to Statistical Inference,* Newbury Park, CA: Sage Publications.

Moss, R.I., and S.H. Schneider. 2000. Uncertainties in the IPCC TAR: Recommendations to Lead Authors for More Consistent Assessment and Reporting. In *IPCC Supporting material, Guidance Papers on the Cross Cutting Issues of the Third Assessment Report of the IPCC,* edited by R. Pachauri, T. Taniguchi, and K. Tanaka. Geneva, Switzerland: World Meteorological Organization, 33–51.

National Research Council. 1983. *Changing Climate.* Washington, DC: National Academy Press.

———. 1984. *Glaciers, Ice Sheets and Sea Level.* Washington, DC: National Academy Press.

———. 1987. *Responding to Changes in Sea Level: Engineering Implications.* Washington, DC: National Academy Press.

Parry, M., and T. Carter. 1984. *Assessing the Impact of Climatic Change in Cold Regions.* Laxenburg, Austria: International Institute for Applied Systems Analysis.

Patt, A.G., and D.P. Schrag. 2003. Using Specific Language to Describe Risk and Probability. *Climatic Change* 61: 17–30.

Renn, O. 1991. Strategies of Risk Communication. In *Communicating Risks to the Public: International Perspectives,* edited by R. Kasperson and P. Stallen. Dordrecht, Netherlands: Kluwer Academic Publishers, 457–481.

Shlyakhter, A., D. Kammen, C. Broido, and R. Wilson. 1994. Quantifying the Credibility of Energy Projections from Trends in Past Data: The US Energy Sector. *Energy Policy* 22(2): 119–130.

Smith, I. 1982. *Carbon Dioxide: Emissions and Effects.* London, United Kingdom: International Energy Agency Coal Research.

Smith, J., and D. Tirpak (eds.). 1989. *The Potential Effects of Global Climate Change on the United States.* Washington, DC: U.S. Environmental Protection Agency.

Tversky, A., and D. Kahneman. 1973. Availability: A Heuristic for Judging Frequency and Probability. *Cognitive Psychology* 5: 207–232.

———. 1988. Rational choice and the framing of decisions. In *Decision Making: Descriptive, Normative, and Prescriptive Interactions,* edited by D. Bell, H. Raiffa, and A. Tversky, Cambridge, MA: Cambridge University Press, 167–192.

UNEP (United Nations Environment Programme). 1989. *The Full Range of Responses to Anticipated Climate Change.* New York: United Nations.

U.S. Department of Energy. 1983. *Carbon Dioxide: Science and Consensus.* Washington, DC: U.S. Department of Energy.

———. 1985. *Information Requirements for Studies of Carbon Dioxide Effects.* Washington, DC: U.S. Department of Energy.

———. 1988. *Workshop Proceedings on Sea Level Rise and Coastal Processes.* Washington, DC: U.S. Department of Energy.

———. 1993. *Climate Change Action Plan: Technical Supplement.* Washington, DC: U.S. Department of Energy.

U.S. EPA (Environmental Protection Agency). 1983. *Projecting Future Sea Level Rise.* Washington, DC: U.S. Environmental Protection Agency.

————. 1987. *Unfinished Business.* Washington, DC: U.S. Government Printing Office.

————. 1988. *Greenhouse Effects, Sea Level Rise and Coastal Wetlands.* Washington, DC: U.S. Environmental Protection Agency.

————. 1993. *Preparing for an Uncertain Climate.* Washington, DC: U.S. Government Printing Office.

————. 1995. *The Probability of Sea Level Rise.* Washington, DC: U.S. Environmental Protection Agency.

U.S. Office of Technology Assessment. 1991. *Changing by Degrees: Steps to Reduce Greenhouse Gases.* Washington, DC: U.S. Government Printing Office.

Van der Sluijs, J. 1997. *Anchoring Amid Uncertainty, on the Management of Uncertainties in Risk Assessment of Anthropogenic Climate Change.* Leiden, Netherlands: Ludy Feyen.

Weber, E. 1997. Perception and Expectation of Climate Change: Precondition for Economic and Technological Adaptation. In *Environment, Ethics, and Behavior: The Psychology of Environmental Valuation and Degradation,* edited by M. Bazerman, D. Messick, A. Tenbrunsel, and K. Wade-Benzoni. San Francisco: New Lexington Press, 314–341.

Zeckhauser, R., and W. Viscusi. 1996. The Risk Management Dilemma. *Annals of the American Academy of Political and Social Science* 545: 144–155.

CHAPTER 7

Limits to Assessment

An Example from Regional Abrupt Climate Change Assessment in the United States

David C. Lund

AS THE NUMBER AND RESOLUTION of paleoclimatic and historical climate observations have improved over the past two decades, so too has our knowledge that climate is capable of changing in surprising ways. Examples include reorganizations in deep ocean circulation, decadal drought in North America, and apparent shifts in the frequency and magnitude of the El Niño Southern Oscillation. In each case, physical change in the climate system is abrupt, occurs over years to decades, is spatially widespread (from continental to global), and involves shifts from one stable state to another, each lasting from years to millennia.

Despite these recent discoveries that climatic changes can be abrupt, global assessments of climate change generally assume there will be a gradual warming of earth's climate over the next several centuries. On the regional level, however, there are examples of abrupt climate change assessments in the United States, including the Colorado River Basin Severe-Sustained Drought Assessment (SSDA), and the Pacific Northwest Regional Assessment, which addresses, among other issues, the Pacific Decadal Oscillation (a type of abrupt change). It is unclear why regionally focused assessments address abrupt climate change more than global efforts, but it may be a function of the difficulty in obtaining consensus on low probability, high impact events in large international assessment projects (Patt 1999, and Chapter 6). The purpose of this chapter is three-fold: (1) to begin to understand how abrupt climate change science can be utilized in integrated assessments by examining the SSDA in the Colorado River Basin, (2) to examine the factors that influenced SSDA effectiveness and their relationship to the issue domain framework presented in Chapter 1, and (3) to compare the SSDA with key elements of the Columbia River Basin experience, which had a different outcome. A key issue that emerges from this study is that even well-designed assessments may lack effectiveness, due to the presence of institutional constraints that limit progressive management options. Paleoclimatic evidence of previous abrupt climate change events appears to be an inadequate catalyst for institutional change, but ongoing crises may provide the necessary motivation for consideration of alternative policy options.

138

Background

A major challenge to global environmental assessments is identifying regional effects of global phenomena. In the case of anthropogenic climate change, for example, current models are not yet capable of reliably resolving regional impacts, and thus the debate has primarily focused on the magnitude and rate of average global temperature change. One of the strengths of the paleoclimatic and historical records is that each is composed of geographically discrete data sets, allowing for regional-level definition of past climate events. While past events do not forecast the future, they can be useful analogies for probing societal vulnerability to, and the level of preparation for, abrupt climate changes.

Tree ring and lake sediment data indicate that during the past 700 years North America likely experienced two long-term droughts, occurring in the late thirteenth and sixteenth centuries, respectively (Fritts 1965; Grissino-Mayer 1996; Meko et al. 1995; Stahle et al. 1985; Woodhouse and Overpeck 1998). These so-called "megadroughts" exhibit the characteristics of abrupt climate change in that they began quickly (over just a few years), covered a large portion of the western United States, and persisted for two decades in some areas. Data coverage is best for the southwestern United States, which experienced a 20-year drought from approximately A.D. 1580 to 1600. While the mechanisms driving decadal drought are unclear, sustained North American droughts of the past 1,500 years correspond to extremes in sea-surface temperature in the Sargasso Sea (Keigwin 1996; Woodhouse and Overpeck 1998), which it turn may be related to the strength of deep Atlantic convection (Bianchi and McCave 1999).

Depending on the region in question, historical observations can provide a detailed record of abrupt climate shifts. For example, the southwestern United States suffered multiyear droughts in the past century (e.g., during the mid-1950s; Cook et al. 1998), but nothing on the order of the late sixteenth century event. In the Pacific Northwest, there is a unique climatic feature known as the Pacific Decadal Oscillation (PDO), which shifts states roughly every 20 years (Mantua et al. 1997). In the warm phase of the PDO, temperatures in Idaho, Oregon, and Washington are significantly warmer than normal, and precipitation is significantly lower (Mote et al. 1999). Salmon mortality off the coasts of Oregon and Washington generally increases during the warm phase of the PDO, when sea-surface temperatures in the Northeast Pacific tend to be warmer and less biologically productive than normal. Salmon survival is further exacerbated by low flow in the Columbia River, which decreases by about 10 percent during the warm PDO phase (Mote et al. 1999). Thus, historical data indicate the Pacific Northwest experiences abrupt shifts in climate, which are analogous to climate events apparent in the paleoclimatic record of the southwestern United States.

Awareness of abrupt climate change in scientific circles has increased substantially in the past five years, as indicated by the large number of recent papers on this topic (e.g., Broecker 1997; Overpeck 1996). Interest is also emerging in the U.S. news media, reflected by stories on deep-ocean circulation shifts in the *New York Times* (Stevens 1998, 1999) and the *Atlantic Monthly* (Calvin 1998), and megadrought coverage in the *Washington Post* (Suplee 1998), the *New York Times* (Stevens 2000), and national news broadcasts. Despite increased attention given to abrupt climate changes, their treatment in global assessment efforts is limited. This is similar to other cases in which consensus-based assessment bodies tend to avoid treatment of low probability, high impact events (see Chapter 6 and Patt 1999).

While abrupt climate change is recognized by the Intergovernmental Panel on Climate Change (IPCC) as an important topic with potentially serious consequences, regionally focused integrated assessments are required to better understand the impact of abrupt climate change on modern socioeconomic systems. One such example is the SSDA, which examined the impacts and mitigation strategies for a 20-year drought in the Colorado Basin (Young 1995). Prior to discussing the details of the SSDA, it is first necessary to outline the context in which the assessment operated. As discussed in subsequent sections, this backdrop is essential to understanding why the SSDA, despite its sophisticated design, had little impact on water management or policy in the Colorado Basin.

The Colorado River Basin

The Colorado River is one of the most highly developed rivers in the world. From its headwaters in the Rocky Mountains to its ephemeral delta in the Gulf of California, the Colorado encounters multiple reservoirs, dams, and diversions. Throughout the Colorado Basin, the river is the primary source of water in an otherwise semi-arid region. Continual conflict over water rights has produced a complex legal structure for water distribution and some of the most notable environmental battles in the history of the United States (McPhee 1971). The Colorado River is subject to two primary demands, consumptive and non-consumptive. The former includes municipal, industrial, and agricultural uses that require taking water from the river, and the latter includes recreational, hydroelectric, and environmental uses that require leaving water in the river. Increasing demand for water in the southwestern United States limits flow to such an extent that the Colorado River often fails to reach the sea (Fradkin 1995).

The Law of the River

Apportionment of Colorado River water between the United States and Mexico, and within the seven U.S. basin states, is governed by a set of rules for water allocation known as the Law of the River (Table 7-1; Getches 1997; MacDonnell et al. 1995). These rules are based on the prior appropriation doctrine, which grants primary water rights to those who first put water to beneficial consumptive use (Wilkinson 1985). Over the past 80 years, U.S. basin states have competed to ensure adequate water supply for their present and perceived future needs and to prevent other states from appropriating water. The result is a highly engineered and regulated system, where consumers have grown to expect a predictable, reliable water supply.

Table 7-1. *Primary Water Allocation Components of the Law of the River*

Law of the River component	Allocation	Method
Colorado Compact, 1922	75 million acre-feet (maf) every 10 years from Upper to Lower Basin	Absolute
Mexican Water Treaty, 1944	1.5 maf per year from Upper and Lower Basin to Mexico	Absolute
Upper Basin Compact, 1948	Colorado 51.75%, Utah 23%, Wyoming 14%, New Mexico 11.25% of available supply	Proportional
Arizona v. California, 1964	California 4.4 maf/yr, Arizona 2.8 maf/yr, Nevada 0.3 maf/yr	Absolute
Colorado River Basin Project Act, 1968	8.23 maf per year release from Glen Canyon Dam	Absolute

Notes: The Colorado River Basin (Figure 7-1) is divided into the Upper Basin (Colorado, Wyoming, Utah, and New Mexico) and the Lower Basin (California, Arizona, and Nevada). The hydrological division between the two sub-basins is Lee's Ferry, Nevada, also known as the Colorado Compact Point.

The five components of the Law of the River listed in Table 7-1 are the primary determinants of water allocation in the Colorado Basin—they dictate which state (or country) receives water, the amount they receive, and how often they receive it. Absolute allocations are independent of total available water, while proportional divisions are based on a percentage of available supply. The Colorado Compact allocates water between the Upper and Lower Basins (Figure 7-1), with the latter guaranteed a volume of 75 million acre-feet (maf) every 10 years, regardless of available flow (MacDonnell et al. 1995). It is now known, however, that the baseline flow rate used to negotiate the Colorado Compact was significantly higher than the twentieth century historical average, implying that in low-flow years, the Upper Basin is more vulnerable to water shortage than is the Lower Basin (Brown 1988). In 1944, an absolute allotment of 1.5 maf/year was established for Mexico, marking the first time total allocation on the Colorado exceeded the historical average flow.[1] During years of average or below-average flow, the Law of the River allocates more water than is actually available, a phenomenon euphemistically known as over-allocation. Today, this is only physically possible because Colorado, Wyoming, Utah, and New Mexico have yet to utilize their entire allocation under the 1922 Colorado Compact (MacDonnell et al. 1995).

In 1948, Colorado, Utah, New Mexico, and Wyoming agreed to share water on a proportional basis under the Upper Basin Compact. This compact was driven by obligations to the Lower Basin and Mexico, as well as the realization that river flows could be much lower than those on which the Colorado Compact was based.[2] To further minimize Upper Basin drought risk, Congress enacted the Colorado River Storage Project Act in 1956, which authorized the construction of several Upper Basin projects, including Glen Canyon Dam (Brown 1988). Glen Canyon Dam is the legacy of flow deliveries required by the 1922 Colorado Compact, and the initial overestimation of available water.

Water allocations guaranteed to the Lower Basin under the Colorado Compact cleared the way for the 1964 *Arizona v. California* Supreme Court decision

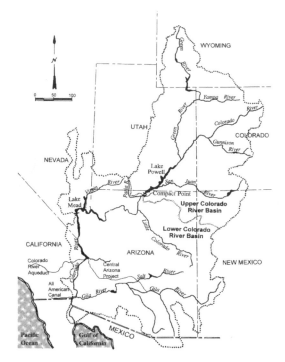

Figure 7-1. *Colorado River Basin*

that designated allotments for Lower Basin states (Table 7-1). With 75 maf/decade from the Upper Basin, the Lower Basin was in the fortunate position of codifying absolute water supply for California, Arizona, and Nevada. In 1968, the Colorado River Basin Project Act established the Long-Range Operating Criteria for Upper and Lower Basin reservoirs, which require that a minimum of 8.23 maf/year be released from Glen Canyon Dam.[3] This absolute allotment removed the limited flexibility available in the Colorado Compact, which originally allowed for an *average* release of 7.5 maf/year over the span of a decade (MacDonnell et al. 1995).

The Colorado Compact set the tone for subsequent components of the Law of the River by encouraging basin states to expect reliable, absolute water deliveries. In essence, it established a feedback mechanism, in which expectations for minimum flows led to the building of massive water storage and delivery structures, which further inflated allocation expectations. This construction/expectation cycle culminated in the annual minimum release from the Upper to Lower Basins of precisely 8.23 million acre-feet.

Accounting for Drought Risk

Drought is accounted for in Colorado River water management in two ways. The first is structural—reservoirs in the Colorado Basin hold a total of approximately four times the historical annual average flow of the river.[4] Most of this water is held in Lake Powell and Lake Mead, massive reservoirs that serve to buffer the basin from short-term drought (and flood) events. Because these reservoirs were near capacity in the 1980s and 1990s, and demand for water continues to increase, much of the controversy in the Colorado Basin focuses on allocation of surplus water, that is, deliveries in addition to the minimum allocations listed in Table 7-1 (Getches 1997). As a result, there was generally little interest in short-term water shortages.

Long-term drought is taken into account primarily through the Long-Range Operating Criteria. The criteria specify reservoir operating rules that depend on projections of water supply and demand. One of the key factors used to determine available supply is the historical critical flow period, which occurred from 1952 to 1964.[5] Simulations of future demand in combination with the 12-year low-flow period indicate the possibility of supply shortages—that is, if the entire Upper Basin storage capacity cannot meet Upper Basin needs plus obligations to the Lower Basin and Mexico. In this scenario, the annual release of water at Glen Canyon Dam is limited to 8.23 maf, limiting the amount of surplus water available to the Lower Basin. There is no provision in the Operating Criteria to authorize lesser flows, even during extreme drought years. In general, the criteria deal primarily with allocating surplus water and guaranteed releases from Glen Canyon Dam (MacDonnell et al. 1995). This optimism is similar to that expressed during the original Colorado Compact negotiations, where state negotiators quickly accepted forecasts of abundant water supply in the hope that it would ease interstate cooperation, at least in the short term (Brown 1988; Hundley 1975).

Basin state representatives, the Bureau of Reclamation, and the Secretary of the Interior review the Long-Range Operating Criteria approximately every five years. Beginning in 1990, participation in the review process expanded to include interested stakeholders such as the Environmental Defense Fund, American Rivers, and the National Park Service (Bureau of Reclamation 1997b). Despite repeated reviews and broadened participation in the review process during the past decade, there have been no changes made in the criteria since their original inception in 1968 (Bureau of Reclamation 1997a). In the words of one Reclamation official, "it's a rubber stamp process," apparently due to the limited flexibility allowed for the criteria by the Colorado River Basin Project Act and to the political difficulty in altering rules that have become accepted operational norms.

The Severe-Sustained Drought Assessment

Tree-ring data reveal the existence of a late sixteenth-century drought event in the Colorado Basin that was both longer and more severe than the critical flow period currently used by the Bureau of Reclamation to determine potential water shortages (Meko et al. 1995; Stockton and Jacoby 1976; Tarboton 1995).[6] How sensitive are water resources in the Colorado Basin to a severe and sustained drought, and what options are available for impact mitigation? To answer these questions, the SSDA project, composed of academically based experts in Colorado River hydrology and policy, tested the capacity of regional reservoirs and the performance of the Law of the River under extreme circumstances similar to the late sixteenth century drought event.

Using a model of basin hydrology, management facilities, and operating rules, the SSDA experts performed a gaming exercise to determine how representatives from Colorado Basin states would respond to the evolution of a 20-year drought event and to the decisions of other participants. While participants in the exercise were able to minimize impacts to consumptive users, nonconsumptive users such as hydropower generators, recreationists, and endangered species were adversely affected (Henderson and Lord 1995). For example, at the peak of the simulated drought, Lake Powell emptied and Lake Mead lowered substantially (Harding et al. 1995), resulting in annual hydropower generating losses of $600 million (Booker 1995) and the local extinction of multiple fish species (Hardy 1995). In general, the Upper Colorado Basin bears a greater drought risk, due primarily to existing compact guidelines that guarantee the Lower Basin minimum flows. As a result, the model showed that deliveries for consumptive use in the Upper Basin fell to about half normal levels, whereas they were relatively unaffected in the Lower Basin (Lord et al. 1995).

To minimize drought impacts, assessment participants recommended the creation of a federal interstate compact that would establish a commission with the technical credibility and political legitimacy to (1) better balance nonconsumptive and consumptive uses, (2) allocate water based on current demands rather than 1922 allocations, (3) establish proportional drought-sharing on a basin-wide scale, and (4) manage interstate water transfers and water banks to minimize the impacts of severe drought (Lord et al. 1995). In other words, the project participants argued for increased diversity of water management techniques to improve system resilience to water shortages. The participants concluded that current allocation rules outlined under the Law of the River lack the flexibility to mitigate drought impacts across the Colorado Basin. Preliminary discussions with SSDA authors, however, indicate that the assessment has had little impact on policy or management strategies in the Colorado Basin. One of the aims of my study was to determine why the assessment was ineffective in changing the status quo.

Understanding the Assessment through Interviews

Semistructured, open-ended interviews were performed with 26 key people in the Colorado Basin, including the principal investigators of the SSDA project and members of the SSDA Advisory Council (Table 7-2). The council included representatives from state engineer offices, state departments of natural resources, the Central Basin Water District (California), the Upper Colorado River Commission, the Western States Water Council, the Colorado River Board of California, the Colorado River Water Conservation District, the Metropolitan Water District of Southern California, and several academic and practicing experts in water law. The council was essentially a "who's who" of interstate water management in the Colorado Basin, established to ensure that results of the SSDA reached the appropriate stakeholders and to provide a venue for feedback regarding study design and recommendations. The council was established in the late 1980s at the end of the initial SSDA scoping stage (Phase 1; Gregg and Getches 1991), with the intent that comments and suggestions could be incorporated into the subsequent stage (Phase 2; Young 1995), the funding for which began in

Table 7-2. *Colorado Basin Interviewee Affiliation*

Professional affiliation		Advisory council representation	
SSDA investigators	9	Upper Basin	4
State engineers	5	Lower Basin	3
Reclamation officials	5	Upper and Lower Basin	1
Miscellaneous organizations	4		
External experts	3		
Total	26		

Notes: The left-hand column includes all interviews performed for the SSDA study. Interviewees included SSD principal investigators from the University of Colorado, University of Arizona, and Utah State University, with expertise in law, economics, hydrology, sociology, and public administration; state engineers or their equivalents from Nevada, Wyoming, Arizona, Utah, and the Metropolitan Water District of Southern California; officials from the Bureau of Reclamation's Lower and Upper Colorado Regional Offices and the Commissioner's Office in Washington, DC; representatives from the Colorado River Board of California, the Colorado River Water Conservation District, the Western States Water Council, and the Upper Colorado River Commission; and water resource policy experts not directly involved in authoring the SSD from the University of Wyoming, and University of California, and the Environmental and Societal Impacts Group at the National Center for Atmospheric Research. Of the individuals in the left-hand column, eight served as members of the Advisory Council, with their geographic allegiance noted.

the early 1990s. Phase 2 was designed to include a more thorough assessment of environmental impacts and a gaming exercise to simulate the interactive decisionmaking process that would occur during an extended drought.

The U.S. Man and the Biosphere Program funded Phase 1, while the U.S. Geological Survey and the U.S. Army Corps of Engineers primarily funded Phase 2. Because these agencies provided financial support through a grant proposal process, they had little vested interest in SSDA results, and they were therefore not interviewed for this study. Additional funds came from the Metropolitan Water District of Southern California, the Upper Colorado River Basin Commission, the University of Arizona's Water Resource Institutes, the University of California, Colorado State University, Utah State University, and the University of Wyoming (Young 1995). Many of these organizations were represented as SSDA principal investigators or members of the SSDA Advisory Council, who were among the people interviewed.

Phases 1 and 2 of the SSDA were the primary documents used for background information for study results and recommendations. Additional written materials covering historical documentation of the Law of the River, environmental controversies in the Colorado Basin, Bureau of Reclamation reservoir operating criteria and hydrological data, impacts of interannual climate variability on Colorado River flow, and the websites of multiple organizations in the basin were essential to formulating interview questions and to providing the historical, legal, and political context of water allocation in the Colorado Basin.

Interview Results and Factors Leading to Ineffectiveness

Environmental assessments can be evaluated both in terms of their effects and their effectiveness. Effects cover the entire range of consequences of an assessment, regardless of original intent, whereas effectiveness is directly related to the intent of the designers and participants in an assessment process. For the purposes of this study, I defined effectiveness as the degree to which each of the following three outcomes occurred (listed with increasing impact on water management and policy in the Colorado Basin):

1. Water managers' framing of extreme drought changed from one based solely on historical experience to one that includes an extreme event from the paleoclimatic record;

2. Methods by which drought risk is determined and included in the long-range management of Colorado Basin reservoirs were expanded to include tree-ring reconstructions of river flow; and

3. A diverse set of water allocation policies (along the lines of those suggested by the SSDA) were created to better cope with the impacts of long-term drought on a basin-wide scale.

Almost without exception, interviewees revealed that the SSDA changed their perception of extreme drought in the Colorado Basin. Many of the SSDA Advisory Council members, while aware of drought reconstructions based on tree rings, were not cognizant of the impacts such a drought would have on modern water uses. In this limited sense, the assessment achieved the first measure of effectiveness listed above—several water managers emerged from the assessment process with an expanded perception of the potential damage of severe drought.

While awareness of prehistoric drought increased as a result of the SSDA study, this awareness appears to have had little impact on Colorado River water management and policy. As discussed in the drought risk section, the Long-Range Operating Criteria for Colorado Basin reservoirs utilize a less extreme historical critical flow period (1952–1964) for long-range supply projections, as opposed to the more severe late sixteenth-century event. Interviews with Bureau of Reclamation officials indicate the choice of using the former flow period is due primarily to bureaucratic inertia and the politically sensitive nature of using a more dire drought event. After using the same critical flow period for the past 30 years, it is now accepted operational practice to make supply projections based on these historical flows. Introducing a low-flow event based on nonstandard hydrologic techniques would likely encounter opposition from state-level engineers. Furthermore, the use of the late sixteenth-century event would have the practical effect of limiting the amount of surplus water available to Lower Basin states, a politically unsavory consequence. In the words of one Bureau of Reclamation official, "[the basin states] don't want to hear a bleak story." Continued review of the Long-Range Operating Criteria may change current decisionmaking patterns, but for the time being, the SSDA appears to have fallen short in modifying the long-range operation of Colorado Basin reservoirs.

The most challenging measure of effectiveness for the SSDA is the third outcome on the list—to what extent did assessment recommendations encourage flexible drought management policies in the Colorado Basin? All of the interviewees agreed the assessment had little or no impact on policies that would

Table 7-3. *Primary Factors Influencing the Effectiveness of the SSDA Policy Recommendations*

Factors	Principal investigators	Advisory committee	Bureau of Reclamation
Political-legal context	X	X	X
Timing*	X	X	X
Choice of drought scenario*	X	X	
Consumptive uses protected	X		
Follow-up, continuity*	X		

Notes: * denotes factors that could potentially be influenced by assessment design—others were beyond the scope of the assessment. Each row of the table represents a key determinant of SSDA effectiveness (see text for details). The interviewees were separated into three groups (principal investigators, advisory committee, and Bureau of Reclamation), each represented by a column. The Xs represent those factors that each group believed were important determinants of effectiveness.

improve resilience to drought on a basin-wide scale. According to interviews with principal investigators, SSDA Advisory Council members, Bureau of Reclamation officials, and other water managers and experts, several factors led to this ineffectiveness (Table 7-3).

Political–Legal Context The highly political nature of water allocation in the Colorado Basin and the issues addressed by the SSDA are critical to explaining both the effects of the assessment and its lack of effectiveness in encouraging alternative water management policies. Without exception, each person interviewed highlighted political barriers as the primary reason that policy recommendations in the SSDA had little impact. In general, decisionmakers have little interest in data that imply the river is over-allocated or that the current allocation rules are too rigid to adequately cope with severe drought. Interviews reveal that a basin-wide interstate compact commission, the primary recommendation of the SSDA participants, is a politically contentious topic, largely due to fears that such an arrangement would erode state-level control of water rights and allocation procedures. SSDA Advisory Council members from California and Arizona expressed little interest in the formation of a basin-wide commission—historically they've been able to successfully address their water needs without one. Indeed, the Law of the River is testament to the decades-long struggle of basin states to earn and retain water rights—it is not surprising that a perceived threat to that allocation would be unpopular. While a basin-wide entity would likely improve equitable sharing of water with Native American and environmental demands (Getches 1997), the idea seems to be viewed by many states as a zero-gain prospect, at best adding a layer of bureaucracy and preserving current allocations, at worst yielding water deliveries lower than historical amounts.

Interview results also indicate that the establishment of interstate water banks between Upper and Lower Basin states is an unpopular idea, particularly in the Upper Basin. This is despite the fact that Colorado, Wyoming, Utah, and New Mexico have yet to utilize their entire allocation under the 1922 Colorado Compact (MacDonnell et al. 1995). Although the banks would in theory act as temporary water transfer devices, most likely from agricultural to municipal uses, states fear that this water would be lost permanently. This fear is not entirely unjustified, given growing demand in Upper Basin states and the continued requests for surplus water from Lower Basin states, particularly California. The Colorado River Water Conservation District, for example, which protects the water rights of 15 counties in west central and northwest Colorado, likens itself to the biblical David, and California to Goliath "casting a covetous eye on currently unused Colorado River water" (CRWCD 2000a).

One effect of the SSDA was to clarify the goals of competing water uses on the Colorado River, particularly between consumptive and nonconsumptive uses and between the Upper and Lower Basins. As highlighted in the summary chapter of the assessment:

> Existing operating rules . . . favor consumptive water uses over such nonconsumptive uses as hydroelectric power generation, environmental protection, salinity control, and recreation. The extent of this favoritism . . . is out of all proportion to what are, arguably, the public values involved (Lord et al. 1995, *942*).

Three of the eight SSDA Advisory Council members interviewed found assessment principal investigators to be predisposed to protecting nonconsumptive uses. In one case, a council member recalled that the study helped him clarify the agenda of nonconsumptive water use advocates. Furthermore, recommendations such as creation of a basin-wide commission and the establishment of interstate water banks had been previously suggested in other venues with limited positive reception from the basin states.

With positions on key issues already established by groups represented in the SSDA Advisory Council, the SSDA recommendations clarified different agendas—in the words of one council member "we went in with stated positions on these issues, and study results weren't likely to change them."

Timing and Crisis The SSDA study was published in 1995. Between 1995 and the time of this study, levels in Lakes Mead and Powell have been near capacity (Figure 7-2; Bureau of Reclamation 2000). Together, these two reservoirs can hold more than three times the annual flow of the Colorado River. When they're full, the concern for drought tends to be low. Rather than focusing on long- or short-term drought, the Colorado Basin states are currently preoccupied with allocating surplus water. Interviews across the three groups listed in Table 7-3 concur that drought isn't an immediate or long-range concern in the Colorado Basin, and that the timing of SSDA during abundant water years seems to have prevented it from having a greater impact.

If a severe drought crisis had coincided with the publication of the study, the result may have been different. Indeed, in "closing water systems" such as the Colorado, crisis appears to be necessary to promote recognition of interdependent uses and negotiation of agreements among basin states (Pulwarty and Melis 2001). As discussed in the next section, the crisis of declining salmon populations in the Columbia River Basin has led to the consideration of alternative management approaches, including recognition of the ocean's role in salmon mortality, and the once-radical idea of dam removal. Given the lack of a crisis event in the

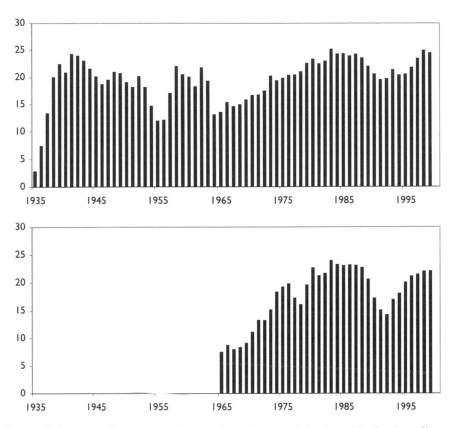

Figure 7-2. *Annual Average Volumes for Lake Mead (top) and Lake Powell (bottom), 1935–1999 (in millions of acre feet)*

Colorado Basin, it may be premature to evaluate the effectiveness of the assessment, as its findings and recommendations could yet be used in future drought scenarios.

Although the timing of publication may be important, another key timing element is that of decisionmaker involvement in the assessment. Much of the West was in the grip of drought during the SSDA Advisory Council meetings in 1991 and 1992. During that time, water levels in Lake Powell were at their lowest since the initial filling of the reservoir, and storage in Lake Mead dropped by nearly 4 maf compared with the wet years of the mid-1980s (Figure 7-2). Both Arizona and California, for example, faced limits to surplus water from the Colorado River, and strict conservation programs were enacted in southern California and Las Vegas (Fradkin 1995). Thus, the SSDA was quite timely from a decisionmaker standpoint because the impacts of drought were occurring in the states represented by council members. If poor timing undermined the effectiveness of the SSDA, then influence on policy appears to require a drought more severe than the 1987–1992 event.

Finally, limited dissemination of study results to the general public precluded the accumulation of political support necessary to change policy. During the early 1990s, the SSDA Advisory Council members were apparently the only stakeholders intimately aware of both the ongoing drought event and the policy implications of the assessment.

Choice of Drought Scenario Assessments of abrupt climate change are faced with the difficult task of addressing low probability, high impact events that adequately represent the range of climate variability in a region, while simultaneously steering clear of examples so extreme that nobody pays attention (i.e., that lack salience). For example, initial discussions with SSDA principal investigators revealed concern that the drought scenario chosen for analysis was too extreme. In particular, the low probability of the event may have prevented the assessment from registering with stakeholders in the Colorado Basin. Further interviews, however, indicate that only a minority of SSDA Advisory Council members and investigators believed that the low probability of the drought prevented its full consideration by water managers. In most cases, interviewees felt that the historical occurrence of severe drought, even if only once in the past 500 years, was adequate justification to consider the implications of such an event for current and future water allocation policies. Thus, despite the limited precedence for analogous drought events, the SSDA participants appear to have chosen an event that was salient to decisionmakers in the region.

The drought scenario used in the assessment was based on water flows estimated using tree ring widths, but with the flows rearranged in time. The rearranged scenario ordered annual values such that each year during the hypothetical 20-year event had progressively lower flow (Tarboton 1995). Although the total volume of Colorado River flow over the entire drought remained the same, the timing of the flows was altered to create an event even more extreme than that in the paleoclimatic record. Despite original intentions to determine the impacts associated with both scenarios, the rearranged drought was the event on which the entire assessment was based. This scenario was used to create a worst-case contingency for testing the ability of the regional reservoirs and the Law of the River to cope with extreme, and perhaps unrealistic, drought.

One of the groups that funded the study—the Metropolitan Water District of Southern California (MWD)—doubted the credibility of the rearranged scenario. Correspondence between the MWD and the study organizers indicate the MWD found the rearranged scenario to have no hydrological basis, and therefore not appropriate for use in the assessment. Referring to the rearranged drought, an engineer from the MWD wrote to SSDA organizers in 1994:

> Considering its nature, it is possible that water resources managers will dismiss the findings, conclusions, and recommendations contained in the report due to the authors' choice of assumption for the representative drought. (Matusak 1994)

The MWD repeatedly made attempts to comment on the SSDA report, but most of the comments, including doubts about the rearranged scenario, were left unaddressed (Matusak 1997, 2000). This apparently is the result of both the linear nature of the assessment process and skepticism of the MWD's motivations. Although the SSDA lasted for nearly 10 years, each stage of the project built on the previous stage—to make the major changes requested by the MWD would have required that the study be completely redone, an unrealistic request given funding constraints. Also, the MWD has expressed serious reservations about the need for a basin-wide commission, because majority-based voting in such an arrangement might have led to reduced water supplies for California (Matusak 1997, 2000). Because establishment of a basin-wide entity was a primary SSDA recommendation, it is possible that the MWD chose to discredit the study on technical grounds. Interviews with assessment organizers indicate that skepticism of the MWD's motivations was a major reason that comments about the rearranged drought scenario were not taken into account.

Consumptive Uses Protected Several of the SSDA investigators indicated that one reason the assessment was ineffective was due to the ability of the established Colorado Basin water distribution system to protect consumptive water uses. With the Law of the River in its current form, the assessment determined that damages to consumptive uses in the Upper Basin during a severe-sustained drought could be significant—up to $350 million per year during the most severe part of the drought—but still minor compared with nonconsumptive use damages. With additional management options, such as water transfers, marketing, and retaining water at high elevations to minimize evaporation, overall damages to consumptive uses could be reduced by greater than 80 percent (Booker 1995). In other words, the assessment found that even when subjected to the most severe drought of the past 500 years, the system of reservoirs, dams, and aqueducts on the Colorado River could sustain most consumptive water uses, at least at current demand levels. In this sense, the assessment may have assuaged fears that long-term drought would have devastating impacts on agricultural, industrial, and municipal uses, thus resulting in little or no change in current policies or management strategies. Curiously, this response was not observed among SSDA Advisory Council members or Bureau of Reclamation officials.

Follow-up and Continuity The SSDA study, which lasted nearly 10 years in total, went through a variety of funding uncertainties and leadership changes. The initial source of funding was through a grant award by the U.S. Man and the Biosphere Program, which in the mid-1980s funded the Phase 1 scoping stage (Gregg and Getches 1991). Similarly, Phase 2, designed to elaborate on the findings of Phase 1 and incorporate stakeholders through the advisory council process, was primarily funded through grant awards by the U.S. Geological Survey and the U.S. Army Corps of Engineers. Several other organizations also provided support for Phase 2, although the contributions were minor in comparison. As the assessment came to a close in 1994–1995, project finances were running low, and it was necessary to cobble together funding from multiple sources to ensure completion of the study.

Partly as a consequence and partly as a cause of its patchy funding history, the SSDA study had several different leaders through its 10-year lifespan. Leadership changed between investigators at the University of Colorado and Arizona, to

Utah State University, back to the University of Arizona, and then finally to Colorado State University. By the time the study had neared completion, funding had run out, and many of the authors, with responsibilities to teach and explore new research, had little incentive to actively pursue publication of the SSDA. Through the exceptional efforts of a few individuals, the study was eventually published in the peer-reviewed *Water Resources Bulletin,* and then republished by the Powell Consortium as a monograph on severe-sustained drought.[7] There were no additional efforts to further publicize the study. As a result, the assessment became an academic document, therefore minimizing public uptake of its results and implications.

Although the SSDA process lasted several years, there was little opportunity for the principal investigators to incorporate significant changes into the assessment design or methodology, particularly once the study had begun. If the funding history had been more stable and the leadership more continuous, it is conceivable that changes could have been fully incorporated as alternative assessment scenarios, allowing for multiple iterations based on continued communication between the principal investigators and stakeholders. In this sense, the assessment is a somewhat traditional, linear assessment; as opposed to the more dynamic and long-term regionally based efforts now emerging in the United States. It is of course impossible to know whether a more responsive assessment would have yielded a different outcome. The Columbia River Basin assessment process, however, appears to have benefited from a more dynamic approach, suggesting that a different outcome may have been possible for the SSDA, had it used a similar methodology.

The Issue Domain Framework and the SSDA

The issue domain is a broad concept, including the actors involved in the issue, their beliefs, and their strategies for dealing with the issue; the actors' institutions; the decisions, policies, and agreements that emerge from the institutions; and the potential impacts of these behaviors on the natural environment itself. The SSDA is an interesting case study because it was well designed in terms of legitimacy, salience, and credibility, yet in the end it was largely ineffective. This case highlights the importance of other issue domain factors, such as institutional and timing constraints, that are key determinants of effectiveness.

Legitimacy

At the outset of this research project, I assumed that SSDA participants were largely from academic institutions, and the users of the study were primarily funding agencies, such as the U.S. Geological Survey and the U.S. Army Corps of Engineers. It was originally hypothesized that the scope of participation in the SSDA study eroded assessment legitimacy among excluded stakeholders—those whose interests were potentially affected by assessment recommendations but who were not directly involved in the assessment. Subsequent research, however, indicates that many key stakeholders were directly involved in the assessment as part of the SSDA Advisory Council. While the actual authors of the assessment were largely from an academic background, the council included representatives from the state engineer's offices, state departments of natural resources, the Upper Colorado River Commission, the Colorado Water Conservation District, and others. Given this broad representation—both between the Upper and Lower Basin and covering consumptive and nonconsumptive water uses—people interviewed for this study generally viewed the assessment as having legitimate representation.

Salience

One of the primary challenges facing assessments of abrupt climate change is issue salience. Abrupt climate changes tend to occur infrequently, often separated in time by decades or centuries, and therefore few stakeholders have experienced the impact of these events. At the beginning of this study, I hypothesized that the low probability of the SSDA scenario prevented it from being salient to stakeholders in the Colorado Basin. Interviews with SSDA Advisory Council members, however, reveal that the drought event was salient to stakeholders. Most of the council members interviewed (six of eight) indicated that the occurrence of such an event in the past implied that similar events could occur in the future, and this was reason enough to better understand the impacts of extreme drought. These sentiments echo the initial approval of the drought scenario by the council during the early 1990s (SSDA 1992). Overall, the interface between science (SSDA principal investigators) and policy (SSDA Advisory Council members) appears to have been sufficiently porous to ensure that the scenario that formed the basis of the assessment was salient to most water managers in the Colorado Basin.

A minority of the stakeholders interviewed indicated that the low probability of the drought scenario used in the assessment did affect their perceptions of assessment results, insofar as they felt that it was highly unlikely that such an event would occur again. In this respect, it may have been helpful to include another drought scenario in the assessment based on a more recent, less extreme event. Of course, this would have required additional investments of time and money, perhaps an unrealistic expectation for a study with such a long and complex funding and leadership history. If time and budgetary constraints had been relaxed, the assessment process may have allowed for the inclusion of an additional drought scenario, thus improving the overall salience of the assessment.

According to interviews with both principal investigators and SSDA Advisory Council members, the assessment never emerged onto a broader political agenda or received significant media attention, despite its credentials of salience and legitimacy. Major environmental issues tend to have attention cycles characterized by (1) a pre-emergent stage, when knowledge and activity related to the issue is concentrated in limited scientific and management circles; (2) an emergent stage when media and political attention dramatically increases, and a much larger group of stakeholders becomes involved in the assessment; and 3) a post-emergent stage, when general public and high-level political interest in the topic wanes, but assessment activity continues with an altered group of stakeholders

(Clark et al. 2001). An assessment that fails to change with the times, altering its participation in concert with the stage of the issue attention cycle, will tend to be ineffective.

The primary reason the SSDA did not make it to the emergent stage was the absence of a crisis event. If an extreme drought had occurred coincidentally during the assessment process, then it is quite likely that the impact of the assessment on water policy in the Colorado Basin would have been different. The primary issue for the assessment is not its adaptation to evolving media and political attention, but the absence of an event required to move the issue of extreme drought onto a broader stage in the first place. This is largely due to the topic the assessment addressed—a low probability event that in all likelihood would not occur during the assessment process. Many of the assessments studied in this book tend to focus on chronic environmental problems such as global warming, ozone depletion, and acid rain. As a result, these matters have a salience advantage over environmental issues more periodic or infrequent in nature (such as extreme drought in a discrete geographic region) because the chronic environmental problem and its assessment tend to occur simultaneously. Thus, the timing of an assessment relative to the impacts on which it focuses is an important factor influencing assessment salience, and hence effectiveness.

Credibility

The drought scenario used in the SSDA was also generally credible to SSDA Advisory Council members. Despite the unusual severity and duration of the drought, most members of the committee felt that it was a reasonable case for testing management of Colorado River water resources under extreme circumstances (see Choice of Drought Scenario section, above). One notable exception was the Southern California Metropolitan Water District. The MWD doubted the validity of the re-arranged drought scenario and suggested the unrealistic flows would preclude adoption of the SSDA recommendations by water resource managers. Given that the drought used in the SSDA was an altered version of the tree-ring record, the MWD criticism seems reasonable. However, the MWD had openly opposed the formation of a basin-wide commission and recognized that the use of a more severe drought scenario in basin reservoir management would result in less frequent deliveries of surplus water to the Lower Basin. While politically motivated questioning of SSDA credibility by the MWD is difficult to document, it is not surprising that an entity tasked with managing the water supply for 18 million people in southern California (MWD 2000) would be skeptical of information that could potentially interfere with its responsibilities.

Evolution of the Issue Domain

One of the primary factors that influenced the effectiveness of the SSDA is the political–legal context of water management in the Colorado Basin. A long history of conflict over water in this semi-arid region has produced an intricate arrangement of interstate compacts, a Supreme Court ruling, an international treaty, and federal statutes that dictate states' water allocations. Previous studies of Colorado River water law (e.g., Brown 1988) and interviews from this study with principal investigators of the assessment, SSDA Advisory Council members, and Bureau of Reclamation officials, indicate the Law of the River places formidable restrictions on water allocation in the Colorado Basin and would likely inhibit flexible basin-wide mitigation of severe drought impacts.

In the Colorado Basin, institutional characteristics inhibiting progressive interstate water management were the primary determinants of SSDA effectiveness. An institution is defined here as the "the sets of rules or conventions that govern the process of decision-making, the people that make and execute these decisions, and the edifices created to carry out results" (Gunderson et al. 1995). In the case of the Colorado Basin, the rules or policies governing decision-making are outlined by the Law of the River, the people who make decisions are primarily those in the Bureau of Reclamation and state engineers' offices, and the edifices include the multitude of water storage and conveyance structures throughout the region. If one were to implement a well-designed assessment in terms of participation, the treatment of uncertainty and dissent, and the science–policy interface, larger political and legal factors may still prevent it from being effective, even if it addresses salient issues, uses credible science, and is politically legitimate. Indeed, the negative influence of political and legal factors appears to be the case for the SSDA, where ineffectiveness was more a function of institutional characteristics than design factors.

Institutions serve as a filter for assessments, either promoting or inhibiting evolution of the issue domain. The term "evolution" is used here not to imply an improvement, but rather a clear change in stakeholder beliefs, strategies, or institutions due to an assessment or an external crisis factor. The SSDA had two primary effects on the issue domain in the Colorado Basin: it raised stakeholder awareness of the potential magnitude of long-term drought and its attendant impacts, and it clarified competing interests among stakeholder groups. In issue domain terms, the beliefs of stakeholders developed, but their strategies and behaviors for coping with long-term drought remained unchanged. As outlined above, this is primarily due to the institutional setting of the issue domain. Given the right circumstances, the issue domain may eventually evolve to formally acknowledge paleoclimatic droughts via altered behaviors and institutional characteristics, but the assessment on its own was unable to catalyze these changes.

The Columbia River Basin

The Columbia River drains an area comparable in size to the Colorado Basin, but the Columbia has nearly 10 times the annual flow of the Colorado River (Skogerboe 1982). As a result, conflict over consumptive water uses in Oregon, Washington, Montana, and Idaho is essentially nonexistent compared to the contentious history of water rights in the Colorado Basin. Nonconsumptive uses are the primary sources of conflict in the Columbia Basin—particularly the balancing of hydropower generation and salmon habitat preservation. A total of 79 different hydropower projects provide nearly 80 percent of the region's electricity (Wilkinson and Conner 1987), making the Columbia one of the most heavily developed and managed river systems in the world. As a result of hydroelectric development, over-fishing, habitat degradation, and hatchery fish production, wild salmon populations in the Columbia Basin declined precipitously during the twentieth century.

In 1980, Congress enacted the Pacific Northwest Electric Power Planning and Conservation Act, which established the Northwest Power Planning Council (NPPC), an interstate body charged with creating and coordinating regional plans for hydropower development and salmon conservation. Using proceeds from the Bonneville Power Administration, which markets hydropower from federal dams, the council funds an ambitious salmon restoration program costing more than $130 million per year (Lee 1995). Engineering efforts of epic proportions have been undertaken for salmon recovery, including fish ladder construction to facilitate upstream migration of adult salmonids, hydropower turbine screens, barging programs, and increased dam spillage to aid downstream migration of juveniles. Despite these efforts, salmon populations continue to decline. In the early 1990s, several salmon stocks in the Snake River, a major tributary to the Columbia, were listed under the Endangered Species Act (Larmer 1999).

Superimposed over this general decline in wild salmon stocks is decadal-scale variability in salmon abundance, which only recently has been linked to periodic shifts in oceanic productivity. The primary mode of decadal climate variability in the Pacific Northwest is the Pacific Decadal Oscillation (PDO)—a shift in North Pacific oceanic and atmospheric conditions that appears to occur roughly every 20 years (Hare and Francis 1995; Mantua et al. 1997). Salmon mortality along the coasts of Oregon and Washington generally increases during the warm phase of the PDO, when sea-surface temperatures in the Northeast Pacific Ocean tend to be warmer and less biologically productive. During the cold phase, climatic effects tend to improve the chances for salmon survival.

The purpose of this section is to determine if and how knowledge of shifts in oceanic productivity is used in the management of salmon in the Columbia Basin. By comparing this case to the SSDA experience in the Colorado Basin, the object is to outline key factors influencing the transfer of abrupt climate change science into regional-level resource management. Based on these two cases, the primary factors motivating application of this information appear to be institutional flexibility and the presence (or absence) of crisis events.

Understanding Adoption of PDO Information through Interviews

Semistructured, open-ended interviews were conducted with climate assessment experts from the Pacific Northwest (Table 7-4). This primarily included members of the Pacific Northwest Regional Assessment, also known as the University of Washington's Joint Institute for Study of the Atmosphere Oceans Climate Impacts Group (JISAO-CIG). Given the time constraints of this study, and given that a primary responsibility of the Pacific Northwest Regional Assessment is to routinely interact with resource managers in the region, I used the JISAO-CIG interviews as a barometer for the adoption of abrupt climate change information in the Pacific Northwest. Discussions with JISAO-CIG members were supple-

Table 7-4. *Pacific Northwest Interviewee Affiliation*

Total number of interviewees	9
Climate impacts experts	4
Fisheries experts	2
Industry representative	1
NPPC representative	1
External climate expert	1

Note: Interviewees included experts on the socioeconomic impacts of climate variability and change from the Joint Institute for the Study of Atmosphere and Ocean-Climate Impacts Group, and the Center for Analysis of Environmental Change, Oregon State University; fisheries experts from the National Marine Fisheries Service and the University of Washington; a representative from the Columbia River Alliance (an industry group); a representative of the Northwest Power Planning Council; and a climate and resource policy expert from the Environmental and Societal Impacts Group at the National Center for Atmospheric Research.

mented with interviews of key people familiar with the impacts of climate variability on salmon populations in the Columbia Basin. These discussions covered academic, industry, and state and federal agency perspectives.

I used the recent report by participants of the Pacific Northwest Regional Assessment (Mote et al. 1999) as a primary background document in this study. Additional written materials, including historical accounts of salmon management in the Columbia Basin, Northwest Power Planning Council documents, studies on the use of interannual climate information in the Pacific Northwest, and the websites of organizations in the basin, were essential to formulating interview questions and providing the historical and political context of salmon management in the Columbia River Basin.

Interview Results and Inter-basin Comparison

The recent scientific connection between variable climatic conditions and salmon populations has been quickly assimilated into the debate on how to restore salmon populations in the Columbia Basin. Over the past five years, the framing of the debate has shifted from one focused on freshwater habitat as the primary zone of salmon mortality to one that more fully recognizes the role of the ocean, where salmon spend most of their lives (Bisbal and McConnaha 1998, 1999; ISAB 1999). A 1996 amendment to the original 1980 Northwest Power Planning Act now mandates the NPPC to "consider the impact of ocean conditions on fish and wildlife populations" when planning restoration efforts.[8] This is a significant change in management of the Columbia River and reflects an awareness of the oceanic component of salmon ecology.

Partly as a result of the 1996 amendment, the NPPC advocates limited hatchery production and hydropower generation to allow salmon populations to better cope with a variable oceanic environment. Advocates of this view, including representatives of the NPPC, argue that management strategies must take into account both freshwater and oceanic habitat and the natural variability in each. For example, competition between hatchery and wild salmon over limited resources could be reduced if hatchery production were reduced during intervals of poor ocean conditions (NPPC 1999). Water spills from reservoirs could also be timed not to maximize the total number of juvenile fish migrating downstream, but to increase their diversity, thus improving chances for oceanic survival (Bisbal and McConnaha 1998). Broad biological diversity is the primary means by which salmon populations adapted to past climatic and oceanic variability—constraints on this diversity from hydropower development, habitat degradation, and hatchery production serve to increase salmon mortality over the long-term. In further recognition of the habitat continuum between the ocean and freshwater environments, the NPPC highlights the need for research into the role of estuarine and oceanic river plume conditions in salmon ecology. By improving conditions in the Columbia River estuary and plume to approximate predevelopment characteristics, overall salmon survival should benefit as a consequence (Bisbal and McConnaha 1998).

While the management strategies advanced by the NPPC are in a formative stage, there is clearly the institutional desire and monetary support to pursue progressive fishery management techniques. On the Snake River, a major tributary of the Columbia, the breaching of four dams has been considered as a viable option by federal agencies in salmon recovery plans.[9] The early 1990s Endangered Species Act listing of the four remaining Snake River salmon stocks and their continued spiral toward extinction spurred this radical approach. The crisis was also partly prompted by the spring 2000 deadline from the Clinton administration for the U.S. Army Corps of Engineers, Bonneville Power Administration, and National Marine Fisheries Service to create a restoration plan for the endangered species (Larmer 1999).

Breaching of Snake River dams was controversial, particularly for those industries that rely on the dams for barge transportation and hydropower. The Columbia River Alliance, which represents aluminum manufacturers, wheat growers, and other industrial interests in the basin, argued that breaching was an "unreasonable course of action" that was "economically harmful and will not help recover salmon" (Lovelin 2000). Industry groups argued that factors other than dams, such as over-fishing and poor oceanic conditions, exerted the dominant control over salmon populations (Barker 2000). Many environmental interests, Native American tribes, and fishery organizations countered that breaching must occur to save the salmon from extinction (Larmer 1999). According to one fisheries expert, "minor modifications in the system aren't working . . . the dams need to be removed." The controversy became part of the U.S. presidential election in 2000, with both candidates offering their opinions on whether the dams should be breached (Mapes 2000; Seelye 2000).

The continued crisis of depleted salmon stocks was a major factor influencing the incorporation of abrupt climate change information into the Columbia Basin salmon recovery debate. Those wishing to minimize freshwater habitat restrictions desired more aggressive salmon conservation measures to improve the diversity and hence resilience of salmon to natural variability, while those benefiting directly from dams used high oceanic salmon mortality to justify status quo measures of juvenile salmon barging, hydropower turbine screens, and the like. In the contentious environment of balancing the needs of economic development with salmon habitat preservation, it appears that knowledge of variable oceanic conditions has been employed by both sides to advance existing political agendas.

Timing and Crisis The frequency of climatic anomalies is a key difference between the Columbia and Colorado cases. Abrupt climate change information is more likely to be assimilated into the salmon management debate in the Columbia Basin because the PDO changes phase approximately every 20 years. The last PDO shift in the mid-1970s provides a historical analog for resource managers to better understand the impacts of future PDO changes on salmon populations (Mantua et al. 1997). A drought like the one outlined by the SSDA study has no adequate twentieth century analog in the Colorado Basin, and thus there is no institutional memory of the magnitude or impacts of such an event. Furthermore, it is likely that a PDO phase shift will occur in the near future, therefore providing motivation for stakeholders in the Pacific Northwest to utilize PDO information. The apparent low probability of the assessment scenario could invite resource managers in the Colorado Basin to write off the possibility of it occurring during their policy or management tenures.

Structural factors may lead to more frequent salmon management crises in the Pacific Northwest. In the Columbia River Basin, the effects of poor oceanic conditions are exacerbated by the phalanx of dams that salmon encounter in their migration to and from the ocean. In the Colorado Basin, on the other hand, the construction of Hoover and Glen Canyon Dams dramatically increased basin states' buffer against multiyear drought.[10] During the 1987–1992 drought, for example, there was enough water in Lake Powell and Lake Mead to prevent major restrictions on the basic allocations outlined in the Law of the River (CRWCD 2000b). Thus, dams facilitate water resource management in the Colorado Basin, mitigating the effects of water shortages; they conversely frustrate salmon management in the Columbia Basin by contributing to habitat degradation, thus increasing the likelihood of crisis and the consideration of alternative policy options.

Institutional Constraints Crisis often clarifies key issues and offers an opportunity to revise entrenched resource management policies, clearing the way for new approaches that were once infeasible for technical or political reasons (Gunderson et al. 1995). In the Columbia Basin, management approaches that take into account recent scientific advances are now being considered. There is a movement away from command-and-control strategies to one that acknowledges uncertainty in oceanic survival and advocates the application of ecological principles. The most noticeable example of this shift is the serious discussion of dam breaching on the Lower Snake River.

The SSDA lacked a crisis event comparable to that of salmon extinction in the Pacific Northwest. Because severe drought has not yet affected the Colorado Basin in its modern structural and institutional state, water managers have little incentive to re-evaluate current interstate water policies. Recent intrastate water marketing and transfers in California and discussion of interstate banking in the Lower Basin indicate a changing political environment, but it is unclear how advances in water allocation flexibility will improve the basin-wide response to a severe sustained drought. For the time being, it appears that constraints imposed on interstate water management by the Law of the River and by the long history of competition over scarce Colorado River water prevent the adoption of the institutional recommendations advocated by the assessment.

In the Columbia Basin, a more open and adaptive management system led by the NPPC and the Pacific Northwest Regional Assessment, seems to facilitate the consideration of progressive management approaches. The NPPC has established itself as an important basin-wide voice in the salmon debate by guiding salmon restoration and hydropower development and acting as an information clearinghouse on related technical and policy issues. In this sense, salmon management in the Columbia Basin is more akin to a distributed assessment system, where "integrated networks of research, assessment, and management bridge numerous levels and include sustained, long-term interactions between scientists, decisionmakers, and stakeholders" (Cash 2000). The NPPC and the Pacific Northwest Regional Assessment have the funding and stability to facilitate long-term interactions between scientists and decisionmakers. These organizations, in addition to the ongoing decline in salmonid populations, played a primary role in the rapid assimilation of abrupt climate change information into the salmon restoration debate. While the use of abrupt climate change information in the Columbia Basin is in its infancy, making it difficult to see concrete examples of new management approaches, there is clearly political and institutional desire to use the information to the greatest extent possible.

Conclusions

The SSDA presents a unique case for the study of environmental assessments. It was well designed in terms of the factors influencing salience, legitimacy, and credibility, yet it had little impact on management techniques or policy. The principal investigators of the assessment were both innovative and sophisticated in their approach, creating a product that was multi-institutional and interdisciplinary in origin, covering everything from tree ring-based river flow reconstructions to sociological analyses of drought mitigation options. It also utilized a spatial scale of analysis that reflects the interconnected nature of water resource management in the Colorado Basin. The assessment involved experts from around the region, and it actively sought the participation of key water managers, offering them the opportunity to guide and provide feedback on assessment structure and process. In many ways, the assessment was an exemplary process, incorporating several of the design factors that have been shown in other chapters to lead to effectiveness.

My point is not to argue that the SSDA represents an ideal assessment, but rather to highlight it as an example of a case in which it is necessary to look beyond design factors under the immediate control of those managing the assessment. Using cases like it we can evaluate the environment in which a well-designed assessment either flourishes and reaches the elusive state of effectiveness or withers into a set of moribund documents that collect more dust than interest from decisionmakers. The assessment had its faults. A drought scenario taken directly from the tree ring record, for example, rather than the re-arranged version adopted by the assessment, would have improved credibility with some stakeholders, and its implications for water policy would have been similar. The assessment did, however, avoid major pitfalls, such as addressing issues based on the presumed interest of decisionmakers, or failing to involve an adequately diverse group of stakeholders. Careful and thoughtful design is crucial to ensure that an assessment can function in a politically contentious environment.

If the SSDA was so well-designed, then why was it ineffective? Largely this is due to the institutional constraints that confounded efforts to create a water management system more resilient to long-term drought. As a result of the semiarid setting of the Colorado Basin and the high demand placed on the river from a variety of uses, conflict over water rights has generally been resolved through formal legal arrangements that create expectations for reliable water flows. Expectations for predictable supply in turn require extraordinary engineering efforts to limit natural hydrological variability. Many of the assessment recommendations, which seem quite reasonable from an academic standpoint, are politically contentious in reality because they challenge the status quo of interstate water policy—policy that has been hammered out over decades of political conflict and negotiation.

Previous studies of natural resource management systems imply that surprise and crisis are the inevitable consequences of command-and-control resource management techniques (Gunderson et al. 1995). In the Columbia River Basin, for example, the continued crisis of potential salmon stock extinctions has driven the new science of estuarine and oceanic salmon ecology to the forefront of the debate on whether dams on the Lower Snake River should be breached. Mitigation options have expanded beyond typical technical fixes, to the once radical realm of decommissioning major structural elements of a water management system. The Colorado Basin, a prime example of command-and-control management, has yet to experience an event to prompt serious reconsideration of current long-term drought contingency plans. While the drought of 1987–1992 led to strict conservation programs in southern California and Las Vegas (Fradkin 1995), the magnitude of this event was inadequate to catalyze a basin-wide crisis.

In two river basin systems with a seemingly infinite number of confounding variables, it is difficult, if not impossible, to determine why one has progressed to the point of considering new management and policy options while the other has not. Nevertheless, speculative comparison of the two cases raises some interesting points. The existence and magnitude of external crisis events is an important factor, but different internal institutional characteristics also play a role. In the Columbia River Basin, the NPPC and the Pacific Northwest Regional Assessment have the mandate and stability to act as long-term basin-wide networks for research, assessment, and resource management. An independent and well-funded interstate body like the NPPC does not exist in the Colorado Basin to coordinate water management. If such an organization did exist, and if it were tasked with balancing consumptive and nonconsumptive uses, as well as Upper and Lower Basin interests, the SSDA would have resonated in the Colorado Basin more than it did. Stakeholders who could benefit from a basin-wide commission are interested in creating such an organization, while those who stand to lose their current water allocation are not. Ironically, the creation of a basin-wide commission that could coordinate severe drought response, a primary recommendation of the assessment, appears necessary to ensure the efficacy of assessments like the SSDA. Therefore, it seems that an external influence, on par with the salmon crisis in the Pacific Northwest, is necessary for the Colorado Basin to construct the flexible institutional framework essential for response to a massive drought event. Hopefully it will not require the actual occurrence of a 20-year drought.

Notes

1. Total allocation had reached 15.75 maf/year at Lee's Ferry, Nevada (7.5 maf for the Upper Basin, 7.5 maf for the Lower Basin, and the Upper Basin's share of the Mexico allotment of 0.75 maf), compared to the 15.1 maf/year historical average flow. The historical average is calculated from the Bureau of Reclamation's natural flow database for the Colorado River at Lee's Ferry, Nevada (1905–2000). Natural flow is calculated by adjusting for the effects of consumptive use withdrawals, reservoirs, and dam releases under the Law of the River. It is an estimate of what flow would have been without human intervention in the river system.

2. Average Colorado River flow at Lee's Ferry from 1930 to 1939 was 13.1 maf (Bureau of Reclamation 1999), well below the total allocation of 15.75 maf/year.

3. This flow is calculated by adding the 7.5 maf allotment for the Lower Basin, plus 0.75 maf for Mexico, minus 0.02 maf tributary inflow between Glen Canyon Dam and Lee's Ferry. The Colorado Basin Storage Act also authorized the Central Arizona Project, an aqueduct to allow Arizona to use its full entitlement of 2.8 maf.

4. Total reservoir capacity in the Colorado Basin is approximately 60 maf, four times the annual average flow of approximately 15 maf (Bureau of Reclamation 2000).

5. As listed in the Criteria for Coordinated Long-Range Operation of Colorado River Reservoirs Pursuant to the Colorado Basin Project Act of September 30, 1968 (P.L. 90-537). During the critical period in the natural flow database, flows averaged 12.2 maf/year at Lee's Ferry, Nevada.

6. The severe drought occurred from 1579 to 1598, with an annual average flow at Lee's Ferry of approximately 11 maf.

7. The Powell Consortium is a collaborative group of water research centers based at universities around the Colorado Basin. For more details, see http://wrri.nmsu.edu/powell/.

8. Northwest Power Planning and Conservation Act. 1996. Section (4)(h)(10)(D).

9. The proposed dam breaching involves removing the earthen portion of four Lower Snake River dams, leaving the hydropower portions intact, but allowing salmon to freely pass without encountering either fish ladders or turbines (Larmer 1999).

10. Recent events indicate that increased resilience to short-term drought has increased the risk of severe flooding in the Colorado Basin. In 1995, the January forecast underestimated spring runoff by 5 maf (Pulwarty and Melis 2001). Fortuitously, reservoirs were low as a result of the 1987–1992 drought, and they could easily absorb the extra inflow. Had the reservoir levels been higher (as was the case in 1983), severe flooding likely would have occurred. In the late spring of 1983, a severe flood led to unusually high water levels in Lake Powell, which in

turn required unprecedented water releases from Glen Canyon Dam. The dam was severely damaged by the high volume flow, to the point that Bureau of Reclamation engineers doubted its structural integrity (Fradkin 1995). The consequences of a Glen Canyon Dam failure and the subsequent draining of Lake Powell would have been catastrophic. This nearly instantaneous release of 23 maf would have caused severe downstream flooding, the potential collapse of Hoover Dam, and a drastic reduction in the system's ability to deliver water to millions of users.

References

Barker, E. 2000. To Breach or Not to Breach, *High Country News,* February 28.

Bianchi, G.G., and I.N. McCave. 1999. Holocene Periodicity in North Atlantic Climate and Deep Ocean Flow South of Iceland. *Nature* 397: 515–517.

Bisbal, G.A., and W.E. McConnaha. 1998. Consideration of Ocean Conditions in the Management of Salmon, *Canadian Journal of Fisheries and Aquatic Sciences* 55: 2178–2186.

———. 1999. Consideration of Ocean Conditions in the Management of Salmon. Background paper for symposium on Ocean Conditions and the Management of Columbia River Salmon, Northwest Power Planning Council, July, 1, 1999, Portland, OR. http://www.nwcouncil.org/library/ocean/ (accessed May 15, 2000).

Booker, J.F. 1995. Hydrologic and Economic Impacts of Drought under Alternative Policy Responses. *Water Resources Bulletin* 31: 889–906.

Broecker, W.S. 1997. Thermohaline Circulation, the Achilles Heel of Our Climate System: Will Man-Made CO_2 Upset the Current Balance? *Science* 278, 1582–1588.

Brown, B.G. 1988. Climatic Variability and the Colorado River Compact: Implications for Responding to Climate Change. In *Societal Responses to Regional Climatic Change: Forecasting by Analogy,* edited by M.H. Glantz. Boulder, CO: Westview Press. 280–305.

Bureau of Reclamation. 1997a. Fact sheet, 1995 Review of the Criteria for Coordinated Long-Range Operation of Colorado River Reservoirs. Boulder City, Nevada: U.S. Department of Interior.

———. 1997b. Public comments matrix, 1995 Review of the Criteria for Coordinated Long-Range Operation of Colorado River Reservoirs. Boulder City, Nevada: U.S. Department of Interior.

———. 2000. Reservoir volumes for Lake Powell at Glen Canyon Dam (1965–2000) and Lake Mead at Hoover Dam (1935–2000). Boulder City, Nevada: U.S. Department of Interior.

Calvin, W.H. 1998. The Great Climate Flip-Flop. *Atlantic Monthly* January: 47–64.

Cash, D. 2000. Distributed Assessment Systems: An Emerging Paradigm of Research, Assessment and Decision-Making for Environmental Change. *Global Environmental Change* 10: 241–244.

Clark, W.C., J. Jäger, J. Cavender-Bares, and N.M. Dickson. 2001. Acid Rain, Ozone Depletion and Climate Change: An Historical Overview. In *Learning to Manage Global Environmental Risks–Volume 1: A Comparative History of Social Responses to Climate Change, Ozone Depletion, and Acid Rain,* edited by W.C. Clark, J. Jäger, J. van Eijndhoven, and N. Dickson. Cambridge, MA: MIT Press.

CRWCD (Colorado River Water Conservation District). 2000a. Colorado River Water Conservation District. http://www.crwcd.gov (accessed February 2, 2000).

———. 2000b. Personal communication between Eric Kuhn, Colorado River Water Conservation District, and the author. February 9, 2000.

Cook, E.R., D.M. Meko, D.W. Stahle, and M.K. Cleaveland. 1998. National Oceanic and Atmospheric Administration—National Environmental Satellite and Data Information Service. North American Drought Variability. http://www.ngdc.noaa.gov/paleo/pdsi.html (accessed October 10, 1999.).

Fradkin, P.L. 1995. *A River No More: The Colorado River and the West.* Berkeley, CA: University of California Press.

Fritts, H.C. 1965. Tree-Ring Evidence for Climatic Changes in Western North America. *Monthly Weather Review* 93: 421–443.

Getches, D.H. 1997. Colorado River Governance: Sharing Federal Authority as an Incentive to Create a New Institution. *University of Colorado Law Review* 68: 573–658.

Gregg, F., and D.H. Getches. 1991. *Severe Sustained Drought in the Southwestern United States: Phase I Completion Report*. National Technical Information Service Document No. PB92-115013. Springfield, VA: National Technical Information Service.

Grissino-Mayer, H.D. 1996. A 2129-Year Reconstruction of Precipitation for Northwestern New Mexico, U.S.A. In *Tree Rings, Environment, and Humanity*, edited by J.S. Dean, D.M. Meko, and T.W. Swetnam. Tucson, AZ: Radiocarbon, 191–204.

Gunderson, L.H., C.S. Holling, and S.S. Light. 1995. Barriers Broken and Bridges Built: A Synthesis. In *Barriers and Bridges to Renewal of Ecosystems and Institutions*, edited by L.H. Gunderson, C.S. Holling, and S.S. Light, New York: Columbia University Press, 489–532.

Harding, B.L., T.B. Sangoyomi, and E.A. Payton. 1995. Impacts of a Severe Sustained Drought on Colorado River Water Resources. *Water Resources Bulletin* 31: 815–824.

Hardy, T. B. 1995. Assessing Environmental Effects of Severe Sustained Drought. *Water Resources Bulletin* 31: 867–875.

Hare, S.R., and R.C. Francis. 1995. Climate Change and Salmon Production in the Northeast Pacific Ocean. *Canada Special Publications of Fisheries and Aquatic Sciences* 121: 357–372.

Henderson, J.L., and W.B. Lord. 1995. A Gaming Evaluation of Colorado River Drought Management Institutional Options. *Water Resources Bulletin* 35: 907–924.

Hundley, N. Jr. 1975. *Water and the West: The Colorado River Compact and the Politics of Water in the American West*. Los Angeles: University of California Press.

ISAB (Independent Scientific Advisory Board). 1999. Looking for Common Ground: Comparison of Recent Reports Pertaining to Salmon Recovery in the Columbia River Basin, Northwest Power Planning Council. http://www.nwcouncil.org/library/isab/isab99-3.htm (accessed March 22, 2000).

Keigwin, L.D. 1996. The Little Ice Age and Medieval Warm Period in the Sargasso Sea. *Science* 274: 1504–1508.

Larmer, P. 1999. Unleashing the Snake. *High Country News*, December 20. http://www.hcn.org/servlets/hcn.Article?article_id=5452 (accessed January 30, 2005).

Lee, K. 1995. Deliberating Seeking Sustainability in the Columbia River Basin. In *Barriers and Bridges to Renewal of Ecosystems and Institutions*, edited by L.H. Gunderson, C.S. Holling, and S.S. Light. New York: Columbia University Press, 214–238.

Lord, W.B., J.F. Booker, D.H. Getches, B.J. Harding, D.S. Kenney, and R.A. Young. 1995. Managing the Colorado River in a Severe Sustained Drought: An Evaluation of Institutional Options. *Water Resources Bulletin* 31: 939–944.

Lovelin, B. 2000. Personal communication between B. Lovelin, executive director, Columbia River Alliance, and the author, March 23.

MacDonnell, L.J., D.H. Getches, and W.C. Hugenberg Jr. 1995. The Law of the Colorado River: Coping with Severe Sustained Drought. *Water Resources Bulletin* 31: 825–836.

Mantua, N., S.R. Hare, Y. Zhang, J.M. Wallace, and R.C. Francis. 1997. A Pacific Interdecadal Oscillation with Impacts on Salmon Production. *Bulletin of the American Meteorological Society* 78: 1069–1079.

Mapes, J. 2000. Gore Says Salmon Must Survive. *Oregonian*, May 13. http://www.oregonlive.com/news/oregonian/index.ssf?/news/oregonian/00/05/lc_72salmn13.frame (accessed January 30, 2005).

Matusak, J. 1994. Letter from J. Matusak to Severe Sustained Drought Assessment organizers. Metropolitan Water District of Southern California, October 31.

———. 1997. Formal comments regarding the Severe Sustained Drought Study methodology, Presented at Symposium on Climate Variability, Climate Change, and Water Resource Management, October 26–29, Colorado Springs, CO.

———. 2000. Personal communication between J. Matusak, Engineer, Metropolitan Water District of Southern California, and author. February 9.

McPhee, J. 1971. *Encounters with the Archdruid*. New York: Noonday Press.

Meko, D., C.W. Stockton, and W.R. Boggess. 1995. The Tree-Ring Record of Severe Sustained Drought, *Water Resources Bulletin* 31: 789–801.

MWD (Metropolitan Water District of Southern California). 2000. http://www.mwdh2o.com/mwdh2o/pages/about/about01.html (accessed May 31, 2000).

Mote, P.W., D.J. Canning, D.L. Fluharty, R.C. Francis, J.F. Franklin, et al. (19 authors). 1999. *Impacts of Climate Variability and Change, Pacific Northwest.* Seattle, WA: National Atmospheric and Oceanic Administration, Office of Global Programs, and Joint Institute for Study of Atmosphere and Ocean Climate Impacts Group.

NPPC (Northwest Power Planning Council). 1999. Symposium on Ocean Conditions and the Management of Columbia River Salmon. http://www.nwcouncil.org/library/ocean/ (accessed May 10, 2000).

Overpeck, J.T. 1996. Warm Climate Surprises. *Science* 271: 1820–1821.

Patt, A.G. 1999. Extreme Outcomes: The Strategic Treatment of Low Probability Events in Scientific Assessments. *Risk Decision and Policy* 4: 1–15.

Pulwarty, R.S., and T. Melis. 2001. Climate Extremes and Adaptive Management on the Colorado River: Lessons from the 1997–1998 ENSO Event. *Journal of Environmental Management* 63(3): 307–24.

Seelye, K. 2000. The 2000 Campaign: The Vice President—Gore Speaks Out on Dams, and Maybe Suicide. *New York Times,* May 13, A12.

SSDA (Severe Sustained Drought Assessment). 1992. Progress report to the U.S. Geological Survey, activities from October 1, 1991, to September 30, 1992.

Skogerboe, G.V. 1982. The Physical Environment of the Colorado Basin. *Water Supply and Management* 6: 221–232.

Stahle, D.W., M.K. Cleaveland, and J.G. Hehr. 1985. A 450-Year Drought Reconstruction for Arkansas, United States. *Nature* 316: 530–532.

Stevens, W.K. 1998. If the Climate Changes, It May Do So Fast, New Data Show. *New York Times,* January 27, F1.

———. 1999. Arctic Thawing May Jolt Sea's Climate Belt. *New York Times,* December 7, F3.

———. 2000. Megadrought Appears to Loom in Africa. *New York Times,* February 8, F3.

Stockton, C.W., and G.C. Jacoby. 1976. *Long-Term Surface Water Supply and Streamflow Levels in the Upper Colorado River Basin.* Lake Powell Research Project, Bulletin No. 18, Institute of Geophysics and Planetary Physics. Los Angeles, CA: University of California.

Suplee, C. 1998. Past Patterns Suggest a Future "Megadrought." *Washington Post,* December 21, A3.

Tarboton, D.G. 1995. Hydrologic Scenarios for Severe Sustained Drought in the Southwestern United States. *Water Resources Bulletin* 35: 803–814.

Wilkinson, C.F. 1985. Western Water Law in Transition. *University of Colorado Law Review* 56: 317–345.

Wilkinson, C.F., and D.K. Conner. 1987. A Great Loneliness of the Spirit. In *Western Water Made Simple,* edited by E. Marston. Washington, DC: Island Press, 54–64.

Woodhouse, C.A., and J.T. Overpeck. 1998. 2,000 Years of Drought Variability in the Central United States, *Bulletin of the American Meteorological Society* 79: 2693–2714.

Young, R.A. 1995. Coping with Severe Sustained Drought on the Colorado River: Introduction and Overview. *Water Resources Bulletin* 31: 779–788.

CHAPTER 8

Can Assessments Learn, and If So, How?

A Study of the IPCC

Bernd Siebenhüner

W HEN CARRIED OUT OVER a longer period of time with several recurrent processes, assessments provide an outstanding opportunity for learning from past experiences and from other assessments. Through learning, assessments could improve their procedures and enhance their effectiveness in supporting issue development. If assessments were considered continuous learning processes, rather than one-time events, they could be organized as constant processes of improvement and reflective change of the institution and, consequently, they might become more powerful institutions in the process of solving environmental problems.

In this context, this chapter addresses the following questions: Did the assessment at hand learn over the years and over different phases of the assessment process? In which design issues did it learn? How could the learning process be characterized—as an adaptation to given targets and belief systems or as a more self-reflective process that even induces changes of the objectives and underlying convictions of the actors involved? How and by whom was the learning initiated? How could it have been done better when compared with insights from literature and with experiences from other assessments? Which internal structural, cultural, and personal factors could facilitate learning by and within assessments?

After clarifying the general notion of learning in relation to assessments in section 2, section 3 addresses these questions, based on a case study of the strategic decisionmaking within the Intergovernmental Panel on Climate Change (IPCC). The IPCC provides an example of a long-term assessment that has gone through various modifications regarding its structure and management and in the way in which the assessments have been designed, evaluated, and redesigned. This chapter does not focus on the level of the IPCC's working groups or their internal dynamics, but rather on the overall framework of the IPCC and the way its assessments are carried out.

Learning in and of Assessments

Across various academic disciplines, learning is a widely discussed topic that has been defined in numerous ways. In the context of scientific assessments, however, learning processes have rarely been studied despite the fact that many long-term assessment processes have changed their procedures and processes, indicating that learning took place. The scientific assessments under the Convention on Long-Range Transboundary Air Pollution provide an excellent example in this respect. The assessment bodies changed their structures over time and were able to incorporate new scientific findings in their conceptual work (Siebenhüner 2002).

In political science, learning is mostly regarded as a process of change over time in the knowledge and the resulting behavior of certain actors or actor groups. For example, Haas (1991, 63) puts it as follows: "I define learning as any change in behavior due to a change in perception about how to solve a problem." Thus, the essence of learning processes is twofold, comprising changes in human cognition and changes in their behavior. Sabatier (1987) challenges this notion by introducing another element in his definition of policy-oriented learning. By referring to the belief system, he incorporates values and emotions that lie behind the actual cognitive parts of knowledge.

For the purposes of this chapter, learning is taken to be *a process of long-lasting change in the behavior or the general ability to behave in a certain way, which is founded on changes in knowledge and beliefs.* This concept follows the notion that learning is not necessarily an absolute increase in knowledge, because there are always losses of knowledge that allow for the acceptance and memorization of new knowledge.

In many models of learning, different kinds of learning processes have been differentiated. With regard to the comparatively broad definition of learning employed in this study further specification will be useful. Therefore, I adopt a typology that is based on Argyris and Schön (1996). The fundamental criterion for this classification is how far the underlying objectives, norms, and beliefs of the assessment bodies have changed during the learning process, according to the perceptions of the individuals involved. This approach distinguishes three forms of learning:

- *Adaptation or single-loop learning:* The simplest form of learning is the adaptation of new knowledge to existing frameworks of objectives and causal beliefs. Based on a simple feedback loop between given expectations and the real outcomes of a process, this instrumental type of learning allows for error correction and leads to the adjustment of results that differ from the preexisting expectations. For example, a product manager may detect unexpectedly high emission rates from one production process, and on his search for the causes he might find a technical flaw that has to be corrected.

- *Double-loop learning:* According to Argyris and Schön (1996), the advanced form of learning could be framed as "double-loop learning," which also includes the underlying objectives, values, norms, and belief structures in the learning process. Thus, there are two feedback loops—an instrumental one of error correction (as above), plus a more fundamental loop that creates changes in the general framework of beliefs, norms, and objectives. For example, new results in environmental research might require a company to overturn existing orientations on a certain product type (e.g., chlorofluorocarbons) and call for a reorientation on substitution technologies with far-reaching consequences for the whole organization.

- *Deutero-learning:* If learning takes place on the meta-level of how to learn, one could speak of deutero-learning. This is a form of learning of the ability to learn itself. For example, an organization might gather experiences with certain approaches to learning and attempt to improve its internal learning system consisting of—among others—communication channels, information systems, training procedures, and routines.

In their studies of business corporations, Argyris and Schön found few examples of deutero-learning. Most learning processes usually remain within the scope of the first two categories of learning. In general, single-loop learning is largely sufficient when limited errors or deviations from goals have to be corrected, but it is no longer sufficient when the underlying norms and belief systems of an organization or other agents conflict with new internal or external developments or requirements.

By way of focusing on what I call *reflective mechanisms,* this study attempts to grasp the learning phenomena taking place in the assessment under examination. These mechanisms should help to make use of past experiences by reflecting on them and turning them into action. Such mechanisms might be very informal, as with personal communication among participants in the assessment, or they might be highly formalized and sophisticated in the form of institutionalized committees with a distinct set of rules of procedure. As part of the learning system of the assessment organization, reflective mechanisms might be able to facilitate either instrumental single-loop learning or more demanding double-loop learning.

Assessments are mostly carried out by individuals in some kind of collective effort. Therefore, the question arises as to who comprises the learning agent or agents. The agents could be simply the individuals by themselves, or the agent could also be the social entity formed by the individuals as a group.

The research reported in this chapter assumes that assessments can be seen as a collective learning endeavor that cannot entirely be reduced to the sum of the individual learning processes, because some knowledge remains with the organization even if the individuals change. This assumption simply states that learning of and in assessments can be more than mere individual learning, because of the existence of internal relationships between the participants of the process.

Aiming at explaining the learning processes themselves, I discuss a number of internal characteristics of the assessment organization. According to organizational-learning literature, relevant organizational features include evaluation procedures, formal and informal communication structures, exchange of experts among different organizational units, and informal networks, as well as shared values among the participants such as openness and flexibility. Moreover, the role of the personalities involved has to be addressed when possible change agents are to be identified. External factors such as political pressures, new scientific findings, and criticisms from nongovernmental organizations (NGOs) or from the media are deliberately excluded from this study, given the complexity of these influences and the constraints of time and space.

The Case of IPCC Assessments

Climate change has been the subject of scientific debate since the nineteenth century when Svante Arrhenius phrased his hypothesis that global temperatures would rise due to manmade emissions (Arrhenius 1896). Interestingly, Arrhenius already served as a scientific adviser to the Swedish government. He stressed the need for international cooperation in this field and was very much concerned about the clear distinction between science and policymaking (Elzinga 1997). Therefore, he pioneered a tradition in policy-oriented climate change research whose latest and most prominent outcome is the IPCC.[1]

The IPCC was established in 1988 as a scientific advisory body to the United Nations Environmental Programme (UNEP) and the World Meteorological Organization (WMO; see Box 1-3 in Chapter 1). Over the years, the IPCC has undergone several changes with regard to its internal structures and procedures. In the next section, I first characterize the IPCC assessments with regard to the critical design elements identified in Chapter 1. I secondly focus on the reflective mechanisms and procedures used in the IPCC processes. Thirdly, I scrutinize some internal factors that might have had an influence on the organization's ability to learn.

Assessment Design Goals and Initiation

When the IPCC was founded in 1988, it was more than the mere addition of another scientific advisory body in the climate change arena. Although the IPCC was not the first international assessment forum for climate change, it had a unique and innovative structure and design. Whereas its precursor in the international arena, the Advisory Group on Greenhouse Gases, established in 1986, consisted of a handful of scientists almost exclusively from northern industrialized countries (Agrawala 1999), the IPCC was based on an intergovernmental approval mechanism and incorporated the expertise of scientists from all over the world. The centerpiece of this intergovernmental mechanism was the involvement of numerous governments in the formulation of the questions addressed and in the approval of the final reports. Thereby, it was different from national assessments, which were hardly noticed or accepted by nations other than those where they originated. The IPCC allowed for a broader acceptance among the large number of governments that were hitherto skeptical or ignorant of climate change. The intergovernmental approval mechanism provided the chance to deal with the inevitably global structure of the problem of climate change and to ensure the governments' ownership of the resulting reports (Bolin 1994a; Watson 2001).

As formulated in its first assessment report (IPCC 1990), the original objectives of the IPCC were threefold. It was charged with

- assessing the state of existing scientific knowledge on climate change;

- examining the environmental, economic, and social impacts of climate change; and

- formulating response strategies.

The IPCC addressed these objectives by setting up three working groups that were more or less charged with one of these three tasks.[2] By and large, the formal objectives remained unchanged over the three subsequent assessment reports (Houghton 2001). Nevertheless, the way that the IPCC pursued these objectives, specified the tasks, and organized the division of labor varied over time and was subject to reflective and adaptive adjustments.

Organizational Structure

On first sight, the organizational structure of the IPCC seems very complex—there are various different functions, bodies, and decisionmaking procedures. But at second glance the organization proves to be rather lean. The chairperson presides over the IPCC Bureau, consisting of five vice-chairs plus the co-chairs and vice-chairs of the working groups. Each working group has two co-chairs and six vice-chairs that are usually given equally to representatives from industrialized and developing countries. The work of the working groups is coordinated and administered by individual technical support units that are mostly located in industrialized countries from which they receive their funding. Only the IPCC secretariat, based in Geneva, is funded by the parent organizations (UNEP and WMO). The IPCC Bureau prepares the decisions to be adopted at the plenary sessions, which are attended by government officials from the member nations of UNEP and WMO. At these regular annual meetings, the IPCC accepts and approves IPCC reports, report structures and outlines, work plans, IPCC rules and procedures, and the budget; it also elects the chairperson and the bureau. It is the responsibility of the co-chairs of the working groups to select the lead authors of the chapters (based on government nominations) and to coordinate their work.[3] Meanwhile the number of authors involved has increased to nearly 1,500 leading and contributing authors over the three working groups.

Participation

One of the key issues of the design of the IPCC was the struggle for balanced participation of scientists from all parts of the world. Because the international set up and the involvement of governments from all over the world was the centerpiece of the IPCC, the participation of experts from all regions of the world was regarded as crucial for the acceptance of the assessment results by policymakers (in the industrialized countries as well as in the developing South).[4] As expressed by WMO Secretary-General Godwin O.P. Obasi, the initial goal of the IPCC was to ensure membership of the major greenhouse gas emitting countries, of all geographic regions, and of those countries with strong scientific expertise in the field. However, because experts from the developing world, in particular, lacked the necessary funding opportunities and a great deal of crucial research capacity, their participation has been a constant subject of debate in the IPCC Bureau (Agrawala 1998b). To deal with this problem, a special committee was established to find ways to increase participation by developing countries. Moreover, quotas were fixed for the composition of the main IPCC committees, and funding opportunities for travel expenses were introduced to allow participants from developing countries to attend the IPCC meetings. Thereby, the segment of experts from these regions increased over time, although the representation of world regions among the IPCC lead authors is still not equally balanced in all working groups (Leary 2001).

On the side of government representatives, a broader and more balanced participation could be observed. The IPCC's plenary sessions, which started out with only 30 participating governments in 1988, now usually attract government officials from more than 100 nations. This number demonstrates not only the improvement of the international balance on the side of the political decisionmakers, but also perhaps a growing interest in the issue of climate change and its ecological, social, economic, and political outcomes in general. Industry groups and environmental organizations had a similar rising interest in the issue, even though nongovernmental organizations do not have an official voice in the sessions.[5] However, some bodies such as the Global Climate Coalition have been approved as reviewers in the revision process of the assessment reports (Franz 1998).

Peer-Review Process

To ensure scientific quality and credibility to both the scientific and political communities, a specific and highly sophisticated type of review procedure has been developed over the three assessments. In the first assessment, each chapter had been reviewed by two or three experts and governmental officials simultaneously, whereas in the second assessment the review process was much more refined (Parry 2001). The review process took place in two rounds. First, the drafts prepared by the lead authors were circulated among specialists in the area at hand, other lead authors, and experts from relevant international organizations. Then in the second round, the revised drafts were distributed among governments, and their comments were solicited. Finally, the lead authors included the governments' comments in a final draft that was submitted for acceptance at the working group plenary meeting. While the lengthy chapters comprising the bulk of the IPCC reports only require acceptance by the working group, the shorter and more focused executive summaries and the summaries for policymakers must be approved line by line by the IPCC Plenary, consisting of all the government officials.

The main intention of this iterative review and approval process has been to "ensure that the reports present a comprehensive, objective, and balanced view of the areas they cover" and not to allow for the intrusion of political or economic interests in the assessment process (IPCC 1995). Although government officials are always tempted to introduce politically biased statements into the reports to promote their national interests, experiences with the intergovernmental approval process have shown that it cannot do major harm to balanced and scientifically solid reports.[6] The third assessment report stuck to these procedures and added so-called "review editors," who were in charge of supervising the process of peer review by tracking the comments from the reviewers and the resulting changes in the drafts prepared by the lead authors (IPCC 1999). Although the review editors added another element in the review process, the additional time requirements were marginal (Watson 2001). Nevertheless, not all authors regarded the review editors a completely helpful improvement of the process, because not all of the review editors were equally diligent in fulfilling their job. While some regarded it as rather trivial, others took the task very seriously—a task that required reading and consideration of the various versions of the chapter drafts and as many as 200 comments (Leary 2001).

Science–Policy Interface

As an organization at the interface between science and policy, the IPCC assessments are expected to fulfill a twofold purpose: They should provide credibility to the scientific community, and they should feed scientific and technical information into the political negotiation and implementation processes (Bolin 2001). Therefore, it is worth questioning how the interaction between scientific experts and the political community is designed. Because of the high level of contestation inherent in the issue of climate change and the political options involved, this interface has been crafted very carefully.

The official interaction between scientists and policymakers is restricted to well-defined stages of the assessment process. Scientists of the IPCC Bureau develop an outline of the report and/or the division of labor among the structure of working groups. They suggest it to the political community at one of the plenary sessions, where a final decision is made. Then they select the authors and reviewers based on government nominations and the principles of scientific expertise and geographic representation. Governments might influence this process by their nomination policies. By contrast, the whole process of preparing the chapters and the first round of peer-review remains exclusively in the scientific realm. Governments come into play once again in the second round of review when their comments are being solicited and they have a major role in the approval of the Summary for Policymakers and the Synthesis Report.[7]

In terms of organizational structure, the science–policy interface is filled with a number of committees (Figure 8-1). From early on, a joint working group between the IPCC and the negotiating bodies was established to facilitate direct communication among the scientific and political committees.[8] On the side of the IPCC, the group consists of the chairperson of the IPCC and several members of the bureau; on the side of the United Nations Framework Convention on Climate Change (UNFCCC) delegates are the director of the UNFCCC secretariat and several of his/her staff members as well as members of the subsidiary bodies under the convention. The group has provided a comparatively informal forum to discuss the projects of the IPCC and the information needs of the negotiation processes. Because the joint working group was established at a rather advanced stage of the second assessment report, its influence in this phase remained limited. During the preparation of the third assessment report, the

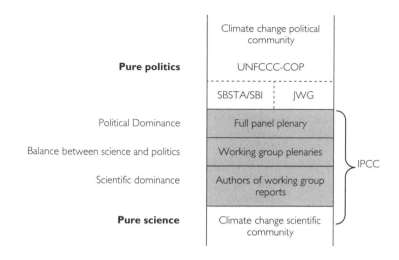

Figure 8-1. *The Science-Policy Interface of the IPCC*
Source: Based on Alfsen and Skodvin (1998).

group became significantly more influential. The group met on a regular basis and had influence especially with regard to the introduction of a new type of specialized IPCC report (Moss 2001).

Apart from the joint working group, the convention established two standing bodies consisting of government delegates: the Subsidiary Body for Scientific and Technological Advice (SBSTA; see Chapter 9 for more on the SBSTA) and the Subsidiary Body for Implementation (SBI). It is the task of the former to advise the negotiating parties on scientific, technological, and methodological matters relating to the convention. Therefore, the SBSTA should link scientific information provided by the IPCC to the policy-oriented needs of the Conference of Parties. In this function, the SBSTA has to cooperate closely with the IPCC and could request specific studies from it. The SBI is charged with the assessment and review of the implementation of the convention. Among others, it has to examine the national emission inventories submitted by the parties and, thereby, has to cooperate especially with the IPCC Task Force on National Greenhouse Gas Inventories.

Concerning the communication processes among the scientific and political communities, it has frequently been stressed by interviewees that—apart from keynote addresses—scientists involved in the IPCC process usually refrain from participating in the negotiation process in order to ensure their scientific neutrality (Houghton 2001; Moss 2001). However, no documented rules exist on this issue. Direct interaction between both spheres is thereby limited to the more formalized fora and processes. Nevertheless, it has been emphasized that the introduction of highly specialized reports in 1994 and so-called "rapid response" technical papers in 1996 was subject to intense discussions in the joint working group. Since then, these reports have significantly facilitated the transfer of scientific and technical information into the political process (Agrawala 1998b).

Treatment of Uncertainty

Given the complexity and interconnectedness of the issue, the treatment of uncertainties has always been a crucial issue to climate change assessments. Within the IPCC, the topic was intensely debated since the beginning, and authors have been using different approaches to estimate and document uncertainties. Many of them even refused to give any estimation on uncertainties because of their scientific ethos to focus exclusively on reliable and certain research findings. Several researchers had to be convinced that the policy-oriented approach of the IPCC urgently required these estimations to advise political judgments on specific risks—even when no scientific certainty was available. Thereby they sometimes left policymakers alone in their judgment of the presented data and research results (Schneider 2001).

To give better advice to the policy community and to achieve more consistency in the use of language on uncertainty, Moss and Schneider (2000) developed a guidance paper in the third assessment that was sent out to all working groups. The guidance paper introduced a qualitative framework to describe "states of knowledge" and suggested a numerical scale of the various degrees of confidence. The objective was to increase consistency in the whole report and to allow authors to make their inevitably partly subjective estimations more transparent and understandable for policymakers.[9]

While most authors adhered to this framework, one group of lead authors from Working Group I refused to do so. Instead of using the suggested scale with intervals of 5 percentiles (5 percent, 10 percent . . . 95 percent confidence) they used likelihood measures of a third or 99 percent and avoided the exact description of uncertainties by referring to the assumptions of various models leaving policymakers without clear guidance how to evaluate the related risks. After lengthy discussions among the authors, and following a request from governments, the authors partly revised the chapter. The other working groups that had to build on the results from Working Group I, however, had hardly any chance to adapt to these changes because of the short time frame between sessions of the working groups to approve the third assessment report (Schneider 2001). Therefore, minor inconsistencies could also be found in this report although the general treatment of uncertainties is significantly more reflective, consistent, and clearer than in the previous assessment reports.

Reflective Mechanisms

Which kind of mechanisms were in place in the IPCC assessments to reflect on past experiences, to make use of experiences from other assessments, and to feed them into changes of future assessment designs? As a long-term recurrent assessment, the IPCC provides an excellent opportunity to learn from the past and to achieve a continuous improvement of its assessment design and its effectiveness.

By contrast, in their comparative study of learning in the early stages of the IPCC, Haas and McCabe (2001) conclude that the "INC [International Negotiation Committee] and the IPCC learned very little."[10] On the basis of this study of the IPCC over the whole period of its existence since 1988, a different conclusion could be drawn given the evidence found in empirical studies of the IPCC processes. Therefore, the following section focuses on the way the IPCC learned on a collective level and how the learning was facilitated. It delineates the mechanisms that have been used for learning over the period of the IPCC's existence. The section addresses structural, cultural, and personal factors that have very likely been influential in learning.

Although there has never been an official systematic evaluation or institutionalized reflection process within or outside of the IPCC to examine the experiences and results of previous work and to draw conclusions about necessary changes in the structures and processes of the assessments, there have been several unofficial ways of doing so. In parallel to the design elements mentioned above, the learning mechanisms of the IPCC also evolved over time. Therefore, I describe them chronologically following the sequence of the assessment reports.

First Assessment and Supplementary Reports (1988–1992)

At the beginning of the IPCC, lesson-drawing from other environmental assessments was the predominant learning mechanism in place. The example of the assessments of the stratospheric ozone layer provided this lesson. Having resulted in strict and effective political regulation, the ozone assessments were handled as success stories to provide a blueprint for the climate assessments of the IPCC.[11] The former co-chair of the stratospheric ozone assessment and the second chair of the IPCC, Robert Watson, makes that clear: "So when we designed the IPCC, we certainly had in mind some of the experiences from the international ozone assessment. . . . We looked at the ozone assessment for lessons learned and how to get the really world-class people involved as the co-chairs, et cetera" (Watson 2001). However, there was only a small number of people who participated in both assessments and who were able to bring in their experiences.

Nevertheless, even within the climate area, experiences with international assessments were available. These assessments had hardly been successful and therefore, threw some doubt on the idea that a simple repetition of the ozone case would be realistic. The Advisory Group on Greenhouse Gases, existing from 1985 until 1992, was by far less effective in terms of political outcome than were any of the ozone assessments, and significant design flaws could be held accountable for that. First, the participating members represented only a very limited number of countries, and therefore the assessment's legitimacy was questioned (Boehmer-Christiansen 1994b). Second, salience for policymakers was rather low, because no serious link to the political process had been institutionalized. Third, the organization lacked financial resources and political support. Last but not least, participants in the advisory group received little scientific credit from their peers and institutions, which reduced members' commitment to the group's activities (Agrawala 1999).

These experiences, among other factors, could explain the call for the intergovernmental mechanism of the IPCC that emerged from intensive consultations between the WMO, UNEP, and several foreign affairs ministries of WMO member countries, especially the United States (Agrawala 1998a). As a means to push the assessment closer to the political process, the consulting bodies set up the intergovernmental approval process—unprecedented in the environmental area. Lessons could only be drawn from other UN processes with which most of the scientists involved were not familiar due to their almost exclusively scientific career paths.[12] Thus, the approval mechanism was new to most of its participants, and they had to learn how to use it effectively and how to develop it further given the little formalization it had in the beginning.[13]

The finalization of the first assessment report in 1990 provided the opportunity to reflect on the experiences of the IPCC processes and on what could be learned for future assessments. This reflection took place in the form of discussions in the IPCC Bureau and the plenary sessions, but no formal document was prepared for these talks. One of the main lessons learned by the participants of the first assessment related to the political aspects of the IPCC process. Scientists had to adapt to the fact that the agenda of the assessment was mostly set by political needs. Major discussions emerged, and significant skepticism was expressed, by some of the scientists involved. Nevertheless, acknowledging the political function of the whole endeavor, a supplementary report was prepared that—together with the first assessment report—was said to be highly influential on the negotiations of and final agreement on the framework convention in 1992 (Agrawala 1998b; Bolin 1994b). Therefore, this reflection process could be seen as helpful to increase the effectiveness of the IPCC assessment by raising salience for political decisionmakers.

Second Assessment Report (1992–1995)

During the time of the preparation of the second assessment report, one main incidence of learning—in the sense of reflection of past experiences—could be detected. At its sixth session, the panel established the special Task Force on the IPCC Structure that reported to the eighth session of the IPCC in 1992. The purpose of the task force was, on one hand, to comply with the requirements of the framework convention introducing several new bodies such as the Conference of Parties. On the other hand, the task force was asked to outline the working group structure of the second assessment and to suggest ways to allow for the participation of NGOs in the IPCC process, which had been demanded by several environmental NGOs. The task force prepared a report that led to the adoption of a four-page document about new rules of procedure of the IPCC at its ninth plenary session (IPCC 1992).

The report of the task force paid attention to several critical issues inherent in the IPCC and its procedures. It addressed the criticism that the process lacked transparency, and it suggested making greater effort to inform nations about ongoing IPCC activities and the decisionmaking processes (Bolin 1994b). The task force also responded to the call for broader participation by developing country experts, which was seen as fundamental for the IPCC process. The task force made recommendations to facilitate scientific exchange between industrialized and developing countries, to foster international cooperation in climate research, and to integrate a greater number of experts from developing countries.

By and large, the report and the resulting decisions during the subsequent plenary sessions reacted to critical comments from governments and NGOs. But while numerous changes resulted mainly from claims from the political world, only a few criticisms from the scientific community were addressed in the report and in the revision of the rules of procedure. Nevertheless, the task force could be seen as a first institutionalized effort to reflect on the experiences of the first period of the IPCC, and the task force was fairly successful in promoting changes of IPCC procedures.

Third Assessment Report (1995–2001)

A somewhat different effort was made to facilitate the learning from the second assessment and to feed these reflections into the design of the third assessment report. In 1996, the first chairman of the IPCC, Bert Bolin, was about to step down, and Robert T. Watson had been elected as the new chairman. In a one-year transition period, both worked closely together to ensure continuation of the work and to allow for a transfer of knowledge from predecessor to successor.

The new chair started out with the attempt to "really consider what was the right structure of the IPCC" (Watson 2001). Instead of establishing a specific task force, he himself prepared a white paper addressing a number of key questions and suggesting new structure and procedural improvements for the third assessment (Watson 1997). The draft was based on critical articles in the scientific literature, on government critiques, and on a number of workshops on specific substantial matters such as mitigation technologies, regional projections of impacts, and uncertainty (Moss 2000, 2001). In a first consultation round, the co-chairs of the working groups and Richard Moss, head of the Technical Support Unit of Working Group II, commented on the paper. A revised version was presented to the IPCC Bureau, which requested several revisions (IPCC Bureau 1997). In a second round, comments on the revised version of the paper were solicited and collected from governments, NGOs, and scientific experts. The chairman received more than 90 responses, and he attempted to incorporate them in the preparation of the final decision paper on the design of the third assessment report, which was adopted in September 1997 by the plenary session. All together, this iterative process took more than a year, and it was certainly a larger effort than was the design of the second assessment when measured by comments considered and by rounds iterated.

A number of suggestions came out of this process that were influential for designing the final structure of the third assessment in 1998. Based on criticisms on the artificial separation of Working Groups II and III in the second assessment, a new working group structure emerged to allow for a better integration of socioeconomic and technical/scientific aspects of adaptation and mitigation (see discussion in endnote 2). Another substantive issue was the treatment of cross-cutting themes such as uncertainty, costing methodologies, and equity issues. These issues had either not been addressed in a consistent manner or had not been addressed at all in the second assessment report (Moss 2000). Therefore, a special subgroup under Working Group III was established that prepared guidance papers on each of these issues in a lengthy review process (Pachauri et al. 2000).

One of the main challenges for the design of the third assessment report was how to deal with the criticisms raised in the so-called "Chapter 8 debate."[14] In the aftermath of the release of the second assessment report, a number of U.S.-based scientists backed by the Global Climate Coalition launched a massive assault against the final version of Chapter 8 of Working Group I, which concluded that "the balance of evidence suggests a discernible human influence on global climate" (Houghton et al. 1996, 4). The accusations published in widely read newspapers like the *Wall Street Journal* were serious; they charged that the lead authors of the chapter had changed the text of the final version after it had already been officially approved by the working group plenary session. The accusers thereby claimed that the chapter authors violated the IPCC's rules of procedure and the fundamental standards of peer review. However, the accused authors and the leaders of the IPCC process successfully proved these accusations wrong, and no IPCC member state government joined in the criticism (Edwards and Schneider 2001). Yet, the debate brought to the surface some deficits to be addressed in future assessments.

One flaw, and the source of the abovementioned criticism, allowed authors to change text in the body of the chapters after the working group plenary session had approved the report. By contrast, the text of the summary for policymakers could not be altered afterward because it has been *accepted*—a line-by-line procedure different from the approval process.[15] The IPCC realized this deficit and reacted by adjusting its rules of procedure enforced in 1999. In the new rules, "changes made after acceptance by the working group shall be those necessary to ensure consistency with the Summary of Policymakers," and lead authors must indicate their changes to the panel (IPCC 1999, 5).

Moreover, in the second assessment report, authors were not accountable to the reviewers or to anybody else when they finally revised their chapters. Thus, they could even ignore reviewer's comments without plausible reason. This deficit in the rules became apparent in subsequent "Chapter 8 controversy" debate and led to another institutional innovation in the preparation of the third assessment. Incoming Chairman Watson suggested in his white paper the introduction of review editors, whose function was to oversee the review process—that is, to ensure that authors appropriately address the comments from expert and government reviewers. Chairman Watson had already gathered some experience with unofficial review editors in the preparation of the Working Group II report of the second assessment and was able to base his suggestions on this history (Watson 2001). Because skeptics were afraid of delays in the timing of the assessment, the number of review editors was limited to two, and they were invited to the author meetings to witness the processes and to give immediate comments and suggestions. Their work was therefore not a blind review, but on the other hand, no significant delays occurred due to review editors. Similar to the recruitment of assessment authors and reviewers, selection of review editors was based on government and NGO nominations.

The introduction of the review editors demonstrated a considerable degree of reflection on the ongoing processes and can be seen as more than single-loop learning, because it surmounts the framework of previous conceptions of the IPCC process. These changes were paralleled by a shift in the general perception of the IPCC process. When IPCC scientists were confronted by legalistic arguments in the debate about their scientific statements and the procedures that led to them, the scientists had to realize that the form of scientific discourse they were used to was not appropriate in the IPCC forum. They encountered procedural arguments from lawyers that could not be addressed with scientific arguments based on a common notion of truth and credibility (Schneider 2001). Therefore, the scientists had to adapt to the conditions of legal discourse, which resulted in the more precise formulations of existing rules and procedures and in procedural innovations. It seems hardly exaggerated to state that the IPCC had to incorporate a new rationale. In so doing, it fulfilled the criteria of double-loop learning, although this is not to say that every individual involved in the IPCC participated in this change—but the organization on a collective level did.

Options for Future Developments

While still concerned with the process of completion of the third assessment report in early 2001, the IPCC Bureau was already reflecting on past assessment processes and developing ideas about future steps. As in the third assessment, the chairman prepared an "Issues Paper on the Future Work Program of the IPCC" and submitted it to the bureau for discussion. Comments were again expected from the vice-chairs, co-chairs, lead authors, and governments but not really from individuals or organizations external to the IPCC (Watson 2001).

Alternative options are available to organize the reflection process, but they have never been thoroughly discussed by officials in the IPCC. One option that has been outlined by Watson (2001) is an evaluation of the IPCC by a special commission to either WMO or UNEP or to both. This commission might consist of members of the central decisionmaking bodies of these organizations (i.e., the Governing Council of UNEP and the WMO Executive Council), and it could also include individuals from the negotiating bodies that are closer to the political processes. In this way, the evaluation might be able to gain a broader perspective of the scientific and political functions of the IPCC.

Moss and Schneider (1996) developed another concept for evaluating the IPCC. They suggested the establishment of a so-called "assessment court" or a "panel of independent judges," which would evaluate each assessment based on observations of the debates, the recruitment of authors, the review procedures, and the outcomes. If highly respected and independent personalities could be recruited for this kind of committee, the credibility of the whole IPCC process could be increased due to this additional and institutionalized reflection process.

However, both concepts have significant shortcomings. In both cases, it seems hardly possible to find a group of individuals who could be agreed upon by all parties involved as equally independent and competent. Given the intensity of political debate on issues of climate change, it would be highly problematic to identify individuals who could claim to be "independent" from both political interests and scientific prejudices or preferences (e.g., for certain approaches, models, or theories). The selection procedures could add another question mark, because it is unlikely that all parties involved would accept the World Health Organization and UNEP as the IPCC's parent organizations as legitimate institutions to select the jurors, especially when politically sensitive issues such as response strategies are being discussed. Moreover, both forms of independent review would require significant financial means to provide the expenses for the committee. As funding structures are usually linked to possible influences and interest structures, funding would have to be designed very carefully and rather indirectly to ensure independence of the committee.

Influences on the Ability to Learn

Which factors have been important for the learning ability of an organization like the IPCC? Following insights from organizational learning literature, I highlight some factors that could be considered as influential in the learning processes of the IPCC, as described above. Nevertheless, due to the complexity of social processes, such as causal relationships, learning ability can be traced back to these factors only with great difficulty.

Communication structures are regarded as crucial for organizational learning, because it is through communication that new knowledge is generated and distributed inside the organization. As far as learning mechanisms and other design issues in the IPCC are concerned, the most important form of communication has been informal communication among a core group of people. These individuals were primarily members of the IPCC bureau who met regularly to discuss and decide about certain issues and about the future design of the IPCC, among other things. This predominance of informal forms of learning mechanisms holds especially for the learning endeavors in the period of Bert Bolin's chairmanship from 1988 to 1997. Conditions changed slightly when Robert Watson took over and kept the reflection mechanisms more under his personal supervision. In addition, informal communication has been given high priority at the meetings of the chapter authors. These meetings allow authors to interact with many of their collaborators and to establish personal relationships with them. Nevertheless, most of the informal communication has remained within the working groups, and exchange among the authors of different working groups has been limited. Therefore, the exchange of new information between the working groups is more limited than the exchange within the working groups (Leary 2001). That might explain the crucial role of the IPCC chairperson to link together the working groups and their results, but it also makes understandable the need for cross-cutting guidance papers to develop common standards and a language for all the working group reports in the third assessment.

New knowledge must be documented and stored in a systematic way to make it accessible to other individuals. Looking at the IPCC, one could find a fairly simple system in place. New knowledge is maintained in the first place by the secretariat in Geneva, where all the official documents are stored. Given the heavy work load of the secretariat, however, its responsiveness to requests for certain materials is limited. The technical support units of each working group hold most of the procedural information and specific information about the individual working groups. The support units provide the necessary basic information for new participants in the process, be they authors, reviewers, or review editors. Except for the technical support unit of Working Group I, which remained in the United Kingdom, the location and staff of the support units changed with each of the three assessment rounds due to changing host countries. The continuation of the work and the ongoing processing of information could suffer from these shifts, especially because there is no advanced system of information storage in place.

In general, the whole IPCC process relies heavily on the individual personalities involved and their contact among each other. It is their constant contribution to and engagement in the IPCC procedures that keeps the process running, especially because the IPCC does not provide financial compensation to participants. Although the whole organization has gone through a process of bureaucratization during the past decade, the important decisions are still prepared by a core group of people who know each other personally. Nevertheless, the rules of procedures in place and the bureaucracy behind it are strong enough to keep individuals from dominating the process (Leary 2001). Fundamental decisions can only be made by the panel itself and not by the chairperson or other bureau members. However, the chairperson is powerful enough to keep certain ideas or concepts out of the process by executing his powers as chair of many sessions. In this respect, learning of the IPCC on a collective level depends on learning by individuals and on their willingness and ability to learn. As described above, this core group of individuals has proven to be flexible and able to learn in cases of several procedural innovations such as attempts to broaden participation, to introduce review editors, and to establish specialized technical papers.

What are the underlying cultural values and norms that the people involved share and how did these values contribute to learning in the IPCC? One basic shared belief among the participants is that science can lead to better decision-making about climate change (Leary 2001). Because scientists prepare the IPCC assessments, basic rules and principles of science also govern most of their discussions, arguments, and ways of interaction. For example, a commonly shared notion of truth often provided a venue to bring discussions to a conclusion (Houghton 2001). According to most scientists involved, truth can be found through repeated testing and empirical research. In the case of the "Chapter 8 debate," the mere focus on truth was not a viable solution once the debate began to include legalistic arguments that were based on a largely different rationale. Here, the basic beliefs of most of the participants had been challenged and had to adapt to new ways of thinking, which in the end led to the adoption of new rules of procedure.

Conclusions

Learning mechanisms in assessments are one part of the design-related decisions that organizers must consider when planning and conducting the assessment. The method by which past experiences are brought into bearing could be viewed as crucial for the credibility, legitimacy, and salience—and thereby the overall effectiveness—of assessments.

Nevertheless, the relationship between learning mechanisms in assessments and the political outcomes of the assessments is difficult to measure. The overall effectiveness of the IPCC assessments is especially complicated to judge, given the various factors affecting the outcomes of negotiations and the highly contested arena of climate change. However, even participants in the IPCC negotiation process claim that without the 1990 and the 1992 reports, there would not be such a thing as the Framework Convention on Climate Change (Cutajar 1999). Other things being equal, advanced learning mechanisms that allow for adaptive (or even more fundamental) changes of the value and belief systems of an assessment organization or of the underlying network can enhance the assessment's overall effectiveness rather than inhibit it.

On the whole, designers of assessments have several options at their disposal to institutionalize learning procedures. The simple adoption of one or a number of these mechanisms will hardly be sufficient for thorough learning processes, however. A number of supporting factors are crucial for successful learning as well, including open and direct communication structures, well-functioning information processing systems, supportive shared values among the participants, and others. Bearing these caveats in mind and attempting to deduce some general recommendations from the case study of the IPCC, I suggest that assessment designers adopt the following components as steps to turn assessments into permanent learning endeavors:

- *External committees* consisting of politically independent personalities whose authority is derived from the institution that requests the evaluation,

- *Workshops* with independent experts on certain issues,

- *Assessment courts* with a system of judges and lawyers to address the contested issues of an assessment,

- *Internal iterative procedures* that lead to internal communication processes and result in a revision of the former assessment process, and

- *Specific task forces* charged with evaluating the assessment from inside.

Acknowledgments

I am especially grateful to Shardul Agrawala, William C. Clark, Nancy Dickson, Robert Frosch, Jill Jäger, Myanna Lahsen, Sandy MacCracken, Richard Moss, and Wendy Torrance for their intellectual, technical, and administrative support of my work. I am also deeply indebted to all the interviewees for their time and cooperation. Finally, I greatly acknowledge funding provided by Deutscher Akademischer Austauschdienst and the National Science Foundation.

Notes

1. The IPCC has been described in its structure and evolution over time by Boehmer-Christiansen (1994a, 1994b); Alfsen and Skodvin (1998); Agrawala (1998a, 1998b); and Franz (1998). An overview can be found at http://www.ipcc.ch.

2. The evolution of the working group structure is demonstrated by the changes in the titles of the working groups over the three assessment reports. Working Group I was called "The Scientific Assessment of Climate Change" in 1990; the title was amended to "The Science of Climate Change" in 1995, and it remained unchanged for the third assessment. Despite the name change, the thematic scope of Working Group I remained largely the same. The other working groups, by contrast, underwent more significant changes in their thematic focus. Working Group II was entitled "Impacts Assessment of Climate Change" in 1990; the title changed to "Impacts, Adaptations and Mitigation of Climate Change: Scientific-Technical Analyses" in 1995 and to "Impacts, Adaptation, and Vulnerability" in the third assessment report. While the issues of mitigation were incorporated in this working group's contribution to the second assessment, the task was given to Working Group III in the third assessment. Working Group III in the first assessment was charged with "Response Strategies" and with the more general "Economic and Social Dimensions of Climate Change" in 1995. Because considerable overlaps had been detected, especially concerning economic issues of adaptation and mitigation (Watson 1997), Working Group III was restructured in the third assessment to focus on nothing else but on mitigation and its scientific, technical, environmental, economic, and social aspects.

3. In particular, in the third assessment the selection of authors was a highly demanding process. In Working Group II, for example, co-chairs had to choose 80 authors from about 1,100 government nominees (McCarthy 2001).

4. Schneider (1991, 25) refers to a discussion with Bert Bolin, the first chairman of the IPCC, in which the latter clearly delineates the international aim of the IPCC work: "Bolin agreed that the diversion of talent and resources was not a trivial cost, but he emphasized the international aspect of the study. 'Right now, many countries, especially developing countries, simply don't trust assessments in which their scientists and policymakers have not participated,' he said. 'Don't you think credibility demands global representation?' "

5. As in other UN processes, nongovernmental organizations and industry representatives are only allowed to attend negotiations under the Framework Convention on Climate Change (FCCC)—the so-called Conferences of Parties—as observers not as negotiating parties.

6. This opinion was expressed by nearly all of the interviewees involved in the IPCC process. Although they admitted to considerable arguments at the plenary sessions over the wording of the summaries for policymakers, they said that the conflict resolution mechanisms in place worked and lead to a neutralization of extreme positions among the government delegations. Due to the consensus principle, all delegates must agree to the final wording. Opposing positions have to be explained in the plenary session, and if no compromise can be found, the discussion will be continued in smaller contact groups. Although this mechanism in most cases delivers acceptable solutions, sometimes certain countries try to push forward their claims even further. If still no compromise can be found, a dissenting vote will be included in the text. Because the dissenter is mentioned in the document, countries usually dislike to fall back on this option—especially because it is mostly the same small number of countries with clear political or economic interests, as with the major oil producing countries, which have tried to weaken certain statements in the report (for examples, see footnotes in IPCC 1995). Experience has thus shown that these procedures could not lead to significant changes or weakening of the final documents (Schneider 2001).

7. Because the approval of the synthesis reports of the first and second assessment led to major discussions that could hardly be consensually concluded, the procedures concerning synthesis reports were changed in the third assessment. First, it addressed a list of key questions that were developed in consultation with officials from the negotiating bodies of the FCCC. Second, the synthesis report was split into a longer document that had to undergo a hitherto unknown section-by-section approval process, whereas the more focused summary for policy-makers of the synthesis report had to be approved line-by-line, which means in practice a word-by-word approval, according to participants in the plenary sessions (IPCC 1999).

8. The group was founded in 1993 based on an initiative by IPCC Chairman Bert Bolin. After the first Conference of Parties under the UNFCCC in 1995, the group acquired its current title as IPCC/UNFCCC Joint Working Group (Agrawala 1998b).

9. The paper was based on a workshop at the Aspen Global Change Institute in 1996 (see Moss and Schneider 1996). For an overview of the related issues, see Moss (2000).

10. The INC was the negotiating body before the UN climate convention entered into force.

11. After the successful closure of the ozone negotiations in 1985 in Vienna and 1987 in Montreal, the then-UNEP director Mustafa Tolba was reported to be outspokenly optimistic to be able to repeat the ozone "miracle" for climate (Agrawala 1998a).

12. For example, co-chairs of the first assessment reportedly "didn't have much of an idea themselves" how to organize the assessment process. Because "not much [was] centrally planned" and hardly any rules were available, one author summarized, "we made them up as we went along" (Parry 2001).

13. Initially, the IPCC took over the general rules of procedure from the WMO. The first set of rules of procedure that were specific to the IPCC were formally approved in 1991 and filled only one page (Bolin 2001). At the end of the third assessment, the document was 16 narrowly printed pages long (IPCC 1999).

14. An extensive analysis of the "Chapter 8 debate" can be found in Lahsen (1998). An in-depth study of the arguments concerning peer review put forward in this discussion is included in Edwards and Schneider (2001).

15. The IPCC distinguishes between the "approval" of a document and its "acceptance." The approval builds on the review procedures and is merely a formal acknowledgment of the main body of the working group report by the working group plenary session. Acceptance, by contrast, requires the line-by-line discussion and agreement from all government delegates (IPCC 1994, 1999).

References

Alfsen, Knut, and Tora Skodvin. 1998. *The Intergovernmental Panel on Climate Change (IPCC) and Scientific Consensus. How Scientists Come to Say What They Say about Climate Change.* CICERO Policy Note 1998:3. Oslo, Norway: Center for International Climate and Environmental Research.

Agrawala, Shardul. 1998a. Context and Early Origins of the Intergovernmental Panel on Climate Change. *Climatic Change* 39: 605–620.

———. 1998b. Structural and Process History of the Intergovernmental Panel on Climate Change. *Climatic Change* 39: 621–642.

———. 1999. Early Science–Policy Interaction in Climate Change: Lessons from the Advisory Group on Greenhouse Gases. *Global Environmental Change* 9: 157–169.

Argyris, Chris, and Donald A. Schön. 1996. *Organizational Learning II. Theory, Method, and Practice.* Reading MA: Addison-Wesley.

Arrhenius, Svante. 1896. On the Influence of Carbonic Acid in the Air upon the Temperature of the Ground. *Philadelphia Magazine* 41: 237–271.

Boehmer-Christiansen, Sonja. 1994a. Global Climate Protection Policy: The Limits of Scientific Advice. Part 1. *Global Environmental Change* 4(2): 140–159.

———. 1994b. Global Climate Protection Policy: The Limits of Scientific Advice. Part 2. *Global Environmental Change* 4(2): 185–200.

Bolin, Bert. 1994a. Science and Policy Making. *Ambio* 23(4): 25–29.

———. 1994b. Next Step for Climate-Change Analysis. *Nature* 368: 94.

———. 2001. Telephone interview with author, March 13.

Cutajar, Michael Zammit. 1999. Statement by the Executive Secretary of the UNFCCC. In *Report of the 15th Session of the IPCC, San José 15–18 April 1999*. Bonn, Germany: United Nations Framework Convention on Climate Change Secretariat.

Edwards, Paul N., and Stephen H. Schneider. 2001. Self-Governance and Peer Review in Science-for-Policy: The Case of the IPCC Second Assessment Report. In *Changing the Atmosphere: Expert Knowledge and Environmental Governance,* edited by Clark Miller and Paul N. Edwards. Cambridge, MA: MIT Press.

Elzinga, Aant. 1997. From Arrhenius to Megascience: Interplay between Science and Public Decisionmaking. *Ambio* 26(1): 72–80.

Franz, Wendy. 1998. *Science, Skeptics, and Non-State Actors in the Greenhouse.* ENRP discussion paper E-98-18. Kennedy School of Government. Cambridge, MA: Harvard University.

Haas, Ernst B. 1991. Collective Learning: Some Theoretical Speculations. In *Learning in U.S. and Soviet Foreign Policy,* edited by George W. Breslauer and Philip E. Tetlock. Boulder, CO: Westview Press.

Haas, Peter, and D. McCabe. 2001. Amplifiers or Dampeners: International Institutions and Social Learning in the Management of Global Environmental Risks. In *Learning to Manage Global Environmental Risks: A Comparative History of Social Responses to Climate Change, Ozone Depletion and Acid Rain,* edited by the Social Learning Group. Cambridge, MA: MIT Press.

Houghton, John. 2001. Personal interview with the author, March 20.

Houghton, J., L.G. Meira Filho, B.A. Callandar, N. Harris, A. Kattenberg, and K. Maskell (eds.). 1996. *Climate Change 1995: The Science of Climate Change. Contribution of Working Group I to the Second Assessment Report of the Intergovernmental Panel on Climate Change.* New York: Cambridge University Press.

IPCC (Intergovernmental Panel on Climate Change). 1990. *Climate Change. The IPCC Scientific Assessment.* New York: Cambridge University Press.

———. 1992. *Report of the 8th Session of the IPCC, Harare, 11–13 November 1992.* Geneva, Switzerland: Intergovernmental Panel on Climate Change Secretariat.

———. 1994. IPCC Procedures for Preparation, Review, Acceptance, Approval and Publication of Its Reports. In *Report of the 10th Session of the IPCC, Nairobi, 10–12 November 1994,* edited by Intergovernmental Panel on Climate Change. Geneva, Switzerland: IPCC Secretariat.

———. 1995. IPCC. *Second Assessment Synthesis of Scientific–Technical Information Relevant to Interpreting Article 2 of the UNFCCC.* New York: Cambridge University Press.

———. 1997. *Report of the 12th Session of the IPCC Bureau, Geneva, 3–5 February.* Geneva, Switzerland: Intergovernmental Panel on Climate Change Secretariat.

———. 1999. Procedures for the Preparation, Review, Acceptance, Adoption, Approval and Publication of IPCC Reports. In *Report on the 15th Session of the IPCC, San José, 15–18 April 1999.* Geneva, Switzerland: Intergovernmental Panel on Climate Change Secretariat.

Lahsen, Myanna. 1998. The Detection and Attribution of Conspiracies: The Controversy over Chapter 8. In *Paranoia within Reason: A Casebook on Conspiracy as Explanation,* edited by George E. Marcus. Chicago: University of Chicago Press.

Leary, Neil. 2001. Personal interview with the author, Washington, DC. February 27.

McCarthy, James. 2001. Personal interview with the author, Cambridge, MA. March 9.

Moss, Richard. 2000. Ready for IPCC-2001: Innovation and Change in Plans for the IPCC Third Assessment Report. *Climatic Change* 45: 459–468.

———. 2001. Personal Interview with the author, Washington, DC. February 27 and 28.

Moss, Richard, and Stephen H. Schneider. 1996. *Characterizing and Communicating Scientific Uncertainty: Building on the IPCC Second Assessment.* Aspen, CO: Aspen Global Institute.

————. 2000. Uncertainties. In *Guidance Papers on the Cross Cutting Issues of the Third Assessment Report of IPCC,* edited by R. Pachauri et al. Geneva, Switzerland: Intergovernmental Panel on Climate Change Secretariat.

Pachauri, R., T. Taniguchi, and K. Tanaka (eds.). 2000. *Guidance Papers on the Cross Cutting Issues of the Third Assessment Report of IPCC.* Geneva, Switzerland: Intergovernmental Panel on Climate Change Secretariat.

Parry, Martin. 2001. Telephone interview with the author, March 5.

Sabatier, Paul. 1987. Knowledge, Policy-Oriented Learning, and Policy Change. *Knowledge: Creation, Diffusion, Utilization* 8: 649–692.

Schneider, Stephen H. 1991. Three Reports of the Intergovernmental Panel on Climate Change. *Environment* 33(1): 25–30.

————. 2001. Telephone interview with author, March 5.

Siebenhüner, B. 2002. How Do Scientific Assessments Learn? Part 2. Case Study of the LRTAP Assessments and Comparative Conclusions. *Environmental Science and Policy* 5: 421–427.

Watson, Robert T. 1997. White Paper on the Third Assessment Report and the IPCC Bureau, Annex to the *Report of the 12th Session of the IPCC Bureau, February 1997*, edited by Intergovernmental Panel on Climate Change, Geneva, Switzerland.

————. 2001. Telephone interview with author, March 8.

CHAPTER 9

The Design and Management of International Scientific Assessments
Lessons from the Climate Regime

Clark A. Miller

IN RECENT DECADES, international scientific assessments have become a prominent and increasingly important feature of world affairs and global civil society. Facing an array of complex global challenges, diplomats and international public managers have turned to scientific assessments as a tool not only for identifying policy problems and potential solutions but also for bridging the cultural gaps that divide nations. Many international organizations now conduct scientific assessments as a regular part of their ongoing activities.[1] International environmental, trade, and security agreements increasingly include provisions that establish expert advisory institutions and mandate assessment procedures.[2] A growing number of independent and quasi-independent scientific assessment organizations also now exist: some issue specific assessments such as those produced by the Intergovernmental Panel on Climate Change (IPCC) and the Millennium Ecosystem Assessment, while others, including the InterAcademy Council, are designed to provide advice and assessments on a wide range of policy issues.[3]

This chapter examines an important but sometimes overlooked aspect of the design and management of international scientific assessments—namely, their need to speak credibly to numerous scientific, policy, and public audiences. Most international scientific assessments fall into the class of what I term *public assessments,* meaning they are conducted by or on behalf of governments or government-like entities, and their efforts are seen as relevant to and meant to influence the formulation, implementation, and/or evaluation of public policy. Hence, businesses, the media, nongovernmental organizations, regulatory officials, scientists, and informed members of the public pay careful attention to their conduct and results. These diverse audiences may hold widely varying norms and expectations regarding best practices for the production and validation of policy-relevant knowledge, a problem that is especially acute in cross-cultural contexts.[4]

At one level, cross-cultural differences create a problem for assessment design, requiring negotiation among competing policy framings, legal requirements, customary practices, and evidentiary standards.[5] U.S. law, for example, requires that scientific advisory committees conduct their business in a manner that is transparent and open to public involvement.[6] Historically, British practice has taken the opposite approach, with most scientific advisory processes operating behind closed doors. Bridging these kinds of divides in international assessment design is not always straightforward. The problem runs more deeply, however, than just finding workable arrangements that can satisfy participants. Audiences trust science advice in part by the degree to which it is generated in accordance with culturally embedded norms and standards of assessment design and conduct. Assessments that do not mirror those expectations may come under public criticism and may ultimately fail to persuade audiences of the veracity of their claims. For example, when the World Trade Organization organized its expert panels behind closed doors, U.S. environmental and consumer organizations complained loudly, calling into question the credibility of knowledge produced in this nontransparent fashion.

In this chapter, I analyze the implications for international assessors of cultural differences in audience expectations regarding the design and management of credible scientific assessments. I begin with a review of the literature on comparative science advice and regulatory politics, which offers a basis for understanding why cultures arrive at divergent standards for conducting public assessments. I then turn to a more specific analysis of science advice in the climate regime, which offers a useful case study of the challenges that divergent cultural standards for assessment practice pose for the design and management of international scientific assessments. I also examine how institutions, in turn, have sought to meet these challenges. I begin by suggesting several instances in which the IPCC and the Subsidiary Body for Scientific and Technological Advice (SBSTA) of the United Nations Framework Convention on Climate Change have run into criticisms that are at least partially grounded in divergent cultural norms and expectations regarding science advice. I then compare the strategies adopted by the IPCC, which focused on tightly defined standards, and by the SBSTA, which adopted a more flexible, open, and deliberative approach, in response to these criticisms.

My analysis of these strategies suggests that the IPCC's effort to standardize a well-defined set of rules for international assessments may be less effective than the more flexible, deliberative, and exploratory approaches adopted by the SBSTA in shoring up assessment credibility vis-à-vis widely divergent audiences. Standards can seem appealing when dealing with cultural heterogeneity. In a global context, however, the absence of an authoritative body empowered to legislate global standards—and also shared norms for warranting policy-relevant truths—challenges the ability of standards to bridge cultural divides. In the past, when few people outside the scientific and diplomatic communities paid much attention to international scientific assessments, negotiators could finesse these differences through ad hoc settlements. Today, however, such ad hoc compromises are increasingly less viable. Contemporary issues assessed by international scientific organizations, such as climate change and the safety of genetically modified organisms, penetrate deeply into people's daily lives and concerns. Consequently, the media and general public now attend closely to the activities and conclusions of the organizations. These outside observers judge the behavior of international scientific organizations according to culturally embedded norms and expectations. Standard-setting processes must therefore acknowledge cultural divides in audience expectations and find ways to build commonly held norms and expectations among divergent and scattered scientific, policy, and public audiences. For this work, the more tentative, experimental, open, and adaptable approaches of the SBSTA seem more likely to achieve greater long-term success.

The Politics of Credibility

To understand why cultures differ in how they design, manage, and evaluate the credibility of scientific assessments, it is necessary to understand that credibility has a political element. Conventionally, the credibility of assessments has been understood in terms of the content and transmission of assessment reports. What does the report say? What doesn't it say? How well are its conclusions supported by evidence? To whom are the report's conclusions communicated? Who reads the report? Who uses the knowledge presented by the report? Were the assessors biased in their presentation of data? Do institutional factors act as barriers to the communication of assessments to appropriate policymakers? Do politics distort policymakers' use of knowledge presented by assessment reports?

These questions are important, but they present only a partial picture of the politics of credibility. In this conventional model of credibility, politics have only a distorting role to play. However, political discourse and institutions can also play an important role in shoring up the credibility of scientific assessments (Miller 2004b; Miller et al. 1997). It may be useful to think about the credibility of assessments the way one might think about the performance of a play. In *Science on Stage,* a recently published study of assessments carried out by the U.S. National Academy of Sciences, Stephen Hilgartner suggests that the process of conducting a scientific assessment and making its conclusions public shares many of the same features as a stage performance (Hilgartner 2000). Much as Shakespeare can be performed well or poorly, so too can assessments be performed well or poorly. Like stage managers, Hilgartner asserts, designers and managers of scientific assessments often divide their performance space into public and private domains so that much of the work that goes on "backstage" is invisible to the audience. Like actors, participants in assessments must skillfully present themselves in character if they are to be believed. A good performance will help strengthen the credibility of the assessment's claims; a bad performance may doom it to obscurity.[7]

Thinking about the metaphor of a performance highlights the importance of the relationship between assessors and their (often multiple) audiences. Studies of scientific controversies indicate that, in practice, credibility must be achieved dynamically; it is not a static feature of a given set of data or factual claims.[8] Moreover, the achievement of credibility often follows different paths in different contexts, and different audiences judge credibility using different criteria. Establishing the credibility of science in a courtroom setting, for example, does not follow the same practice one scientist would use to convince another of a recent finding. Instead, lawyers use different forms of performance to strengthen (or weaken) the credibility of expert witnesses in jurors' minds, such as high-lighting the witnesses' credentials or business ties, presenting visual demonstra-tions of data, and carefully scripting the messages that jurors hear (Jasanoff 1998a). Just as stagecraft follows different rules in the Kennedy Center than it does in a small experimental theater to account for differences in audience expectation, so too is the performance of credibility different at a scientific con-ference, in the courtroom, and in a congressional hearing.[9]

To be sure, many assessments are intended for a single audience and so can afford to ignore how other audiences may view their work. For public assess-ments, however, differences in audience expectations matter enormously. Public assessments, as I use the phrase, are those that are carried out by or on behalf of one or more governments for the specific purpose of generating knowledge as an input to public policy decisions. The ability of public assessments to speak credibly to multiple audiences is important precisely because many different groups typically care about the policy choices being made on the basis of the claims made in the assessment. As a result, these same groups also tend to care whether assessments that inform those policies provide credible knowledge.

To cope with this problem, most governments have developed standards for the design and management of public assessments. Some standards are established by legal requirements, as in the case of scientific advisory committees in the United States whose work falls under the 1972 Federal Advisory Committee Act. Other legal requirements in the United States can be found in the 1946 Administrative Procedure Act and in subsequent legal interpretations of both acts by U.S. courts. Other standards are not legally binding but nonetheless acquire solidity and often a degree of permanence through custom and tradition. There is no legal requirement in the United States, for example, that U.S. Environmen-tal Protection Agency (EPA) advisory committees contain a balanced member-ship of nongovernmental organization (NGO), industry, and university partici-pants. Indeed, the courts have ruled that a formal requirement to that effect is not permissible. Nevertheless, EPA administrators have come to recognize the importance of balancing membership in this way if the committee's work is to be seen as credible among all parties (Jasanoff 1990).

Importantly, however, the standards governments have developed for the conduct of credible scientific assessments often differ markedly from country to country. Even among "modern" democracies, scientific assessments are often conducted according to widely varying standards. For example, although the concept of risk is a universally recognized feature of contemporary environ-mental politics, risk assessments vary systematically in Europe and North Amer-ica along a range of axes, including who counts as an expert, what evidentiary standards are used, how uncertainty is treated, how much weight is given to sci-entific knowledge in comparison to other kinds of knowledge, who has access to assessment meetings and documents, how the review of assessments is han-dled, and how assessments relate to other political and scientific institutions.[10] These differences reflect deeply held norms and values regarding the conduct of government and the warranting of policy-relevant truths, often reinforced by distinct constitutional infrastructures—both elements of what social scientists term *political culture*.

Two examples illustrate my point. U.S. political culture, which is highly adversarial and transparent, strongly favors toxicological assessments of chemical carcinogenicity derived from animal models over alternative methods of risk analysis. These studies, which are highly standardized and methodologically rigorous, hold up better in the face of public scrutiny than, for example, epidemiological studies, which are the other frequently used source of scientific knowledge about cancer causation. By contrast, British expert advisory committees, who judge evidence in closed meetings, tend to give more weight to epidemiological evidence, whose flaws can be evaluated and weighed more readily by experts and which give more direct knowledge of human carcinogenicity. One result of these differences is a marked difference in the ratio of toxicologists and epidemiologists working in the two countries (Brickman et al. 1985).

A second example of differences between the two cultures can be found in their distinct notions of expertise. In the United States, membership on advisory committees is determined by possession of specialized, often disciplinary-specific knowledge and skills. In the United Kingdom, by contrast, citizens expect their government officials to have broad experience as well as narrow technical expertise. As a consequence, the public servants who head scientific advisory committees usually embody the public trust in some easily recognized manner, often through previous demonstrations of leadership on controversial public issues or formal acknowledgment and recognition of their service to the country through such mechanisms as knighthood or appointment to the House of Lords.[11] Sir John Houghton, who was the head of the UK Meteorological Office and representative to the IPCC, offers an apt example of both. He has both led expert assessments of other weighty environmental issues and been formally recognized for his public service.

The Cultural Credibility of International Scientific Assessments

When attention shifts from domestic to international policy, the importance of political culture as a factor in the credibility of scientific assessments takes on added importance. Across a range of policy issues, disagreements over the interpretation and application of knowledge and evidence have become increasingly central to international policy controversies. Consider a few examples: U.S.–Europe differences over the safety of genetically modified organisms and the use of growth hormones in milk production, conflicts between the World Health Organization and the city of Toronto over severe acute respiratory syndrome quarantine, and evidentiary disputes surrounding the alleged presence of weapons of mass destruction in Iraq. Each case brought culturally embedded policy framings, notions of expertise, treatment of uncertainty, and evidentiary standards into conflict.

Such conflicts have also plagued the climate negotiations, up to and including the Bush administration's rejection in 2001 of the science of global warming. Examining the history of the IPCC and the SBSTA, the two primary international scientific advisory institutions in the climate regime, it is possible to identify three stages in the design and conduct of international scientific assessments at which cultural variations may matter significantly. The first stage is the design phase, when participants from different countries may disagree with one another about how best to design and manage international scientific advisory processes. The second is when audiences evaluate the credibility of the assessment product (e.g., the report), which may occur after the assessment is completed but also may occur earlier—during final stages of the assessment process itself. The third stage is when the assessors undertake to evaluate and review the assessment process, often for the purpose of revising the rules and procedures for subsequent assessments.

At stage one, variations in the political culture of participants can significantly affect an assessment, including leading to disputes over design parameters that can seriously reduce the credibility of the assessment in the eyes of some audiences or even lead to deadlocks that prevent the assessment from taking place at all. One amusing but serious anecdote illustrates how variations in political culture can influence even the participation of countries in the design process: in 1988, nations were first invited to send representatives to the IPCC to establish its basic rules and procedures. These invitations went out under the auspices of the World Meteorological Organization to national meteorological offices. In the United States, as a result of extensive intergovernmental coordination on international issues, a delegation of 24 officials and scientists, under a delegation leader from the State Department, was sent to Geneva. By contrast, the German meteorological office, which, unlike its U.S. counterpart, has no responsibility for climate research, forwarded the invitation at the last minute to the appropriate German research organization (a semi-independent scientific organization), which sent one scientist, Hartmut Grassl, then chair of the Climate Advisory Board. When he discovered the nature of the meeting, he quickly left the room to call the German embassy in Switzerland to ask the embassy to send a diplomatic representative, too (Grassl 1997).

Such seemingly minor issues can have serious long-term consequences. Many developing countries opted not to send delegates to the initial IPCC meeting at all. Although the IPCC quickly recognized the problem this posed for the global credibility of its results and immediately took steps to increase the participation of developing country scientists, the organization was unable to overcome its initial credibility deficit, and developing countries ultimately refused to allow the IPCC to become a formal player in the climate negotiations until the late 1990s. At stake were two principal issues: the framing of climate change as a problem of greenhouse gas emissions, which developing countries saw as unfairly shifting the burden of responsibility from rich consumers (whose profligate energy consumption led to high carbon dioxide emissions) to poor communities (whose agricultural activities were essential to survival; Agarwal and Narain 1991); and the framing of the IPCC assessment as a scientific study of the climate system, rather than as an economic and political assessment of the need to transfer resources and technologies to poor countries to enable them to pursue sustainable, climate-friendly development (Ripert 1991).

At stages two and three (the publication and evaluation of assessment outcomes), cultural variations also matter. For example, in part as a result of disagreements about the initial design of the assessment, and in part as a result of its ongoing inability to attract developing country participants, the IPCC encountered subsequent difficulties with developing country audiences after it had published its first report. In late 1990, at the close of the IPCC's first round of assessment, developing country negotiators rejected an effort by the United Nations to establish the IPCC as the host institution for diplomatic efforts to craft a climate convention in 1991–1992. Instead, they established a special committee of the UN General Assembly to carry out that task. Developing countries subsequently refused to allow the IPCC to be included as a formal part of the treaty organization created by the 1992 Framework Convention on Climate Change. Consequently, negotiators were forced to establish the SBSTA as a second scientific advisory institution within the regime.

Nor have culturally grounded disagreements over the credibility of the IPCC solely focused on North–South issues. Bert Bolin and Sir John Houghton—two prominent European scientists with long experience in advising their respective governments—initially designed the IPCC to function as a closed, relatively informal body whose members had close ties to public officials. This organizational form paralleled that of European scientific advisory bodies. Over time, however, the IPCC has adopted a much more open style of deliberation as well as a much tighter set of rules governing its operational practices. These changes have taken place in a series of stage-three institutional reforms (the first took place in 1991, after the end of the first IPCC assessment, and continued subsequently at the close of each of its major assessments) prompted primarily by criticisms of the organization in the U.S. media.[12] The result is that a number of the IPCC's rules of operation now bear a much greater similarity to those of the U.S. Federal Advisory Committee Act.

Cultural disagreements over the design and management of assessments can even go so far as to give rise to deadlocks that prevent assessments from happening in the first place. Consider an early experience in the SBSTA, which first met in 1995: Initially imagined as an alternative to the IPCC, the SBSTA has evolved into roughly the equivalent of a legislative subcommittee with authority over the design and management of processes for providing expert input into the decisionmaking processes of the climate regime. Members of the SBSTA are typically the same diplomatic representatives of signatory countries to the UN Framework Convention on Climate Change who regularly attend the convention's Conference of Parties. In addition, representatives of international organizations and NGOs participate as observers. The SBSTA works closely with the IPCC to help set the latter's agenda and to facilitate the uptake of its reports among diplomats working on the Kyoto Protocol and other international treaties dealing with global warming. The SBSTA has also established several other expert committees, whose purpose is often to address issues that the IPCC either has chosen not to address or has been prevented politically from assessing—such as the transfer of green technologies to poor countries, and the development of standard methods for accounting for greenhouse gas emissions and for assessing vulnerability to climate impacts.[13]

Many proposals put forward through the SBSTA for new expert advisory and assessment processes have proven controversial. One notable set of wide-ranging deliberations occurred in 1996–1998 over an effort to establish two new technical assessment panels (TAPs) to supplement the IPCC. Over the course of several SBSTA meetings, deep-seated divisions emerged among participants over how to organize the additional advisory panels. Participants differed over a number of issues:

- *Expert affiliation.* Would experts on the panels be invited from governments, the private sector, NGOs, international organizations, or universities?

- *Method of appointment.* Would experts be appointed by governments, nominated by governments and appointed by the Framework Convention secretariat, or nominated by governments and appointed by the SBSTA?

- *Balance of experts.* Would experts from any country be allowed to participate (in an open-ended structure), or would there be a regional/geographic balance, a balance between Annex I (industrialized) and non-Annex I (developing) countries, or a balance of disciplines?

- *Method of review.* Would the TAPs use formal scientific peer review, formal review by the SBSTA, or no review at all?

- *Committee structure.* Would the TAPs have a fixed number of members or a fixed steering committee with flexible ability to create subpanels? Or would the number of TAPs members flexibly accommodate new needs that might arise? Furthermore, how many members would the TAPs have?

- *Duration.* Would the TAPs be permanent, have a fixed duration, or be of contingent duration depending on periodic SBSTA review?

- *Line of authority.* Would the TAPs report to the SBSTA through the IPCC, or would they report directly to the SBSTA?

- *Terms of reference.* Would the SBSTA establish a fixed terms of reference or would the TAPs determine their own terms of reference?

Parties to the SBSTA put forward a number of compromise proposals designed to address these disagreements and to allow TAPs to come into existence. However, all of the proposals were ultimately rejected. Disagreements proved too deep to overcome. Many of the disputes reflected differences in national political culture. U.S. negotiators persistently called for experts to be selected solely on the basis of their claims to specialized expertise in a particular discipline or field, reflecting a common feature of U.S. advisory committees. By contrast, other countries proposed additional criteria for selecting experts, including selection on the basis of national or regional identity, recognized international experience, governmental affiliation, and others. For both sides, this dispute, which was a major cause of the ultimate breakdown, reflected assumptions about what would make the panels both credible and legitimate in the eyes of scientists, political officials, and lay publics in their country. Ultimately the SBSTA shelved the idea of establishing the TAPs, as a consequence of its failure to find a compromise that all sides could accept.

More generally, differing expectations regarding the proper conduct and organization of scientific assessments, rooted in national political cultures, have helped prevent the emergence of formal or informal standards regarding the way in which to conduct international scientific assessments. Although international diplomats share certain expectations regarding the conduct of international public affairs, those expectations are weak compared with their domestic counterparts, and they are frequently not shared more widely among scientists, public officials, or individual citizens. To date, the United Nations has not sought to standardize the design or management of international scientific assessments, and it would likely encounter considerable difficulty if it tried. The absence of a global state with the authority to set such standards has meant, in practice, that each international scientific assessment has had to construct ad hoc rules and procedures to govern its own activities.

Assessment Design in Multinational Contexts

Faced with widespread disagreements over scientific advisory processes based in competing national perspectives, what kinds of strategies have participants in the SBSTA and the IPCC adopted for designing public assessments of climate change? Taken independently, each of the decisions made by these organizations regarding an individual choice of design or management often appears as an ad hoc compromise. Taken collectively, however, several patterns emerge that suggest broader lessons that could be applied to other assessment processes in multilateral contexts. In this final section, I discuss the advantages and disadvantages of four strategies adopted by the IPCC and the SBSTA to deal with the challenges posed by political culture to the design and conduct of international scientific assessments. I conclude that the first strategy (international standardization), while intuitively appealing to many scientists and diplomats, actually has less to offer than the latter three in terms of its ability to manage the widely varying norms and expectations of differing national audiences.

Formal Standards

In its efforts to assess the risks of climate change, the IPCC has sought to set formal standards for the rules and procedures followed by its working groups. The impetus for this standardization stems from early criticism of the IPCC during and after the completion of its first assessment report in 1990. At the time, the IPCC's three working groups operated under different rules: one adopted a form of peer review, asking scientists outside of the organization to comment on its initial draft, while the other two did not. Subsequently, the latter two were characterized as, in effect, diplomatic prenegotiation sessions, rather than scientific assessments. To shore up its credibility, the IPCC reformed its rules of operation in 1991, requiring all three working groups to adopt peer review procedures. This action helped clarify the distinction between the IPCC, which now appeared to be a scientific assessment organization, and the newly established Intergovernmental Negotiating Committee, which housed the diplomatic negotiations that led to the 1992 UN Framework Convention on Climate Change. It also set the tenor for subsequent reforms of IPCC rules and procedures after the second and third assessments, each of which has further specified exactly how the IPCC will operate, including how experts can be nominated to the panel, who will have an opportunity to review the panel's documents, and what procedures authors must follow in responding to reviewers' comments.

The perceived advantage of formal standards is that by applying them rigorously the IPCC inoculates itself against criticisms that it is biasing its scientific work. Observers of U.S. politics have long argued that bureaucratic officials in U.S. agencies often turn to impersonal rules to help defuse claims that they have allowed subjective judgment to cloud their decisions (see especially Porter 1995). The IPCC seems to be following a similar logic. By setting formal standards of assessment design and management, the hope is that assessors will be able to strengthen their ability to perform credibly by demonstrating that they followed the rules laid out in advance. This is perhaps most clear in the case of peer review rules. Following (hotly disputed) criticisms that authors failed to take reviewer comments sufficiently into account during the IPCC's second assessment, the organization established new peer review rules for the third assessment that instituted a new editorial review process. In this process, authors must document their responses to each reviewer comment and persuade an independent editor that their responses are adequate (Edwards and Schneider 2001).

The IPCC's increasingly formalized rules of procedure, however, have not silenced its critics or strengthened its credibility among audiences that are skeptical of its claims—in the United States or in developing countries. One problem is that once rules are set, perceived deviations from required practices can offer critics of the assessment new ammunition to challenge its credibility. Formal and informal standards thus become fodder in disputes over whether to give credence to the conclusions reached by an assessment. Despite its 1996 reforms, after the completion of its third assessment, IPCC scientists were again criticized in 2001 for altering the text of their reports in violation of the rules governing the organization. A well-known feature of legal scholarship is that no set of rules can fully specify a domain of activity. The proliferation of rules has, in some sense, made it easier for political opponents of the IPCC to find instances in which IPCC scientists can be made to appear as having failed to follow the rules. Even where such failures are obviously minor and clearly accidental, the appearance of impropriety can damage the institution's credibility in a critical public environment.

Perhaps more importantly, deriving an intricate set of rules risks creating rules that differ markedly from the expectations of various audiences regarding what makes for good science advice. Formalized rules may limit the ability of the organization to respond to distinct cultural perspectives or new political demands arising out of the rapidly changing character of global governance. Such rules may also demand a great deal of organizational and bureaucratic oversight to maintain, uphold, and adapt—while leaving the organization blind to other problems that may place its credibility at risk. For example, the peer review standards adopted by the IPCC in 1991, which focused attention on the need to strengthen external review *after* chapters had been written, offered little help to authors in identifying potential biases or problematic assumptions built into the existing literature *at the beginning* of the work. As a result, some problems, which might have been easily resolved in early framing discussions, turned into significant political liabilities in those instances in which reviews uncovered problems very late in the day. Perhaps the most illustrative example of this occurred during the second IPCC assessment, when IPCC economists used a much higher statistical value of life for the United States, Europe, and Japan than they did for developing countries when evaluating the potential costs of climate change. This was consistent with the economic literature on valuation, and the assessment authors were therefore unwilling to change the results during the review process, despite significant pressure from developing country scientists and officials to do so. At stake was what reviewers and developing country officials perceived to be a radical underestimation of the costs of climate change to poor countries (Meyer and Cooper 1994). In response, the IPCC altered its summary for policymakers to reflect a more equal valuation but did not alter the underlying chapter—a result that settled the immediate controversy but ultimately did little to avert damage to the IPCC's credibility among many economists and developing country audiences.

Finally, in considering international standards for assessment design and management, it is important to go beyond questions of their strategic use in criticizing or defending the credibility of assessments. When they function well, standards help to deepen the credibility and legitimacy of assessments as public policymaking tools by ensuring that they conform to norms of good governance. In this sense, the development of global standards offers a valuable opportunity to encourage different parts of global society to learn to "reason together" about global risks (Jasanoff 1998b). This can only happen, however, if wider policy and public audiences are actively engaged in the standard-setting process, such that it alters their perceptions of what makes for a credible assessment. If audiences beyond the community of scientists and diplomats actively involved in the assessment do not ultimately arrive at a shared understanding of good practice in assessment design and practice, arbitrary standards are likely to detract from rather than add to the credibility of the assessment. Arguably, in this regard, the approaches to assessment design adopted by the SBSTA, while slower and more cumbersome at the outset, may ultimately arrive at more widely satisfactory assessment processes.

Deconstruction and Deliberation

Like the IPCC, the SBSTA began its existence with considerable confusion and disagreement over its purpose. Some saw it as replacing the IPCC; others saw it as merely a rubber stamp for IPCC products. Still others saw it as a supplement to the IPCC, taking on tasks the IPCC had ignored but leaving risk assessments to the IPCC. Unlike the IPCC, however, SBSTA participants were never able to completely sort out their disagreements and agree on a coherent program of action with clearly defined rules and procedures. Instead, the organization ultimately coalesced around a process in which nations (and to a lesser extent NGOs, business groups, and international organizations) with an interest in climate change could deliberate over and establish short-term scientific and technological advisory committees to explore and advise them on predetermined questions. This process, which involves regular, semiannual meetings of all SBSTA participants (the signatories of the Framework Convention and any other groups that have been formally acknowledged as observers by the Framework Convention secretariat), implicitly combines three strategies for managing the multiplicity of differing national perspectives on international expertise and assessment.

One key feature of the SBSTA's ongoing processes is the opportunity it provides participants to deconstruct assessment designs so as to make visible and open up for deliberation the tacit, value-laden assumptions embedded in design choices. Within a given national political culture, value choices that go into the design and conduct of public assessments frequently become taken for granted by those who carry out assessments and those who rely on the knowledge they generate. Only when U.S. assessments are compared with British assessments, for example, does the commitment of U.S. political culture to transparency and openness become apparent for what it is: a feature of U.S. democratic values. In international contexts, the choice of whether to hold open or closed meetings thus constitutes a normative choice about what kind of institutional arrangements are going to characterize international governance.

The institutional structure of the SBSTA helps guarantee that participants who see important value choices at stake in particular assessment designs have an opportunity to voice their concerns. In debates over the priorities of the SBSTA working group on methodologies, for example, members of the Association of Small Island States were able to successfully argue for the inclusion of an initiative to develop methods for evaluating the potential impact of rising sea levels on coastal regions. Prior to the association's intervention, discussions had focused primarily on methods for measuring national emissions of greenhouse gases. Similarly, representatives of India, China, and the Philippines were able to ensure that the IPCC—an institution they and other developing country diplomats criticized for offering a too narrowly Western framing of the climate issue—reported first to the SBSTA, and that the SBSTA passed IPCC recommendations on to the Framework Convention Conference of Parties. From a scientific standpoint, this change has little effect on the knowledge produced, because the IPCC carries out and publishes its assessments independently of the Framework Convention institutions. From a political standpoint, however, the reporting requirement helps to reassure many governments who do not quite trust the IPCC. The debate about TAPs, discussed earlier, also illustrates the ability of participants in SBSTA to become involved in the deconstruction and deliberation of competing models of how to design and conduct international science advice.

The most important institutional feature of the SBSTA that facilitates this process of deliberation and deconstruction is its voting rule. Rules of procedure in the SBSTA require unanimity among members to adopt decisions, thus affording any member's objection substantial weight. Although this effective veto power is used only infrequently, its existence helps to ensure that it is at least possible to voice perspectives that might not otherwise be heard. To be sure, this procedure frustrates many participants who would like to see more rapid action. When what is at stake are the basic constitutional norms of global governance, however, it seems reasonable to proceed at a more staid pace and to ensure that the assessment designs arrived at by the SBSTA incorporate the concerns of a wide array of countries.

Flexibility and Control

A second important feature of SBSTA deliberations is its combination of flexibility and accountability. As noted above, SBSTA delegates have been unable to reach agreement on the creation of new, permanent expert advisory institutions. Aside from the IPCC, which existed prior to the Framework Convention and with which many governments agreed only reluctantly to work, SBSTA members have avoided establishing any kind of long-term institutional arrangements. Instead, they have established new institutions only on an ad hoc basis. For example, the SBSTA has established a "Roster of Experts," consisting of individuals nominated by governments. With unanimous agreement, the SBSTA can request that the Framework Convention secretariat select a group of experts from this roster to provide an answer to a specific question or set of questions, after which the group is dissolved. The SBSTA also has replaced the idea of constituting two new IPCC-like technical advisory panels with the informal constitution of small working groups of its own members to address the same topic. Any member who wishes may participate in these working groups.

The effect of these decisions is that SBSTA delegates collectively retain a great deal of flexibility regarding the design and conduct of expert assessments and that the design and management of the assessment remains tightly accountable to the SBSTA. Panels are one-shot affairs constituted to address only specified questions. They do not acquire independent authority to address questions other than those directed by the political process. Panels may be composed differently depending on their subject matter, which enables the secretariat to select appropriate panelists to fit the circumstances of a given assessment. Thus, for example, one panel may consist of specialists from particular disciplines while another consists of specialists recruited from a particular group of countries, as in the case of the panel established to report on financial transfers to developing countries. That panel was composed entirely of experts from developing countries to ensure credibility to that particular audience. Another important degree of flexibility, which relates also to the third strategy adopted by the SBSTA (described in the next section), is that governments nominate experts to the Roster of Experts on the basis of their own criteria for who constitutes an expert. There is therefore no need for universal or even majority consensus on criteria for determining what counts as expertise.

The informal working groups on methodologies and technology transfer also reflect an important degree of flexibility and accountability. Participating delegates can quickly respond to new issues as they arise with concrete plans to address them, in part because they have tight political connections to other parts of the Framework Convention process and in part because they have the political authority to act. They do not have to ask permission from another body first. They are also in a position to deliberate and resolve many of the tough political and value-laden choices that suffuse methodology work. As the IPCC discovered in its early efforts to standardize methods for measuring greenhouse gas emissions, selecting standards inevitably involves trade-offs with political consequences. The IPCC found that getting another body to quickly deliberate and resolve those choices—so their own detailed, technical work could continue—was very difficult. Now that such methodology work is accountable to the SBSTA, it is at least clear who is responsible for making the difficult decisions.

This combination of flexibility and accountability helps address the challenges of multinational contexts in two important ways. First, it enables a kind of confidence-building to occur. Initially, individual SBSTA members expressed considerable uncertainty about whether particular expert advisory processes would operate consistent with their notions of how global governing arrangements should function. They thus chose to retain tight control over the subject matter and design of assessment processes, reassuring themselves that these processes conform to their expectations. Over time, as expert advisory processes worked as expected, members have acquired greater confidence in their positive contribution. Thus, the IPCC has acquired greater credibility today among many representatives of developing country governments as a result of its willingness to work with the SBSTA to set its agenda and to respond to concerns expressed by SBSTA members on an ongoing basis.

Second, the combination of flexibility and accountability allows a process of learning to take place. Institutional arrangements that seem to work (such as the Roster of Experts and the informal working groups on methodologies and technology transfer) can be relied on to perform more extensive assessments, while others that don't seem to work (according to the criteria of any member) can be quickly discarded. In international contexts, where the norms and functions of regulatory policymaking rapidly change and models for public assessments that can achieve broad credibility among the world's publics are scarce, this kind of opportunity for learning and dynamic readjustment offers important advantages over alternative approaches that rely on fixed practices. The IPCC, for example, has undergone several difficult periods of institutional reform that might have been made more easily and more effectively had its initial design been more flexible and accountable to the wide range of audiences who care deeply about climate change.

National Adaptability

A third important aspect of many of the SBSTA's approaches to organizing public assessments is its adaptability to the characteristics of individual political cultures. Arguably, the most important form of public assessment in the climate regime is the calculation of national inventories of greenhouse gas emissions. These assessments form the basis of the core regulatory function of the climate treaties—the future limitation of greenhouse gas concentrations in the atmosphere. The assessments distribute responsibility and blame for degrading the atmosphere among countries, and they suggest appropriate policy strategies for individual countries in meeting their targets under the Framework Convention, the Kyoto Protocol, and future agreements. Not surprisingly, the assessments are deeply controversial and are likely to become more so now that the Kyoto Protocol has come into force and the regulatory framework on climate change has acquired a more binding character.

The basic guidelines underlying the calculation of national inventories of greenhouse gas emissions are set in the Framework Convention and the Kyoto Protocol. For example, the Framework Convention requires that inventories include sources *and* sinks as well as all greenhouse gases, not just carbon dioxide. However, the precise methods of calculation are being developed by a series of working groups of the IPCC and Organisation for Economic Co-operation and Development (OECD) for final approval by the SBSTA. Rather than require countries to use identical methods, however, the climate regime has focused instead on establishing a set of default methods that countries may use if they wish, but they may also supplement or replace these methods with nationally specific approaches. If they choose to use their own methods, countries must specify how they differ from the default IPCC/OECD standards.

Countries may wish to adopt their own methods because they have available or require considerably more accurate methods for certain gases and sources than for others. Methane emissions from rice paddies, for example, constitute a significant source of greenhouse gases in India and China. Few other countries need to assess their rice paddies as accurately, however, and it doesn't make sense to require other countries to make extensive measurements for an essentially irrelevant source. Countries also are likely to have very different methods for establishing credible inventories. Different agencies, interested parties, and political cultures are likely to come into play. It makes sense, therefore, to allow countries sufficient flexibility to follow their own processes and use their own institutions for producing policy-relevant knowledge.

To be sure, this approach generates occasional conflicts when national methods are questioned as to their propriety, as happened when several European countries sought to adjust their inventories for annual fluctuations in temperature. Working out such conflicts has become one of the primary responsibilities of the SBSTA. This process seems preferable, however, to a centralized institution with responsibility for conducting all greenhouse gas emission inventories. Few countries would likely trust its claims, regardless of the method used. By distributing responsibility for developing methods for and calculating emission inventories to individual countries, international assessors avoid the appearance of abridging national sovereignties (by taking away the important function of sovereign states to produce policy-relevant indices) and take advantage of well-established systems for rhetorically presenting and defending public knowledge that each national political culture has already worked out.

Conclusions

Over the course of the twentieth century, industrial democracies responded to the growing complexity and scale of human affairs by bringing experts directly into the formulation and implementation of public policy. Scientists, lawyers, engineers, economists, and others with specialized expertise increasingly occupied positions of power and influence throughout government bureaucracies. These individuals not only offered a broad array of skills and knowledge useful for understanding public affairs, but also came to play essential roles in legitimizing the state's power and authority to manage important aspects of economic and social life. Today, few areas of policymaking exist in which experts do not participate.

The incorporation of experts into national governance resulted in the evolution of unique scientific advisory arrangements in each country consonant with the characteristics of its political culture and institutions. Differences in the norms and practices of these arrangements, however, complicate the organization of science advice in contemporary global affairs. With the growth of worldwide concern over the environment, biotechnology, health risks, and other issues of global concern, science carries increasingly greater weight in world affairs. To date, however, global governance has not yet reached the point at which well-established norms and practices for conducting public assessments exist. As a result, and as with other aspects of global governance, the conduct and design of public assessments remains ad hoc and negotiated, often on a case-by-case basis.

While ad hoc arrangements may have worked fairly well in the past, they seem less likely to perform successfully in the future. Small, relatively insular diplomatic and expert communities were often able to reach workable settlements among themselves. However, as processes of globalization draw civil societies more deeply into the management and regulation of global economies, public expectations about the proper relationship between science and governance will carry greater weight in the evaluation of international institutions. Because these expectations vary across cultures, in accordance with the models developed over time for national science advice, assessors face a dilemma of how to speak credibly to heterogeneous global audiences.

Assessors facing this dilemma may be helped by an approach to the design and conduct of international public assessments that includes (1) diverse voices empowered to speak to how public assessments should be organized, carried out, and evaluated; (2) processes that remain flexible, responsive, and accountable to those who will use or will be affected by the use of the assessment's knowledge claims; and (3) adaptable elements that enable nations to produce requisite knowledge according to their own culturally-specific methods and approaches. This approach will inevitably slow the process and incorporate overtly political issues from the start. The experience of the SBSTA suggests, however, that the result can enhance the credibility and legitimacy of global regulatory arrangements among the many diverse public, expert, and policy communities that increasingly care about how we manage the earth. Global governance today remains tentative, fragile, and worthy of our most careful and considered attention.

Notes

1. A list of organizations pursuing frequent scientific assessments include the World Bank, the United Nations Environment Programme, the World Health Organization, the World Meteorological Organization, and numerous others.

2. This trend is particularly evident in environmental governance. International treaties in the 1980s and 1990s addressing such problems as ozone depletion, climate change, desertification, deforestation, and hazardous waste have all mandated the establishment of expert advisory committees. The World Trade Organization and North American Free Trade Agreement have also authorized scientific assessment panels to rule on conflicts between free trade principles and environment, health, and consumer safety provisions of national legislation. In the security arena, China and the United States formed an expert panel in April 2001 to assess the cause of a mid-air collision between a Chinese fighter and a U.S. spy plane in international waters off China's coast.

3. Building on the perceived successes of the IPCC, ecologists and conservation biologists launched the Millennium Ecosystem Assessment in 2000. For example, ongoing controversy between the United States and Europe over trade in genetically modified organisms (GMOs) has given rise to calls for an IPCC-like institution to settle public debates by resolving scientific disputes about the risks of GMOs. Partially in response to the GMO debates, Bruce Alberts, president of the U.S. National Academy of Sciences, announced in 2000 that the world's scientific academies had jointly created the InterAcademy Council to provide global policymakers with scientific assessments on topics of international concern.

4. Comparative studies of regulatory science and politics illustrate this point nicely. See, especially, Daemmrich 2002, 2004; Parthasarathy 2003, 2004; Daemmrich and Krucken 2000; Jasanoff 1986, 1991, 1995, 2000; Farrell and Keating 1998; Brickman et al. 1985.

5. Recent social science work refers to these factors in the aggregate using the concept of *civic epistemology*—the practices and institutions a particular community or country uses to generate, validate, and apply policy-relevant knowledge. See Miller 2005, 2004a.

6. Of particular note are the U.S. Federal Advisory Committee Act (1972) and Administrative Procedures Act (1946).

7. Adopting a literal interpretation of this idea, the Millennium Ecosystem Assessment is developing dramatic performances of its conclusions to try to reach audiences that are unlikely to read a thousand-page document—or even the summary for policymakers.

8. For an overview of several case studies, see Collins 1985; and Collins and Pinch 1982, 1983. For a more general treatment of credibility, see Shapin 1996.

9. For a particularly lucid elaboration of the general subject of credibility, including differences between scientific, public, and policy audiences, see Shapin 1994.

10. This argument has been made previously by several authors. See Jasanoff 1986, 1991; Shackley and Wynne 1994; and Wynne 1990.

11. For a comparison of the relationship between scientific credibility and political culture in the United States and United Kingdom, see Jasanoff 1986.

12. For a discussion of these reforms, see Miller 2004b; for a U.S. view of the need for even further reforms to establish yet tighter rules of operation, see Edwards and Schneider 2001.

13. For more detailed histories of the SBSTA, see Miller 2001a, 2001b. Much of this account is drawn from the former.

References

Agarwal, Anil, and Sunita Narain. 1991. *Global Warming in an Unequal World.* New Delhi, India: Center for Science and the Environment.

Brickman, Ronald, Sheila Jasanoff, and Thomas Ilgen. 1985. *Controlling Chemicals: The Politics of Regulation in Europe and the United States.* Ithaca, NY: Cornell University Press.

Collins, H.M. 1985. *Changing Order: Replication and Induction in Science.* London, United Kingdom: Sage Publications.

Collins, H.M., and T.J. Pinch. 1982. *Frames of Meaning: The Social Construction of Extraordinary*

Science. London, United Kingdom: Routledge and Kegan Paul.

————. 1983. *The Golem: What Everyone Should Know about Science.* Cambridge, United Kingdom: Cambridge University Press.

Daemmrich, Arthur. 2002. A Tale of Two Experts: Thalidomide and Political Engagement in the United States and West Germany. *Social History of Medicine* 15(1):137–158.

————. 2004. *Pharmacopolitics: Drug Regulation in the United States and Germany.* Philadelphia: Chemical Heritage Foundation.

Daemmrich, Arthur, and Georg Krucken. 2000. Risk versus Risk: Decision-Making Dilemmas of Drug Regulation in the United States and Germany. *Science as Culture* 9(4): 505–533.

Edwards, Paul, and Stephen Schneider. 2001. Self-Governance and Peer Review in Science-for-Policy. In *Changing the Atmosphere: Expert Knowledge and Environmental Governance,* edited by Clark A. Miller and Paul N. Edwards. Cambridge, MA: MIT Press.

Farrell, Alex, and Terry J. Keating. 1998. Multi-Jurisdictional Air Pollution Assessment: A Comparison of the Eastern United States and Western Europe. Belfer Center for Science and International Affairs (BCSIA) Discussion Paper E-98-12. Cambridge, MA: Environment and Natural Resources Program, Kennedy School of Government, Harvard University.

Grassl, Herman. 1997. Personal communication with the author.

Hilgartner, Stephen. 2000. *Science on Stage: Expert Advice as Public Drama.* Palo Alto, CA: Stanford University Press.

Jasanoff, Sheila. 1986. *Risk Management and Political Culture.* New York: Russell Sage Foundation.

————. 1990. *The Fifth Branch: Science Advisers as Policymakers.* Cambridge, MA: Harvard University Press.

————. 1991. Acceptable Evidence in a Pluralistic Society. In *Acceptable Evidence,* edited by Deborah Mayo and Rachelle Hollander. Oxford, United Kingdom: Oxford University Press.

————. 1995. Product, Process, or Programme: Three Cultures and the Regulation of Biotechnology. In *Resistance to New Technology,* edited by M. Bauer. Cambridge, United Kingdom: Cambridge University Press.

————. 1998a. The Eye of Everyman: Witnessing DNA in the Simpson Trial. *Social Studies of Science* 28(5/6): 713–740.

————. 1998b. Reasoning Together: The Politics of Harmonization. In *The Politics of Chemical Risk,* edited by R. Bal and W. Halfmann. Dordrecht, Netherlands: Kluwer Academic Publishers.

————. 2000. Technological Risk and Cultures of Rationality. In *Incorporating Science, Economics, and Sociology in Developing Sanitary and Phytosanitary Standards in International Trade: Proceedings of a Conference,* edited by National Research Council. Washington, DC: National Academy Press, pp. 65–84.

Meyer, Aubrey, and Tony Cooper. 1994. Ten to One Against: Costing People's Lives for Climate Change. *The Ecologist* 24(6): 204–6.

Miller, Clark A. 2001a. Challenges to the Application of Science to Global Affairs: Contingency, Trust, and Moral Order. In *Changing the Atmosphere: Expert Knowledge and Environmental Governance,* edited by Clark A. Miller and Paul N. Edwards. Cambridge, MA: MIT Press.

————. 2001b. Hybrid Management: Boundary Organizations, Science Policy, and Environmental Governance in the Climate Regime. *Science, Technology and Human Values* 26(4): 478–500.

————. 2004a. Resisting Empire: Globalism, Relocalization, and the Politics of Knowledge. In *Earthly Politics: Local and Global in Environmental Governance,* edited by S. Jasanoff and M.L. Martello. Cambridge, MA: MIT Press.

————. 2004b. Climate Science and the Reconstruction of Global Political Order. In *States of Knowledge: The Co-Production of Science and Social Order,* edited by S. Jasanoff. London, United Kingdom: Routledge.

————. 2005. New Civic Epistemologies of Quantification: Making Sense of Local and Global Indicators of Sustainability. *Science, Technology and Human Values* 30(2).

Miller, Clark, Sheila Jasanoff, Marybeth Long, William Clark, Nancy Dickson, Alastair Iles, and

Thomas Parris. 1997. Shaping Knowledge, Defining Uncertainty: The Dynamic Role of Assessments. In *A Critical Evaluation of Global Environmental Assessments: The Climate Experience,* edited by GEA Project. Calverton, MD: Center for the Application of Research on the Environment (CARE).

Pathasarathy, Shobita. 2003. Knowledge Is Power: Genetic Testing for Breast Cancer and Patient Activism in the United States and Britain. In *How Users Matter,* edited by T. Pinch and N. Oudshoorn. Cambridge, MA: MIT Press.

———. 2004. Regulating Risk: Defining Genetic Privacy in the U.S. and Britain. *Science, Technology and Human Values* 9(3): 332–352.

Porter, Theodore. 1995. *Trust in Numbers: The Pursuit of Objectivity in Science and Public Life.* Princeton, NJ: Princeton University Press.

Ripert, Jean. 1991. Report on the IPCC Special Task Force on Developing Countries. In *Climate Change: Science, Impacts and Policy. Proceedings of the Second World Climate Conference,* edited by J. Jaeger and H. L. Ferguson, 3–4. Cambridge, UK: Cambridge University Press.

Shackley, S., and B. Wynne. 1994. Climatic Reductionism: The British Character and the Greenhouse Effect. *Weather* 49(3): 110–111.

Shapin, Steven. 1994. *A Social History of Truth: Civility and Science in Seventeenth Century England.* Chicago, IL: University of Chicago Press.

———. 1996. Cordelia's Love: Credibility and the Social Studies of Science. *Perspectives on Science: Historical, Philosophical, Social* 3(3): 255–275.

Wynne, B. 1990. Risk Communication for Chemical Plant Hazards in the European Community Seveso Directive. In *Corporate Disclosure of Environmental Risks,* edited by M.S. Baram and D.G. Partan. Salem, NH: Butterworth Press.

CHAPTER 10

Designing Better Environmental Assessments for Developing Countries

Lessons From the U.S. Country Studies Program

Oladele A. Ogunseitan

CLIMATE CHANGE IS framed very differently by scientists and policymakers in industrialized countries of the Northern Hemisphere compared with their counterparts in developing countries of the Southern Hemisphere (see, e.g. Ogunseitan 2003). The discussion has focused largely on normative issues surrounding equitable assignment of causes, consequences, and costs according to particular economic, technological, or political boundaries (e.g., Baer et al. 2000). Relatively meager attention has been dedicated to understanding assessment processes involving scientists and policymakers in developing countries where disagreements in the negotiations over normative issues have either been settled satisfactorily or set aside for a variety of reasons (Anya 1993; Miller 1998). A retrospective analysis of such assessments can provide valuable insight into design options for framing, participation, and the science–policy interface that may influence the course and outcome of international environmental cooperation. The outcome of such analysis can reveal design parameters that distinguish among alternative outcomes of assessments in terms of their effectiveness, ultimately determined by whether specific program objectives were achieved by participating countries.

This chapter presents a case study of the U.S. Country Studies Program (USCSP) and the participation of nine developing and transition countries selected from each of the eligible continental regions. The countries are Bolivia, China, Czech Republic, Estonia, Hungary, Mauritius, Nigeria, Oman, and Ukraine. The analysis presented here is not intended to provoke judgments on the effectiveness of assessments sponsored by the USCSP in specific countries. Rather, the goal is to explore the USCSP experience as a primer for understanding design elements that may influence the process of assessments. The specific characteristics of potentially influential design elements—including stakeholder participation and the dynamics of salience, credibility, and legitimacy of international cooperative agreements—have been explored by other investigators (see, e.g. Carr and Mpande 1996; Clark 1987; Jasanoff 1996; Jaeger et al. 2001; Kates et al. 1985; Long and Iles 1997; Stern et al. 1992; Turner et al. 1991).

The Context of Effective Environmental Assessments

The effectiveness of environmental assessments in accomplishing the goals set by participants is, expectedly, influenced by several factors. Farrell and colleagues have described many of these factors in detail (Farrell et al. 2002). Factors that are particularly germane to this case study include, first, the framing of assessments and the impact of framing on issue salience and participation, and second, the trade-offs between perceptions of assessment credibility and legitimacy. This second factor is particularly important when assessments must be conducted cooperatively between nations with different technical and diplomatic capacities. The level of credibility attributed to an assessment by end-users may be influenced by perceptions of motives for the assessment, or by independent evaluations regarding the level of scientific rigor embedded in the assessment process. Therefore, it is possible for two different stakeholders to attribute low credibility ratings to an assessment process for entirely different reasons.

The generally accepted idea that developing and transition countries are characterized by underdeveloped scientific expertise suggests that questions about assessment credibility arising from these countries are likely to be based on motives as opposed to scientific rigor. However, the need to include developing country scientists in the debate on global climate change has rested on the argument involving the sensitivity of their respective regions to climate impacts and the necessity for integrating multiple regional or national assessments into a meaningful global outlook. Therefore, problems associated with scientific uncertainties inherent in projections across spatial and time scales have been unavoidable in attempts to frame a global perspective on climate change. The believability of such projections and the scenarios or assumptions used in their construction has been debated extensively in both industrialized and developing countries (Apuuli et al. 2000; Jasanoff 1996; Pittock and Jones 2000). The perception of credibility of assessments by developing country participants may rely on the flexibility of traditional scientific methods and models to incorporate local perspectives, analogies, and historical information.

The salience of an international environmental assessment or the perceived relevance of likely outcomes to national concerns is also expected to exert strong influences early in the decisions of nations exercising the option to participate in the assessment process. However, issue salience is sensitive to many factors external to the assessment process. Therefore, stakeholders may prematurely withdraw from the assessment process or ignore the outcome of the completed assessment if the issue domain is no longer judged to be relevant to pressing national development goals. Issue salience varies among stakeholders and participants, and the fact that an assessment is termed "global" does not necessarily guarantee its relevance to every nation state. It has been argued that labeling an issue domain as global is one way to reframe environmental problems that are rooted in local causes in order to share the burden of finding solutions. Such reframing can broaden participation, but it may come at some cost to the level of credibility assigned to the assessment because of questionable motives (Carr and Mpande 1996).

The perception of legitimacy attributed to international environmental assessments by end-users at the national or international level has also been assumed to exert major influence on the effectiveness of assessments, although there are few reports on what incidences or pronouncements are used to assign the status of legitimacy to specific assessments, particularly within the discourse on global environmental change in developing countries. Resistance to the idea of hegemony in the framing of environmental assessments by powerful industrialized countries is a possible reason why scientists and policymakers in developing countries emphasize the adoption of clearly stated criteria for assessment legitimacy, although this dimension has not been investigated seriously. For many developing and transition countries trying desperately to be visible by securing a place at the international economic and diplomatic tables, assigning national scientists and policymakers to the numerous, long-term, simultaneous global assessments and negotiations may place considerable strain on national personnel and financial resources.

Within the climate change regime, participation in the scientific assessment process is separated from participation in the negotiations in order to render international agreements legitimate. Thus, developing country scientists who participate in the scientific assessment process, either through authorship of chapters in the reports of the Intergovernmental Panel on Climate Change (IPCC) or through their status as lead investigators under the Country Studies Program, are not guaranteed national delegates to the Conference of Parties to the United Nations Framework Convention on Climate Change (UNFCCC). It is possible that initial participation in the international scientific assessment process increased the salience of global climate change at the national level, thereby encouraging participation in the legitimization process. However, legitimizing assessments at the national level may also be influenced by the action of other nations or events external to the assessment domain.

The practical design of global environmental assessments can also exert influence on the legitimization process within participating nation states. For example, climate change assessments fall into three major categories: scientific evidence for global warming, vulnerability and adaptation, and mitigation. These three parts form the basis for dividing the IPCC into three working groups and for distributing financial support to develop technical capacities needed for environmental assessments in developing and transition countries through Country Study Programs.

Country Study Programs as Transboundary Institutions for Assessments

Several multilateral and bilateral initiatives were launched in 1992 to support national climate change assessments in developing countries. These initiatives focused to different extents on the three parts of the IPCC's report. For example, the German Association for Technical Cooperation (GTZ) worked on behalf of the German Federal Ministry for Economic Cooperation and Development (BMZ) to support developing countries in preparing their communication to the Conference of Parties to the UNFCCC. In general, GTZ supplemented assessments already supported by the Global Environment Facility (GEF), and only eight countries participated—namely China, Colombia, Indonesia, Pakistan, Philippines, Tanzania, Thailand, and Zambia. In 2000, the GEF budget for supporting environmental assessments in developing and transition countries totaled US$2.15 billion, out of which US$600 million went to support climate change assessments. The GEF emphasized capacity-building programs for "climate change enabling activities" in developing countries, but the predominant framing reflected in the program topics was inventories and decarbonization strategies. In general, vulnerability assessments were not supported (GEF 2000).

The GTZ provided technical support only for greenhouse gas inventories of sources and sinks, and for technological and political options for mitigating these emissions, emphasizing the reduction of carbon dioxide (GTZ 2000). Vulnerability and adaptation assessments were not supported, and this partial framing of the assessments may have discouraged participation in the program by certain countries. For example, the GTZ supported environmental and development initiatives in 36 African countries with estimated annual costs of US$286 million. However, only two African countries opted to participate in GTZ-sponsored climate change assessments.

This approach differs substantially from the U.S. Country Studies Program assessments in both geographic scope and the extent to which assessment design issues were formalized. The USCSP supported assessments in developing countries under the rubric of the three-way IPCC division described above. Certain qualitative and quantitative distinctions made USCSP assessments particularly potent (among the possible alternatives) for understanding the linkages between the process and outcome of assessments in developing countries and for drawing lessons toward generalizable concepts of design in international environmental assessments. To better understand the institutional structure of the USCSP, a site visit was conducted in April 2000. Formal interviews were conducted with the program director (Jack Fitzgerald), the regional coordinator for assessments being conducted in Africa and the Middle East (Sandra Guill), and the regional coordinator for assessments in Asia and small island states (Joseph Huang). Information derived from these interviews was supplemented with several documents, including formal reports on completed assessments, progress reports on ongoing assessments, and program guidelines and protocols. The following sections in this chapter begin with background on the history and mandate of the USCSP, followed by specific learning experiences according to realms of influence, including issue framing, cross-scale dimensions of the interface between science and policy, and capacity development for environmental assessments, which has become a perpetual feature of collaborations between industrialized nations and developing countries.

Case Study:
U.S. Country Studies Program and Associated Assessments

The UNFCCC was designed to deal with projections of the likelihood of occurrence of human-induced climate change and the possible impacts on different countries (UNFCCC 2005). The third session of the Conference of Parties (COP) to the UNFCCC adopted the text of the Protocol on 11 December 1997 in Kyoto, Japan. The Protocol entered into force on 16 February 2005— the 90th day after at least 55 parties to the Convention, including Annex I Parties accounting for 55 percent of carbon dioxide emissions from that group in 1990, deposited their instruments of ratification, acceptance, approval, or accession (UNFCCC 2005). Article 4 of the UNFCCC requires parties to the convention to conduct inventories of national sources and sinks of greenhouse gases and to develop response strategies for global climate change (United Nations 1992). A logical conclusion of the equity debate within the climate change issue was that industrialized countries would assist developing countries and countries with economies in transition in meeting their assessment obligations to the UNFCCC.

The USCSP was created by executive action following negotiations at the 1992 United Nations Conference on Environment and Development (UNCED) in Rio de Janeiro (USCSP 1999). In that year, the United States committed US$35 million to the USCSP in support of national assessments of climate change (USCSP 1999). The USCSP was mandated to offer technical and financial assistance to participating countries through the operations of a multi-agency Country Studies Management Team. The participating agencies and the hierarchical structure of the program are presented in Figure 10-1. The program exists as a multi-agency initiative with the mandate to assist developing and transition countries in meeting their obligations to the UNFCCC.

The USCSP was initiated with three major goals. The first was to "enhance the capabilities of countries and/or regions to inventory their net emissions of greenhouse gases, assess their vulnerabilities to climate change, and evaluate the options available to them to mitigate and adapt to climate change." The second goal was to "support countries' efforts to establish a process for developing and implementing national policies and measures to deal with climate change over time." The third goal of the USCSP was to "develop data and information that can be used at the national, regional, and global levels; to assess current and future trends in net anthropogenic emissions of greenhouse gases; and to further national and international discussions of climate change issues" (USCSP 1999).

Between 1992 and 2000, the USCSP designed and assisted the implementation of environmental assessments under the broad categories of *greenhouse gas inventories, climate change mitigation,* and *vulnerability and adaptation* in at least 56 countries worldwide (Benioff and Warren 1996; Benioff et al. 1996). In some cases, the assessments formed the basis of "national action plans" for use by local policymakers, or "communications" to the international secretariat of the UNFCCC (Feenstra et al. 1998). Thus, USCSP assessments focused on the *product* of the assessment. However, in the case of using the assessment to inform national action plans, design strategies were implemented to optimize the product's credibility by securing diplomatic consent before initiating the assessments and by maximizing the involvement of domestic expert scientists. However, the need to use the assessment product at the international level for producing synthesis reports from various national communications placed some constraints on the freedom allowed in the national assessments. These constraints were designed to enhance the legitimacy of positions adopted by developing countries partici-

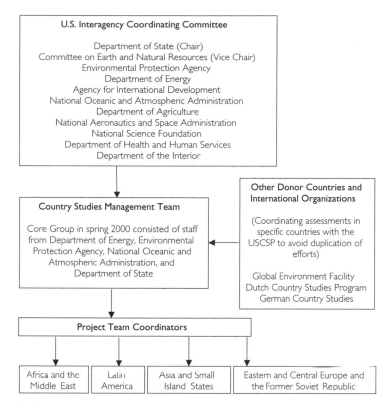

Figure 10-1. *Organization of the U.S. Country Studies Program*
Sources: Adapted from Benioff and Warren (1996); Benioff et al. (1996); and USCSP (1999).

pating in the international agenda to define policy for climate change at the Conference of Parties to the UNFCCC.

Critical evaluation of the *process* of the USCSP assessments occurred within the context of attempts to integrate conceptual and tangible aspects of different phases of the assessment in participating countries (Pittock and Jones 2000; Smith et al. 1996; USCSP 1999; Usher 2000). Evaluation of the assessment process involved a retrospective diagnosis of what went wrong with a particular assessment in a particular country, sector, or scale of analysis. Problems were solved by supplementing methods and models, or by organizing training workshops for lead investigators. The stipulation of procedures was necessary for maintaining cross-national consistency in soliciting assessment proposals, securing diplomatic consent in participating countries, and maintaining communications between participating scientists/diplomats in developing countries and the USCSP management team, including a cohort of U.S.-based consulting scientists and experts (USCSP 1999). Differences were recognized between users of the assessment product at the national level, whose responsibilities include the formulation of local response policies, and users at the international level, who produced synthesis reports for global coordination of mitigation efforts through various political and economic initiatives.

To collect the perspectives of assessment practitioners in developing countries, a site visit was conducted to interview lead investigators for USCSP assessments in Nigeria. A standardized 16-item questionnaire was also used to gather information from each of the lead investigators in other developing and transition countries located in every major geopolitical region containing developing and/or transition countries, namely South America (Bolivia), Asia (China), Eastern Europe (Czech Republic, Estonia, Hungary), Africa (Mauritius), Middle East (Oman), and Former Soviet Republics (Ukraine). The questionnaire was designed to elicit responses to questions on factors that were hypothesized to influence participation, design, and effectiveness criteria of USCSP assessments. Specifically, the questions focused on the rationale for participation; the predominant national framing of climate change; the salience of climate change among national scientists, policymakers, and the public; the rating of credibility attributed to the assessments; and the legitimacy of the assessments in terms of government support and the way in which the assessment product was used. In addition, the questions also focused on capacity development accomplishments and needs identified through participation in the USCSP assessments.

The questionnaire responses received from each of the nine countries indicate that most of the activities of developing and transition country scientists who participated in the USCSP assessments were confined to the first stated goal of the USCSP—specifically the inventory of net emissions of greenhouse gases, assessment of vulnerabilities to climate change, and evaluation of options available to mitigate and adapt to climate change. However, it is difficult to separate participation in achieving the first USCSP goal from contributions to the rather qualitative second and third goals. Within the first explicit goal of facilitating climate change assessments, funding priorities of the USCSP program were, in decreasing order of emphasis, (1) to support national inventory of sources and sinks of greenhouse gases, (2) to evaluate options to mitigate net emissions of greenhouse gases, and (3) to conduct an assessment of vulnerabilities to the impacts of climate change and an evaluation of options to adapt to these potential impacts (USCSP 1999). The results are presented according to three major elements of assessment design, namely participation and issue framing, negotiation of the science–policy interface, and development of scientific capacity.

Participation and Issue Framing

The motives for participating in international environmental assessments are as varied as the participants. To identify potential biases in the criteria for participation in the USCSP, it is important to determine if there are apparent clusters of participants on the basis of geographical distribution, level of economic development, or convergence of policy frameworks on environmental issues. For example, the geographical distribution of countries in which assessments of climate change were supported by the USCSP suggests the existence of political and institutional constraints on participation. The 56 countries that participated in USCSP assessments span the geographical range of continents harboring developing and transition countries, but there are some notable exceptions. The absence of India from the list of participating countries is interesting because

India commands considerable clout within the G-77 coalition of developing economies, and India's challenge to the predominant framing of climate change at the global level is well documented (Agarwal 1990; Agarwal and Narain 1991). The USCSP management team tried to recruit India's participation in the USCSP, and despite the eagerness of several scientists of Indian origin working in India and within the United States to submit assessment proposals to the USCSP for funding, political opposition within India was insurmountable. The political opposition was couched in views on equity and climate change mitigation that have been expressed repeatedly by India's Ministers of Environment and Forestry, beginning with Kamal Nath at the 1992 United Nations Conference on Environment and Development (UNCED) (Huang 2000).

Kandlikar and Sagar (1997) explored the political and scientific intricacies of climate change research in India during the period covered by USCSP assessments. These investigators claim that India's response to the UNCED and the early deliberations on the economic and equity implications of the UNFCCC can only be fully appreciated within the context of radical economic "liberalization" policies instituted as a consequence of structural adjustment conditions imposed by agreements with the International Monetary Fund in 1991. The resulting austerity measures apparently generated substantial political rearrangements, which in turn created new doubts regarding the consequences of accepting foreign financial assistance for programs without critical analysis of the implications for national policy. Nevertheless, Indian scientists coordinated a climate change research agenda nucleated by long-term programs dedicated to the investigation of monsoons and their links to synoptic scale atmospheric circulation patterns. Scientists from India have participated in programs supported by the World Climate Research Program, including the Tropical Ocean Global Atmosphere Program, and the International Research Program on Climate Variability and Predictability. Participation in these international research programs was credited for the establishment of the Indian Climate Research Program with a US$2.5 million five-year budget (1997–2001). The GEF has also provided additional funding to the Indian research program for climate assessments that are directly relevant to international policy, including greenhouse gas inventories. It is noteworthy that similar studies were supported by the USCSP in other Asian countries such as China, Indonesia, and Sri Lanka.

The influence of economic interests on participation in the USCSP is best explored by looking at countries within the Organization of the Petroleum Exporting Countries (OPEC). Through views expressed as an observer organization at the Conference of Parties to the UNFCCC, OPEC has not been particularly sympathetic to the framing of energy decarbonization as the only viable strategy for mitigating climate change (Lukman 2000). In the Middle East region, only Oman participated in USCSP assessments. Nigeria and Venezuela are the only other OPEC member countries that participated in the USCSP process, despite OPEC's explicit opposition to mitigation strategies adopted by the international framework (Lukman 2000). Nigeria's participation in the USCSP assessments of greenhouse gas inventories and mitigation was driven in part by scientists working at the National Center for Energy Research and Development, which advocated the need to engage indigenous scientists in international environmental assessments as a strategy for the government to communicate its willingness to control the troubled petrochemical industry and its role in extensive pollution in the Niger Delta region (Adegbulugbe 2000). Notably, deliberate natural gas combustion in the Niger Delta oil fields contributes as much as 30 percent of global greenhouse gas emissions within the category of flaring activities worldwide (Adegbulugbe et al. 1997). These national contingencies contribute to the general salience of climate change issues in Nigeria, consequently ameliorating Nigeria's policy of typically subscribing to OPEC's positions.

The divergence of frameworks on environmental policy adopted by different countries also has influenced participation in the USCSP. By 1996 when several countries had completed the first phase of climate change inventory assessments, the U.S. government honored requests from 40 developing and transition countries to extend the USCSP by providing financial and technical assistance for national climate change action plans (Benioff et al. 1996). The process of developing an action plan was defined as "a national, regional, or sectoral exercise designed to integrate climate change mitigation and adaptation measures into overall development objectives of a country that produces a framework for intervention and a time-table for action" (Benioff et al. 1996).

Based on the previous four years of experience with designing assessments for developing and transition countries, the USCSP management team composed a set of key challenges and lessons that faced participating countries (Benioff et al. 1996). A major challenge was integrating climate change concerns with other policy priorities, given that climate change is not a priority for most developing and transition countries. According to Benioff and colleagues (1996), the effective mitigation of negative impacts of climate change may require the deployment of actions that were initiated to address other development issues. This constituted a reframing of climate change assessments to address "no-cost adaptation" scenarios that link concepts of national economic development to the international priority of mitigating global climate change. A second challenge was broadening local participation by involving a diverse array of relevant ministries and departments, provincial and local governments, regional organizations, nongovernmental organizations, the private sector, and the public in the planning process.

Although more than 40 countries agitated for technical and financial support for assessments leading to the formulation of national action plans, only 18 countries (Egypt, Mauritius, Tanzania, Bolivia, Mexico, Uruguay, Venezuela, Bulgaria, Czech Republic, Hungary, Kazakhstan, Russian Federation, Ukraine, China, Indonesia, Micronesia, Philippines, and Thailand) actually engaged in the process (USCSP 1999). There are several possible reasons for the relatively low participation rate in this phase of USCSP assessments. One reason is that countries decided to participate based on how well various framings of the rationale for conducting the assessment fit with other priorities. Examples include, but are not limited to, establishment of implementation plans for priority response measures; integration of climate change concerns into other development plans (no cost adaptation); and preparation of communications required under the UNFCCC.

Negotiation of the Science–Policy Interface

The UNFCCC required earlier deadlines and more comprehensive requirements for the official communication of assessment results to the Conference of Parties from industrialized and transition countries (Annex 1 countries) than from developing countries. Therefore, many developing countries prioritized communications over assessments for climate change action plans. Even in developing countries in which salience is relatively high (such as small island states and countries of Central America), participation required extensive negotiations because of structural and functional impediments in the science–policy interface across national boundaries. The structure and function of institutions that bridge the science–policy interface varies among developing and transition countries, and it is difficult to generalize the impact of characteristic interfaces on the design of international environment assessments. Nevertheless, the selected cases described in the following paragraphs illustrate ways in which assessment participants in different countries navigated the local configuration of the science–policy interface, and they also detail the potential impact of such maneuvers on the process of USCSP assessments.

The simultaneous salience of global climate change in both the scientific and political communities of developing countries was necessary to initiate participation in USCSP assessments, but this quality alone was not sufficient to sustain participation in the process (Benioff and Warren 1996; Benioff et al. 1996). Similarly, simultaneous commitment of both the scientific and political systems within the United States was necessary to maintain financial and technical support for the activities of the USCSP. Although the USCSP was established as a multi-agency initiative, sources of permanent staff and annual funding were subject to constant negotiations. The resulting level of institutional uncertainty may have influenced the range of activities to which the USCSP management team was willing to commit (Corell 2000). Therefore, integration of three levels of salience and legitimacy issues were required to initiate and sustain USCSP assessments for realistic effectiveness in meeting the outlined goals. The first level involved integrating the salience and legitimacy of the funding program within the United States, the second level was between the United States and each of the participating countries, and finally the third involved integration within the participating countries. For example, diplomatic letters of support were necessary for scientist-initiated proposals to the USCSP, and in certain cases, low salience of global climate change was detrimental to participation, as illustrated by the following quotation from an interview with Sandra Guill:

> I received a lot of proposals, which were not followed up by diplomatic letters from the ministry of foreign affairs, for unknown political reasons. Those proposals were not funded, and technical assistance in climate change assessment was not possible. . . . We could have had [an] additional 15–20 more Country Study Program participants in Africa, if it were not for the lack of ministerial support. This explains the lack of full coverage in West and Central Africa. I begged, contacted, sent faxes, called, and did everything, but the diplomatic notes were not forthcoming. I told the

principal investigators to go back to their countries and push for the diplomatic note. Examples of countries that applied but whose study could not commence are Mali and Niger. We tried to work with Algeria, but their political situation was not conducive.

—Sandra Guill, USCSP Regional Coordinator
for Africa and the Middle East (interviewed April 5, 2000)

In cases where scientific and political interest in the issue domain was assured, a major impediment to initial participation was eliminated, but differences in the framing of the issue by the USCSP management team and the national community of scientists and policymakers constituted another level of impediment (see, e.g., Dixon et al. 1996; Magazda 2000). Impediments at this level, however, appear to have been easily resolved through negotiations between the parties. For example, when asked to explain the nature and dynamics of the negotiations that preceded country proposals accepted or rejected by the USCSP, Jack Fitzgerald responded as follows:

We [USCSP] imposed some general constraints. We emphasized the nexus between science and policy-relevant questions within the context of the [UNFCCC] convention itself. Inventories must be done first. [We] encouraged proposals to do mitigation assessments. Then came vulnerability and adaptation. If a country needed money for tidal gauges, well, that is not what we do. You [countries] did not have to do mitigation or vulnerability analysis. The only thing you had to do, at least in the second round, was inventory. Certain proposals were cut back because they included aspects that did not quite fit into our agenda. We funded only one regional study (in Central America) where the countries came together at the ministerial level and they wanted to study only vulnerability. They did not want to investigate mitigation, because in their opinion, there will be no political support for mitigation unless they believe that vulnerability is an issue.

—Jack Fitzgerald, USCSP Acting Director (interviewed April 6, 2000)

The coalition of Central American countries (referred to in the above interview) created the Central America Council on Climate Change to negotiate and coordinate scientific and policy assessments for responding to the international agenda on climate change. This coalition, comprising Belize, Costa Rica, El Salvador, Guatemala, Honduras, Nicaragua, and Panama, enabled the prioritization of assessment goals through negotiation with the USCSP, as described above. No other regional coalition was able to negotiate the procedure of assessments with the USCSP management team. Consequently, assessments that should cut across national boundaries were otherwise not included in the process. Examples of regional proposals not funded include regional vulnerability assessments for the Zambezi River basin and the Nile River basin (Guill 2000).

On the other hand, the USCSP process did include multinational coalitions—not organized primarily by geographical proximity, but by interest groups that called for modifications in the scope of USCSP assessments. In one such example, Sri Lanka, Egypt, and South Africa were influential in advocating the inclusion of the health sector in vulnerability assessments in the USCSP agenda for technical and financial assistance. Such coalitions became possible in part through the desire to increase the national salience of global climate change in the advocating countries. The USCSP concurred despite the sense by the management team at the time (circa 1995) that the credibility of training programs, including technical models necessary for climate change assessments in the health sector, could suffer from premature scientific development and a high level of uncertainty. Such uncertainties would make policy recommendations difficult (Fitzgerald 2000; Huang 2000; Guill 2000). The first USCSP-organized international workshop on assessing human health impacts of climate change occurred in 1995 in Harare, Zimbabwe. Notably, the workshop included as presenters scientific experts who had formulated the World Health Organization (WHO) agenda on climate change impacts as early as 1990 (Dixon et al. 1996; McMichael and Kovats 2000; McMichael and Martens 1995; WHO 1990).

Development of Scientific Capacity

Despite the success of coalitions in expanding the range of USCSP assessments, the specific sectors included in vulnerability assessments conducted by each of the participating countries depended in large part on the existing capacity for assessments in particular sectors. Simulation models, methods, and training workshops were provided for assessments in eight sectors, namely agriculture, coastal resources, grassland and livestock, fisheries, forestry, human health, water resources, and wildlife. The representation of these sectors in national action plans was uneven, and the most significant predictors of sector coverage were the presence of relevant experts on the assessment team and the technical capacity to conduct assessments (Ogunseitan 2003).

Technical assistance was the core form of the assistance that the USCSP provided for climate change assessment in developing and transition countries (USCSP 1999). The type, scope, and intent of the assistance offered depended on negotiations with individual countries with respect to the development of national action plans. For example, the objective of technical assistance under the vulnerability assessment process was to "build the technical and institutional capacity within countries to assess their vulnerability to climate change and to evaluate appropriate adaptive responses" (Benioff et al. 1996). The USCSP technical assistance protocol also included guidelines for countries to achieve the most effective approaches (within resource constraints) for conducting assessments, training, providing analytical tools, and promoting information exchange.

Participating countries negotiated the scope of the technical assistance with the USCSP management team regarding their capacities to (1) prepare climate change and socioeconomic scenarios, including access to general circulation model data and guidance on their use and interpretation; (2) evaluate potential biophysical impacts of the most sensitive and important resources in each country; (3) assess the effectiveness of adaptive adjustments to policy and management practices toward reducing the impacts; (4) evaluate economic impacts of climate change; (5) integrate sector-specific results; and (6) identify and assess adaptation policies designed to implement the adaptive adjustments to reduce vulnerability to climate change (Benioff et al. 1996).

International approaches to capacity development in developing and transition countries differ across institutions that engage in the process. The different approaches have become the focal point upon which success or failure of environment and development assistance programs in developing countries is judged. For example, in recognizing the limitations of current capacity-building programs, a strategic partnership was formed between the United Nations Development Program (UNDP) and the GEF to produce a comprehensive approach for developing the capacities needed at the national level to meet the challenges of global climate change (UNDP-GEF 1999). The new partnership distinguishes "capacity development" from "capacity building" in that the former shifts emphasis from a process that is externally driven to one that is self-motivating and takes place gradually from within. In this context, capacity development is defined at three levels: individual, institutional or entity, and systems (the overall system in which individuals and entities interact and operate; UNDP-GEF 1999). Capacity retention at the systems level is perceived as more critical than development at the individual and entity levels. Consequently, many international development organizations have adopted programmatic approaches that focus on the interrelationships between the individual, entity, and system levels. This approach encourages the sense of capacity ownership by developing countries, but short-term grants and workshops are insufficient means of sustaining cross-national capacity development initiatives, particularly where the expectation is the contribution to fast-paced international environmental assessments.

According to Sandra Guill, the approach to capacity building is a distinguishing characteristic of the USCSP in the climate change assessment process:

> We [USCSP] worked closely with other donors so that we do not duplicate efforts. For example in Tanzania, Zambia, and South Africa, the GTZ funded half, and we funded the other half. We combined our work plans. The Dutch have also funded a few countries. However, the U.S. program is much larger and much better defined than the other country study programs elsewhere—from the perspective of participating countries. For example, what they liked about our program was that (1) we did true capacity building—we did none of the work for the countries. We gave them models, and computers, we taught them how to analyze their data, but they were the ones who actually went out and did the field work. (2) UNEP and GEF developed a training manual, but they simply gave the countries half a million dollars for the study and left it up to the investigators to select the model to be used. This probably led to a lot of confusion on the part of some investigators in developing countries. They had no idea what to do. The money disappeared, and no results emerged. It also made it difficult to have a cross-sectional comparison of inventories, mitigation, and vulnerability/adaptation events among the countries.
>
> —Sandra Guill, USCSP Regional Coordinator
> for Africa and the Middle East (interviewed April 5, 2000)

According to responses from USCSP participants in Bolivia, China, Czech Republic, Estonia, Hungary, Nigeria, Mauritius, Oman, and Ukraine, sustaining the capacity-building aspects of the USCSP assessments proved difficult because of the relatively short period in which assessment teams were engaged in the process. The integration of various capacity development strategies from multiple international agencies into environmental assessments having similar goals also proved difficult. There is limited communication among the three levels of actors involved, namely the research scientists who conducted the assessments, the institutions that supported the assessments (including universities, research centers, and regional offices of international organizations such as the WHO and the International Institute for Tropical Agriculture), and the government officials who had to endorse the participation. In a broader view of capacity development, the focus on the ability of individuals and entities is transformed into an enabling environment at the local, national, and regional levels consisting of regulatory frameworks, information, and knowledge technologies that ultimately feed into the improvement of global environmental management (UNDP-GEF 1999). The "self-interest" appeal of open-ended environmental assessments could provide the right incentive for the acceptance of such capacity development initiatives.

Perception of credibility by indigenous scientists and policymakers has been found to be sensitive to the flexibility of capacity development initiatives with respect to the integration of local methodological approaches such as alternative combinations of case studies, and analogical narratives with numerical models (Feenstra et al. 1998; Ominde and Juma 1991; Watson et al. 1997). However, such methodological flexibility has not been compatible with the desire of the USCSP management team to produce integrated assessments with coherent synthesis reports, as noted in this comment:

> What we did was to say: there are several models out there, but we want you to use these specific ones (e.g., COPATH and COMAP for forestry and carbon sinks; and TARGETS for health impacts). By having every one of our countries use the same model (and we are not saying that they are the best or most salient), it allowed us to synthesize the results of the studies into a coherent global picture of climate change assessment and impacts.
> —Sandra Guill, USCSP Regional Coordinator
> for Africa and the Middle East (interviewed April 5, 2000)

A summary of results from a segment of the questionnaire administered to lead investigators involved in USCSP assessments in nine developing and transition countries are presented in Table 10-1. Respondents ranked assessments conducted under the auspices of the USCSP according to salience of the issue domain; credibility of methods, models, and rationale for the assessments; and the legitimate use of the assessment product. Respondents also identified areas of capacity development in which participation in the USCSP contributed to or highlighted national needs. Except for Czech Republic, all participating countries identified technical capacity-building needs that appear not to have been satisfied or sustained through participation in the USCSP. A common rationale for capacity development is to increase the salience of climate change research, particularly in countries such as China, where policymakers have identified the

Table 10-1. *Evaluation of USCSP Assessments by Practitioners in Developing and Transition Countries*

Country	Ranking of USCSP assessments within each country			
	Salience (Sectors identified as national priorities)	Credibility	Legitimacy (Government use or perception of USCSP assessments)	Capacity development (Credit to participation in USCSP assessments)
Bolivia	5 Food security, human health	4	5 UNFCCC communication	5 Establishment of Climate Change Research Institute
China	1 Vulnerable coastal zones	3	1 Scientific uncertainties have limited actions by policymakers	3 Training and professional development, establishment of facilities to support international collaboration
Czech Republic	1 Water management	2	4 Assessment served as background for Czech national climate change strategy	0 Country has sufficient research capacity
Estonia	5 Vulnerable coastal zones	5	5 Participation increased issue salience for Estonian society and government	3 Maintenance of international network
Hungary	1 Flood control	2	4 Assessment was used as basis for comprehensive GHG reduction policy	4 Establishment of Climate Change Focal Point Institute
Mauritius	5 Coastal management	Not ranked	5 Assessment was used as basis for National Action Plan	5 Establishment of institutional structure for funding expert advice, project implementation
Nigeria	3 Economic impact of decarbonization, management of coastal resources	2	3 Assessment served as the basis for government establishment of climate variability impacts center	3 Assessment enhanced capacity to utilize numerical models
Oman	5 Water resources	3	5 Assessment helped consolidate commitment of sultanate to UNFCCC	3 Establishment of environmental monitoring and control programs
Ukraine	2 Coastal resources	3	4 Assessment served as basis for UNFCCC communication and National Action Plan	5 Establishment of rational legislation and standard setting

Notes: Responses on salience, credibility, legitimacy, and capacity needs were ranked on a scale of 1 to 5, representing weak to strong, respectively. Surveys were conducted between December and May 2000.

large degree of scientific uncertainty as the major barrier preventing aggressive formulation of policy and effective participation in international agreements on climate change mitigation. Among the group of participating countries surveyed for this study, only Estonia expressed the maximum degree of confidence in the credibility of international environmental organizations to address issues that are relevant to developing and transition countries. This observation supports the view that sustainable capacity development may serve the important purpose of neutralizing perceptions of intellectual hegemony by creating a sense of ownership of assessments (UNDP–GEF 1999).

Prospects for Improvement

The USCSP management team identified various challenges that hindered environment assessments in developing countries, but they also recognized eight positive lessons from previous assessments that could serve as a basis for improving the design, and consequently the effectiveness, of newly proposed assessments by participating nations. Some of these lessons are paraphrased with italicized comments (Benioff et al. 1996):

1. Planning for the assessment should emphasize integration with other development plans, programs, and measures that have multiple benefits. *In other words, build redundancy into the framing of the benefits to be derived from the assessment. This strategy builds legitimacy into the assessment process.*

2. A diverse group of government agencies and departments should be actively involved in development of the assessment plan.

3. Nongovernmental stakeholders should be involved to secure support for the assessment.

4. The process should identify and maintain a focus on well-defined objectives.

5. The assessment should have a practical (utilitarian) orientation that emphasizes implementation. *This lesson concerns targeting the anticipated users of assessments to help scientific information inform environmental policy.*

6. Assessments should be viewed as living documents that are regularly updated to reflect changing circumstances. *This lesson takes the view of assessments as products but recognizes the advantage of open-ended assessments where the process reflects the flexibility of influential factors such as issue framing, stakeholder participation, and salience of the assessment.*

7. Planning should be under government control, and not driven by the priorities of donors. *This lesson addresses the question of hegemonic framing and control of assessment rationale and process.*

8. A high level of awareness among policymakers and stakeholders regarding climate change issues may be required before an assessment process can gain momentum. *This lesson addresses the concern with salience of the issue domain.*

Because the international assessment of climate change, particularly the aspects pertaining to societal impacts, is an intrinsically open-ended process, it remains to be seen whether the implementation of these lessons has contributed substantially to the effectiveness of assessments in developing and transition countries. The decision of governments in these countries to dedicate scientific, diplomatic, and financial resources to global environmental assessments is complicated by the lack of data showing tangible benefits for national development. Factors that contribute to the decisionmaking process include perceptions of societal vulnerability, the level of conviction that scientific assessments can provide viable options for policy, and the existence of institutional arrangements that can facilitate the integration of national concerns into the agenda of the international network of scientists working on global change. However, the verdict is mixed regarding the degree of success that internationally coordinated assessments have had toward solving environmental problems that exist at the national or regional scale in developing countries (Carr and Mpande 1996; Lopez 1999; MacLean 1990; Shue 1999). The mixed verdict has created the need to understand characteristic features and generalizable principles that define effective environmental assessments, particularly where international negotiations are inevitable.

Transboundary programs such as the USCSP occupy a special position in the communication pathways through which global frameworks are translated into local action. The participation of nations in USCSP assessments required the commitment of local scientists and policymakers to the goals set by a foreign agency regarding capacity development, policy advice, and provision of specific technical expertise according to the requirements of the UNFCCC. These goals were set to maintain acceptable levels of scientific credibility in the assessment process through the provision of specific numerical models and expert training, and to guarantee the legitimacy of the assessment product through its endorsement and use by government officials for communicating national action plans to the international community.

One of the important factors that constrain assessments in developing and transition countries is that financial resources supporting the assessments generally flow only in one direction: from industrialized countries to developing countries. In the case study presented here, this imbalance raised the question of hegemony in the process of environmental assessments. In the absence of matching or supplemental funds provided by countries that host the assessments, there is little room to negotiate important conditions such as framing, participation, methods, and models used to gain scientific credibility and political legitimacy. This absence is particularly relevant in the case of global climate change, where there has been substantial debate about normative aspects of the issue among scientists and policymakers in industrialized countries and their counterparts in developing countries. In the case of the USCSP, resistance to hegemonic framing and implementation of assessment parameters may have reduced the number of participating countries, particularly as discussed in the case of India. According to Jack Fitzgerald, concerns about restrictions imposed by the funding source on the freedom to design and implement assessments may also have played a role in delaying participation of Brazilian scientists and policymakers in the USCSP

(Fitzgerald 2000). There are few strategies for eliminating, or at least reducing, concerns about intellectual hegemony in the design of environmental assessments. The USCSP's approach to the problem was to ensure that participation in scientific assessments within countries was limited to scientists primarily affiliated with specific national institutions that host the assessments, although publications resulting from the assessments were in books edited by experts from other countries who may have served as consultants to the assessment team (see, e.g., Adegbulugbe et al. 1997; Dixon et al. 1996).

The strategy of isolating assessment teams from international hegemony associated with funding sources was interpreted as effective capacity building because it created a sense of ownership of the assessment within the host country. In turn, the sense of ownership legitimized the assessments for national action plans and national communication to the UNFCCC secretariat. However, to maintain a certain level of scientific credibility in the assessment process, the USCSP ensured that similar assessments in all participating countries employed exactly the same scientific models and methodologies, which were provided to the participants by experts from industrialized countries (e.g., Benioff and Warren 1996; Benioff et al. 1996). A wide diversity of methods, models, and scenarios is used in climate assessments, and the choice of one simulation model, method, or scenario over another is often hotly debated. Strict stipulation of methodologies may also have compromised the potential for integration of local perspectives into the assessment process, again raising the issue of overarching influence of funding sources and assessment sponsors. Therefore, trade-offs among strategies to encourage participation, creating a sense of ownership of assessments, and strategies to maintain scientific credibility appear to be inevitable.

Learning from the experience of the USCSP to improve the roles played by cross-boundary initiatives such as the USCSP and the GEF in international environmental assessments will require continuous evaluation of how the framing of environmental change impacts is translated across geographical, political, and cultural boundaries, particularly under constraints imposed by the structure of funding for global environmental assessments. For progress to occur in this direction, additional research is needed on various fronts. There is much to be learned about the configuration of science–policy interfaces within each country and between countries, and how these interfaces and the international negotiation forums influence the attributes of assessments such as salience, participation, credibility, and legitimacy. Several attempts are being made to apply equitable criteria for evaluating the credibility of information submitted by various countries with diverse interests. This line of research has become particularly urgent as the international scientific and diplomatic communities approach the synthesis stage of "national communications" mandated by the UNFCCC.

Acknowledgments

Awards from the Global Environmental Assessment Project (Harvard University), the Global Forum for Health Research (Geneva, Switzerland), the Center for Global Peace and Conflict Studies (University of California, Irvine), and the

Program in Industrial Ecology (University of California, Irvine) supported this work. Special gratitude is due to James McCarthy at the Harvard Museum of Comparative Zoology for helpful discussions. This study would not have been possible without the generous cooperation of Sandra Guill, Jack Fitzgerald, and Joseph Huang at the Washington, D.C., headquarters of the U.S. Country Studies Program.

References

Adegbulugbe, A.O. 2000. Personal communication with A.O. Adegbulugbe, Professor and Director, Center for Energy and Development, Obafemi Awolowo University, Ile-Ife, Nigeria, and the author. January 10.

Adegbulugbe, A.O., I.F. Ibitoye, W.O. Siyanbola, G.A. Oladosu, J.F.K. Akinbami, D.A. Pelemo, F.A. Adesina, and F.O. Oketola. 1997. Greenhouse Gas Emission Mitigation in Nigeria. In *Global Climate Change Mitigation Assessment: Results for 14 Transition and Developing Countries,* edited by S. Meyer, B. Goldberg, and J. Sathaye. Washington, DC: U.S. Country Studies Program, 221–241.

Agarwal, A. 1990. The North–South Perspective: Alienation or Interdependence. *Ambio* 19: 94–96.

Agarwal, A, and S. Narain. 1991. *Global Warming in an Unequal World.* Delhi, India: Center for Science and the Environment.

Anya, A.O. 1993. Science and the Crisis in African Development. Lagos, Nigeria: Ida-Ivory Press.

Apuuli, B., J. Wright, C. Elias, and I. Burton. 2000. Reconciling National and Global Priorities in Adaptation to Climate Change: With an Illustration from Uganda. *Environmental Monitoring and Assessment* 61: 145–159.

Baer, P., J. Harte, B. Haya, A.V. Herzog, J. Holdren, N.E. Hultman, D.M. Kammen, R.B. Norgaard, and L. Raymond. 2000. Equity and Greenhouse Responsibility. *Science* 289: 2287.

Benioff, R., and J. Warren (eds.). 1996. *Steps in Preparing Climate Change Action Plans: A Handbook.* Washington, DC: U.S. Country Studies Program.

Benioff, R., S. Guill, and S. Lee (eds.). 1996. *Vulnerability and Adaptation Assessments: An International Handbook.* U.S. Country Studies Program. Dordrecht, Netherlands: Kluwer Academic Publishers.

Carr, S., and R. Mpande. 1996. Does the Definition of the Issue Matter? NGO Influence and the International Convention to Combat Desertification in Africa. *Journal of Commonwealth and Comparative Politics* 34: 143–165.

Clark, W.C. 1987. Scale Relationships in the Interactions of Climate, Ecosystems, and Societies. In *Forecasting in the Social and Natural Sciences,* edited by K.C. Land and S.H. Schneider. Dordrecht, Netherlands: Reidel, 337–378.

Corell, R. 2000. Personal communication between R. Corell, Senior Research Fellow, Belfer Center for Science and International Affairs, Kennedy School of Government, Harvard University, and the author. May 5.

Dixon, R.K., S. Guill, F.X. Mkanda, and I. Hlohowskyj, I. (eds.). 1996. *Vulnerability and Adaptation of African Ecosystems to Global Climate Change.* Oldendorf/Luhe, Germany: Inter-Research. Also published in *Climate Research* 6: 1–201.

Farrell, Alex, Stacy D. VanDeveer, and Jill Jaeger. 2002. Environmental Assessments: Four Under-Appreciated Elements of Design. *Global Environmental Change* 11(4): 311–333.

Feenstra, J., I. Burton, J.B. Smith, and S.J.T. Richard (eds.). 1998. *Handbook on Methods for Climate Change Impact Assessment and Adaptation Strategies.* Version 2.0. Vrije Universiteit, Amsterdam: United Nations Environment Program and Institute for Environmental Studies.

Fitzgerald, J. 2000. Personal communication between J. Fitzgerald, Director of the U.S. Country Studies Program, and the author. April 6.

GEF (Global Environment Facility). 2000. *The GEF Programmatic Approach: Criteria and Processes for its Implementation.* Washington, DC: Global Environment Facility.

GTZ (Deutsche Gesellschaft für Technische Zusammenarbeit). 2000. CaPP – Climate Protection Programme for Developing Countries. http://www2.gtz.de/climate/download/specials/CaPP-Info_eng.pdf (accessed January 12, 2005).

Guill, S. 2000. Personal communication between S. Guill, Regional Coordinator for Africa and the Middle East, U.S. Country Studies Program, and the author. April 6.

Huang, J. 2000. Personal communication between J. Huang, Regional Coordinator for Assessments in Asia and Small Island States, U.S. Country Studies Program, and the author. April 5.

Jaeger, J., J. van Eijndhoven, and W.C. Clark. 2001. Knowledge and Action: An Analysis of Linkages among Management Functions for Global Environmental Risks. In *Learning to Manage Global Environmental Risks: A Comparative History of Social Responses to Climate Change, Ozone Depletion and Acid Rain,* edited by W.C. Clark, J. Jaeger, J. van Eijndhoven, and N.M. Dickson. Cambridge, MA: MIT Press.

Jasanoff, S. 1996. Science and Norms in Global Environmental Regimes. In *Earthly Goods, Environmental Change and Social Justice,* edited by F.O. Hampson and J. Reppy. Ithaca, NY: Cornell University Press, 173–197.

Kandlikar, M., and A. Sagar. 1997. Climate Change Science and Policy: Lessons from India. Environment and Natural Resources Program discussion paper E-97-08, Kennedy School of Government and International Institute for System Analysis interim report IR-97-035/August. Cambridge, MA: Kennedy School of Government, Harvard University.

Kates, R.W., J.H. Ausubel, and M. Berberian. (eds.). 1985. *Climate Impact Assessment: Studies of the Interaction of Climate and Society.* ICSU/SCOPE report no. 27. Chichester, United Kingdom: John Wiley.

Long, M., and A. Iles. 1997. Assessing Climate Change Impacts: Co-evolution of Knowledge, Communities, and Methodologies. ENRP discussion paper E-97-09. Cambridge, MA: Kennedy School of Government, Harvard University.

Lopez, R. 1999. *Incorporating Developing Countries into Global Efforts for Greenhouse Gas Reduction.* Resources for the Future Climate Issue Brief no. 16. Washington, DC: Resources for the Future.

Lukman, R. 2000. OPEC Statement to the Sixth Conference of Parties (COP6) to the United Nations Framework Convention on Climate Change (UNFCCC). http://www.opec.org/NewsInfo/COP_Statements/COP6_2000.htm (accessed January 12, 2005).

MacLean, D.E. 1990. Comparing Values in Environmental Policies: Moral Issues and Moral Arguments. In *Valuing Health Risks, Costs, and Benefits for Environmental Decision-Making,* edited by P.B. Hammond and R. Coppock. Washington, DC: National Academy Press.

Magazda, C.H.D. 2000. Climate Change Impacts and Human Settlements in Africa: Prospects for Adaptation. *Environmental Monitoring and Assessment* 61: 193–205.

McMichael, A.J., and R. Sari Kovats. 2000. Climate Change and Climate Variability: Adaptations to Reduce Adverse Health Impacts. *Environmental Monitoring and Assessment* 61: 49–64.

McMichael, A.J., and W.J.M. Martens. 1995. The Health Impacts of Global Climate Change: Grappling with Scenarios, Predictive Models, and Multiple Uncertainties. *Ecosystem Health* 1: 23–33.

Miller, C. 1998. Extending Assessment Communities to Developing Countries. ENRP discussion paper E-98-15. Cambridge, MA: Kennedy School of Government, Harvard University.

Ogunseitan, O.A. 2003. Framing Environmental Change in Africa: Cross-scale Institutional Constraints on Progressing from Rhetoric to Action Against Vulnerability. *Global Environmental Change* 13: 101–111.

Ominde, S.H., and C. Juma (eds.). 1991. *A Change in the Weather: African Perspectives on Climatic Change.* African Center for Technology Studies Environmental Policy Series no. 1. Nairobi, Kenya: African Center for Technology Studies Press.

Pittock, A.B., and R.N. Jones. 2000. Adaptation to What and Why? *Environmental Monitoring and Assessment* 61: 9–35.

Shue, H. 1999. Global Environment and International Inequality. *International Affairs* 75: 531–545.

Smith, J.B., S. Huq, S. Lenhart, L.J. Mata, I. Nemesova, and S. Toure (eds.). 1996. *Vulnerability and Adaptation to Climate Change: Interim Results from the U.S. Country Studies Program.* Dordrecht, Netherlands: Kluwer Academic Publishers.

Stern, P.C., O.R. Young, and D. Druckman (eds.). 1992. *Global Environmental Change: Understanding the Human Dimensions.* National Research Council. Washington, DC: National Academy Press.

Turner, B.L. II, R.E. Kasperson, W.B. Meyer, K. Dow, D. Golding, J.S. Kasperson, R.C. Mitchell, and S.J. Ratick. 1991. Two Types of Global Environmental Change: Definitional and Spatial Scale Issues in their Human Dimensions. *Global Environmental Change* 1: 14–22.

UNDP-GEF (United Nations Development Program and the Global Environment Facility). 1999. Capacity Development Initiative. UNDP-GEF Strategic Partnership Terms of Reference. Washington, DC: Global Environment Facility.

UNFCCC (United Nations Framework Convention on Climate Change). 2005. New York: United Nations. http://unfecc.int/essential_background/Kyoto_Protocol/status_of_ratification/items/2613.php (accessed July 26, 2005).

———. 1992. http://unfccc.int/resource/docs/convkp/conveng.pdf (accessed January 12, 2005).

USCSP (U.S. Country Studies Program). 1999. *Climate Change: Mitigation, Vulnerability, and Adaptation in Developing and Transition Countries.* Washington, DC: U.S. Country Studies Program.

Usher, P. 2000. Integrating Impacts into Adaptation Measures. *Environmental Monitoring and Assessment* 61: 37–48.

Watson, R.T., M.C. Zinyowera, R.H. Moss, and D.J. Dokken (eds.). 1997. *The Regional Impacts of Climate Change: An Assessment of Vulnerability.* Geneva, Switzerland: World Meteorological Organization and United Nations Environment Program.

WHO (World Health Organization). 1990. *Potential Health Effects of Climatic Change.* Geneva, Switzerland: World Health Organization.

CHAPTER 11

Grounds for Hope
Assessing Technological Options to Manage Ozone Depletion

Edward A. Parson

DEBATE OVER GLOBAL environmental issues is typically dominated by questions of the reality, degree, and character of the environmental risk, and the associated scientific evidence and uncertainties. Scientific assessments that seek to advance understanding of these risks are the most frequently undertaken assessments, and they feature most prominently in subsequent policy debates. But these questions of environmental risk address only one side of the judgment required to decide how to respond to a risk. Such judgments must also consider the means available to mitigate the risk and their feasibility, cost, and consequences.

This second type of question, concerning technological options to deal with environmental problems, also depends in diverse ways on scientific and technical expertise and so can usefully be informed by expert assessments. These assessments are typically called "technology assessments" or "option assessments." They can address various questions, ranging from simply identifying potential technological, managerial, or policy options for managing risk, to characterizing a specific option's feasibility, state of development, effectiveness, cost, and other consequences with varying degrees of detail and specificity. Although both types of assessment depend on expert judgments, technical assessments of options differ from scientific assessments of environmental risks in multiple ways. They require different types of expertise, including practical engineering and managerial judgments as well as scientific knowledge and skill. They typically are established by different actors, employ different participation and procedures, address different questions in pursuit of different goals, and face different challenges. They often have more direct implications for action and more direct commercial consequences. They are most appropriately evaluated by different criteria, with technical and political aspects intertwining more closely than is the case in scientific assessments.

In general, the effectiveness of technological option assessments for global environmental issues has been low. One major recent study concluded that they have generally "failed to generate a cumulative body of reliable knowledge concerning alternative responses to global environmental risks" (Clark et al. 2001,

70),[1] and have had little or no tangible influence on decisions. Even by the more lenient standard of influencing the agenda for decisions, the occasions in which option assessments have either placed or sustained potentially worthy but unpopular options on the agenda—or effected the removal from consideration of options definitively shown to be inferior—have also been rare (Clark et al. 2001, 72).

This chapter discusses one striking exception to this general pattern: technological option assessments under the international regime to protect the stratospheric ozone layer. While option assessments in the early stages of the ozone issue were as ineffective as option assessments usually are, a series of assessments conducted under the international ozone regime since 1989 has departed sharply from this general pattern, achieving high levels of technical quality, practical utility, and influence that have not been equaled or even approached by any other option-assessment process for any global environmental issue. This case is unique, but it indicates a more general possibility that has not been exploited. Indeed, although technical option assessments have been less frequently undertaken, less frequently effective, and less prominent in policy debate than scientific assessments of environmental risk, this case suggests that they may hold far greater prospect for exercising decisive influence on policy debate and action to manage environmental risks—if the factors contributing to their strong influence in this case can be replicated elsewhere.

Stratospheric Ozone Policy and Option Assessment, 1974–1985[2]

In 1974 two scientists suggested that various halogenated industrial chemicals, principally the chlorofluorocarbons (CFCs), risked destroying the stratospheric ozone layer that shields life on the earth's surface from harmful solar ultraviolet radiation (Molina and Rowland 1974). The publication of this claim was followed by an intense policy debate, which culminated three years later in decisions in the United States and three other countries to ban the use of CFCs as propellants in aerosol spray cans—constituting slightly more than half of worldwide CFC use at the time. Although it was widely recognized from the outset that the problem's scope was global, and consequently global action was needed to address it, several attempts over 10 years to control ozone-depleting chemicals internationally all ended in failure. One attempt in 1980 to broaden U.S. domestic controls to apply to all CFC uses, rather than to aerosols alone, also failed.

In these early policy debates and their outcomes, assessments of technological responses were sometimes conducted but never influential. They played little role in the decisions to ban CFC aerosols, because questions of the cost and difficulty of reducing aerosol CFCs figured little in policy debates. Rather, the availability, low cost, and relative ease of adopting other ways to package products—either nonaerosol formulations or aerosols using non-CFC propellants—were widely known. A few half-hearted attempts were made to argue that eliminating CFCs in aerosols would be costly or difficult, but in the face of evidence that alternatives were both technically feasible and commercially viable, these objections did not meet minimal standards of credibility.

In contrast to aerosol CFC uses, questions of technical feasibility and cost were both important and contentious in debates over controls on nonaerosol uses, and several attempts were made to inform these questions and delimit policy conflict by conducting assessments of potential technological and other options to reduce use. These attempts all failed, however.

The most important of these attempts were two assessments commissioned by the U.S. Environmental Protection Agency in 1979 to inform its consideration of comprehensive CFC controls. The assessments were conducted by the Rand Corporation and by a committee of the National Academy of Sciences (NAS). The two efforts were quite similar; indeed, the NAS assessment relied in part on data and analysis from an early draft of the Rand assessment. Both assessments tried to characterize present (at that time) uses of CFCs, to identify potential technological alternatives to reduce use and emissions, and to characterize the extent of feasible reductions. Both reached extremely pessimistic conclusions about the extent to which CFC use could be reduced or substituted with non-CFC alternatives. The Rand study constructed a marginal-cost curve for CFC reductions attainable at various price increments. It concluded that a tax of $1 per pound (a tripling of the market price) would reduce use by only 20 percent, and that reductions beyond 25 percent were technically infeasible "at any price." Conducting a purely technical analysis, the National Academy of Sciences study concluded that the maximum feasible reduction was 50 percent, and that even modest cuts would be highly costly.[3]

The conclusion that large reductions would be costly and difficult does not by itself indicate an inadequate study, although the extremity of both studies' conclusions gives grounds for skepticism. But the failure of these studies is directly evident in the details of their analysis. Both considered an extremely limited set of potential alternatives, which were almost entirely restricted to alternatives previously identified or used and rejected in favor of CFCs. New chemical alternatives being pursued by CFC manufacturers were not considered because their development was not complete. Therefore, the only substitution considered for most uses was a hydrochlorofluorocarbon (HCFC) called HCFC-22. Both studies adopted a presumption that only commercially available alternatives should be considered, that no adjustment of equipment or manufacturing process could be assumed, and that no significant degradation of product performance was acceptable. This approach created a huge bias toward the conclusion that nothing could be done. Although the National Academy of Sciences assessment identified widespread disagreement among technical experts over what reductions were feasible, it did nothing to reduce this disagreement or to explicate its foundations. The failure of these assessments contributed to the failure to enact the seemingly modest and reasonable 1980 U.S. proposal for comprehensive CFC controls,[4] because proponents were unable to make the case that CFC limits were feasible and their cost acceptable. Industry assertions that significant cuts would be difficult, disruptive, and costly met no effective response (DuPont 1980; *International Environment Reporter* 1980, *401*).

While the causes of failure in option assessments in general may be diverse, the causes of failure in these two cases are clearly related to the assessment bodies' inability to make technical judgments that were independent of industry-held information and industry biases. Authoritative technical information about potential alternatives was overwhelmingly held by the CFC manufacturers and a few major user firms. These parties were inclined to doubt that CFC limits were feasible at reasonable cost, and they had no interest in helping officials or independent assessment bodies make the opposite case. The assessment bodies, forced to rely almost entirely on industry sources for technical information about the availability, development status, performance, and costs of alternatives, were in effect asking industry experts how easily their firms could give up CFCs. Unable to develop the knowledge necessary to conduct an independent critical assessment, the assessment bodies adopted a framework and a set of assumptions that strongly biased their conclusions toward the status quo and the interests of the industries producing and using CFCs.

Subsequent attempts to assess prospects for CFC reductions through the mid-1980s, at both national and international levels, foundered on the same obstacles as the two earlier assessments. A brief attempt to assess CFC-reduction options under the Organization for Economic Cooperation and Development, as part of a broad integrated assessment of the ozone issue in 1981, met sharp opposition and was effectively reduced to a reprise of earlier assessments (U.S. EPA 1982; *International Environment Reporter* 1981, *826*; OECD 1981). After international negotiations on the ozone layer were first convened in January 1982, delegations could not agree how—or even whether—to convene technical discussions of CFC alternatives and controls. One existing international assessment body had these questions within its formal mandate, but the group repeatedly refused to address them out of concern that they lacked necessary expertise and that the questions risked politicizing the committee and damaging its scientific credibility (UNEP 1981, 1982). In the only further attempt to assess CFC reduction options prior to 1987, an ostensibly informal international workshop of experts, held in Rome in 1986, precisely mirrored the lines of conflict that prevailed in official international negotiations at the time. Participants could not even agree as to whether the cost of the original U.S. aerosol ban—by that time in effect for seven years—had been low or high (UNEP 1986).

Establishment of the Ozone Regime and Its Assessment Bodies, 1986–1988

After several years of largely deadlocked negotiations, international management of ozone depletion advanced rapidly between 1986 and 1988, culminating in the signing and entry into force of the Montreal Protocol—the first international agreement with concrete action to protect the ozone layer. This rapid progress was driven by unique factors unrelated to the perceived ease of reducing CFCs. A group of activist officials gained control of the U.S. negotiating agenda and succeeded in sustaining an extreme international negotiating position (CFC reductions by 95 percent) against substantial domestic and international opposition. A shocking report of ozone loss in the Antarctic and claims that global ozone losses could be detected in a satellite record also helped strengthen the activists in their insistence on deep cuts. With a broad international consensus emerging that the appropriate treatment of CFCs was to freeze usage near then-current levels, the activists' persistence was largely responsible for a negotiated agreement to reduce CFCs by half.

Technical assessment of CFC reduction options had little if anything to do with this outcome. In fact, the technical basis for confidence that either 95 or 50 percent cuts were technically achievable at acceptable cost was extremely thin for countries such as the United States that had already eliminated aerosols. An unguarded industry revelation in 1986 had suggested that chemical CFC alternatives could be developed in 5 to 10 years, but serious problems were evident in applying these new chemicals to existing CFC uses (Alliance for Responsible CFC Policy 1986). Uncertainty over the feasibility of these cuts—including significant risks of disruption, premature capital write-off, loss of amenities, and bankruptcies in some usage sectors—was at least as serious as uncertainty over the character of the environmental risk.

But while progress in technical knowledge and technical assessment contributed little to the rapid formation of an international ozone regime, the new regime transformed the subsequent conduct of technical assessment and the significance of technical information. The Montreal Protocol required that parties periodically review the adequacy of the Protocol's control measures in view of advances in knowledge and capability, and it required parties to consider modifying the measures based on advice from expert assessment panels. Although the Protocol's adoption of concrete international CFC controls represented an important first step, these provisions for repeated review and modification of its control measures represented the most central contribution to the ozone-reduction regime's subsequent adaptation and ultimate success.

Panels were initially established in four areas: atmospheric science, ultraviolet effects, technology, and economics. These panels were organized, chairs were identified, and tentative designs and mandates were established at a series of informal consultations between key delegations and United Nations Environment Programme Director Mustafa Tolba, culminating in a series of workshops in The Hague in late 1988 (*International Environment Reporter* 1988, *210*). A series of design decisions made in these initial consultations were decisive for the subsequent effectiveness of the panels. Most importantly, organizational decisions made in the interests of fast work had the effect of substantially reducing the political control over the panels from what was originally envisioned in Protocol negotiations. Rather than authorizing a political body to supervise and integrate the work of four "reporting groups" of independent experts, each of these four groups operated with substantial independence under its chair. The chairs coordinated among themselves to synthesize and publish the four group's work, under minimal oversight by the main political negotiating body. In addition, the organizers of the technology panel made a decisive early choice that the body would rely principally on knowledge and participation from industry experts—although their concern about potential bias or capture is evident from their decision to exclude experts from the CFC producers from the panel, relying instead on experts from user industries and industry associations, as well as government, university, and nongovernmental organization (NGO) experts.[5] This controversial decision, made at the initiative of Mustafa Tolba with the support of several major delegations, reflected negotiators' mistrust of the CFC producers for their long history of obstruction, and it reflected their concern that these firms were too committed to their own chemical alternatives to assess other potential alternatives objectively.

Technology Assessment under the Protocol, 1989–1999

The Montreal Protocol's technology panel, along with the other assessment panels, has conducted four full assessments since it was initially established—to advise renegotiations of the Protocol in 1990, 1992, 1995, and 1999. Although many aspects of the panel's operations have adjusted over time to meet the evolving needs of the Protocol, certain core elements of its organization and the questions it has addressed have remained constant. From the outset, the technology panel has repeatedly addressed central questions of what reductions in ozone-depleting chemicals are feasible in particular uses and sectors, and by what time. Feasibility has been defined, following initial guidance from the parties, as "the possibility to provide substitutes or alternative processes without substantially affecting properties, performance or reliability of goods and services from a technical and environmental point of view" (UNEP 1989b). Although the definition of feasibility was modified after 1990 to include economic as well as technical feasibility, the extent of feasible reductions has remained a single estimate, defined without explicit reference to the cost of alternatives (UNEP 1989c, 9). The panel has addressed these questions by critically examining specific alternatives available and under development, including new production technologies, process changes, changes in product characteristics, and changes in management practices. To answer the questions of feasibility, the panel has relied on working groups assembling the focused knowledge of 15 to 50 experts, principally via a set of Technical Options Committees (TOCs). The committees examine alternatives for each major usage sector, such as refrigeration, foams, and solvents. Although participating experts have consistently included individuals from universities and governments (and government officials have provided leadership and administration), most participating experts have been from private industry, principally from user firms, engineering and consulting firms, and industry associations. Parties to the Montreal Protocol nominate experts to participate, but in practice the chairs of the technology panel and its TOCs have exercised substantial control over participation.[6] They have used this authority to identify expert and energetic people committed to solving problems, with sufficiently wide representation from affected industries in each usage sector to produce high-quality and credible results.

The first decisive phase of the panel's work took place in 1989, when it advised the 1989–1990 negotiations to revise the Protocol. In this round, only the central questions of the degree of feasible reductions in each sector were considered. The conclusions were shocking, in that they stated that at least 95 percent of CFC and halon consumption could be eliminated by 2000. The separate reports of each TOC provided backup, elaboration, and qualifications for this aggregate conclusion, and they illustrated the different degrees of difficulty identified in each sector. At one extreme, the reports found that aerosols could be eliminated immediately except for a few small medical uses; at the other, some long-lived refrigeration equipment would require continued servicing with CFCs to the end of its life, accounting for the few percent of CFC use that the panel judged might be needed beyond 2000. Only the halon TOC failed to reach complete consensus, with the committee splitting over the feasibility of further reductions beyond the 60 percent cuts they agreed were available from reducing nonessential uses and better managing existing stocks (Mauzerall 1990; UNEP 1989a, 3; UNEP 1989b, iv).

This technology panel report, together with that of the atmospheric-science panel, strongly conditioned the negotiations to revise the Protocol in 1989 and 1990. Even before negotiations resumed, these results had prompted many governments and industry actors to endorse strengthening the Protocol to essentially eliminate the original five CFCs by 2000, while several had proposed even earlier phase-outs. The panels did not, however, eliminate all disagreement, but rather channeled discussion into second-order matters such as the precise dates of phase-outs, interim reduction schedules, and the need for a continued small CFC stock for servicing existing equipment. Even on the question of extending controls to new ozone-depleting chemicals, particularly the solvent methyl chloroform—a controversial point, which brought new industry actors into the negotiations, and on which the panel's work had been more hasty and less well-grounded than for other sectors—the panel's judgment of feasible reductions prevailed over contrary claims by industry, and it was subsequently shown to have been correct by the reductions actually achieved (ENDS 1989; UNEP 1989b, v).

After the striking success of its first assessment, the technology panel was reorganized in 1991 to absorb the economics assessment panel—whose work in 1989 had been unsuccessful, in large part because it was divorced from technology—to form the new Technology and Economics Assessment Panel (TEAP). In the next major assessment, conducted to support negotiations for further Protocol revisions in 1992, TEAP once again addressed the central questions of what further reductions in ozone-depleting chemicals were feasible—now with experts from CFC manufacturers participating, and with broadened participation of developing-country experts. In addition, TEAP addressed other specific questions at parties' requests, including the earliest possible date to eliminate methyl chloroform, the likely need and availability of ozone-depleting chemicals for developing countries, and the extent to which eliminating CFCs would require use of HCFCs—transitional chemicals marketed as CFC alternatives that also depleted ozone, but by only a few percent to 15 percent as much as CFCs. As in 1989, the panel reached strong conclusions about the feasibility of further reductions. Noting that progress in reducing ozone-depleting chemicals had been more rapid than anticipated two years earlier, they concluded that substantial further tightening of targets was feasible, eliminating virtually all CFCs, halons, and carbon tetrachloride by 1995 to 1997, and methyl chloroform by 1995 to 2000. Achieving the accelerated phase-outs would also depend on several conditions being met, including increased short-term use of HCFCs (UNEP 1991).

This assessment also saw the panel begin to address operational questions of the management of the ozone regime, reminding parties of concrete steps they would need to take to accomplish phase-outs by 1997. In addition, following the 1990 decision to eliminate halons with an exemption for essential uses, parties delegated to TEAP the task of evaluating proposed essential uses. With no specific guidance from the parties, TEAP and the halon TOC developed criteria to define essential uses. On that basis, they expressed a "qualified opinion" that all essential uses could be supplied until at least 2000 by redeploying existing stocks, and they recommended that parties reject all essential-use applications, subject to periodic re-assessment. Although this conclusion was carefully expressed in purely advisory terms, parties' deference to the panel's findings and subsequent similar decisions represented a substantial delegation of operational responsibility for managing the regime to TEAP.

Through 1991 and 1992, Protocol negotiators experienced substantial conflict in three areas: the relationship between developing-country commitments and associated financial assistance; how sharply to restrict HCFCs in view of their transitional character; and whether to extend controls to methyl bromide, a major agricultural pesticide, on the basis of new suggestions that it was an important contributor to ozone depletion. On each of these issues, parties attempted to limit political conflict by seeking related technical information from TEAP. In carefully worded instructions, TEAP was asked to study the implications of advancing all phase-outs, with specific reference to developing countries; to identify specific uses in which a rapid CFC phase-out required HCFCs and a feasible timetable to eliminate them; and to review uses and alternatives for methyl bromide. Parties' responses to the answers TEAP provided to these questions were mixed. TEAP's conclusion that HCFCs could be reduced but remained essential for eliminating CFCs in some uses was attacked by some delegations, but it provided the basis for negotiating only limited HCFC restrictions. TEAP's conclusion that a substantial advance of cuts in developing countries was feasible with proper financial support did not support any concrete decisions, however, in the face of strong disagreement over how much financial support to provide.

After the 1992 Protocol revisions, the tasks delegated to TEAP continued to expand, and its work was more closely integrated with parties' negotiations. TEAP once again assessed the extent of feasible opportunities for further reductions in ozone-depleting chemicals, now with specific charges to assess alternatives to HCFCs in the uses most dependent on them (refrigeration and insulating foams) and to conduct the first full-scale assessment of alternatives to methyl bromide (UNEP 1991, 1994). In addition, TEAP conducted further essential-use evaluations for all controlled chemicals, presenting recommendations that were implemented by the parties with only small modifications despite substantial political controversy; recommended a strategy for managing the stock of halons; assessed the feasibility of implementing a particularly expansive provision in the Protocol's trade restrictions; and evaluated technologies for recovering and recycling ozone-depleting chemicals. As in 1992, parties repeatedly asked TEAP to address additional questions related to points of particularly sharp conflict in negotiations, some of them new and some of them reconsiderations or elaborations of questions already addressed.

In a series of reports presented through 1994 and 1995, TEAP largely reaffirmed its conclusions of 1992: Further reductions in HCFCs were judged feasible, although these remained necessary for some applications. Accelerating developing-country phase-outs was feasible, but only with adequate financial support and timely implementation of sponsored projects. In addition, a newly established 65-member TOC conducted the first full assessment of methyl bromide, concluding that at least 90 percent of use could be eliminated. Industry representatives, including some who served on the TOC, sharply attacked this conclusion. They claimed that only much smaller reductions, perhaps as little as a few percent, were feasible. By this time, TEAP's influence over policy negotiations was coming to be widely recognized; delegations responded by beginning to oppose proposals to pose questions to TEAP, while TEAP began declining or avoiding parties' questions when they judged them not sufficiently technical that they could provide a helpful resolution. In these negotiations, TEAP's conclusions on the continuing need for HCFCs helped resist calls for rapid cuts, but north–south political disagreement prevented delegates from acting on the large reduction opportunities identified for methyl bromide.

Following the 1995 negotiations, the operations of TEAP and the tasks assigned to it continued to evolve. After the panel estimated the level of funding required for continuing phase-out programs in developing countries, parties implemented its recommended level in the fund re-authorization decision of 1996 (ENDS 1996, 36). As phase-outs of ozone-depleting chemicals in industrialized countries approached, industrialized-country firms became less willing to bear the substantial costs of participating, while developing-country needs for technical assessment increased as their targets approached. In response to these changes, TEAP was reorganized to reduce the number of separate bodies, increase participation of developing-country experts, and increase reliance on ad hoc teams to address questions requiring highly specific expertise. This reorganization provided the opportunity as well to re-constitute the methyl bromide TOC in order to reduce participation by those with commercial stakes in methyl bromide but who offered no alternatives. In its next assessment, this newly reconstituted methyl bromide body reported that it could not find a crop that needed methyl bromide and increased its estimated feasible near-term reduction to more than 95 percent. With the assessment body moving in this direction, the promoters of methyl bromide increasingly operated directly through political channels outside the technology assessment process. Delegations finally broke their deadlock on methyl bromide in 1997 and agreed to a worldwide phase-out, although with certain crudely defined and potentially large exemptions (UNEP 1997).

In sum, the striking success of the Montreal Protocol's technology assessment process over its 10 years of operation is evident in the huge number of specific technical judgments it provided, which were with few exceptions persuasive, technically supported, consensual, and found to be accurate or moderately conservative when tested by subsequent events. The success is also evident in the substantial influence TEAP exercised over parties' decisions, even while carefully avoiding usurping their authority. TEAP's strong, specific, carefully delimited statements of feasible reductions have been disputed by policy actors on very few occasions, even when they have not translated into parties' decisions. The TEAP process itself spurred many innovations and succeeded in keeping top industry expertise engaged through the 10-year process of moving to full phase-outs. One measure of their approval of TEAP's performance is that parties repeatedly asked them to take on new and expanded jobs, even delegating significant de facto operational responsibility in the case of the essential-use process.

Explaining the Success

Before the enactment of the 1987 Montreal Protocol, technological knowledge about the feasibility, performance, and cost of potential chemical alternatives to CFCs was held nearly exclusively by the CFC manufacturers. These firms could withhold technical knowledge about alternatives from other actors, and could also control how much knowledge they themselves possessed, because it was their choice how far to pursue the development of alternatives. These firms had no interest in helping the proponents of CFC controls make them do something costly, risky, and inconvenient, and they successfully promoted an environment of widespread pessimism about the viability of alternatives. Other policy actors had no equivalently authoritative technical information and therefore could not rebut claims by the CFC manufacturers to demonstrate that significant reductions were technically feasible. Unable to effectively engage industry expertise, the few attempts to conduct independent assessments of CFC alternatives either echoed the pessimistic public stance of industry, or could not be undertaken at all. The result was a low-confidence equilibrium, in which the actors who wanted CFC controls could not make the case that they were technically feasible, while those with the best knowledge of technical feasibility would not reveal it. Sustaining the widespread belief that CFC alternatives were infeasible or unacceptably costly counts as a great strategic success of industry organizations through the early 1980s.

The shock of the 50 percent cuts enacted in the Montreal Protocol, together with growing alarm through 1988 over the severity of the ozone-depletion risk and widespread calls to eliminate CFCs entirely, began a sharp shift from the prior low-confidence equilibrium to a high-confidence equilibrium that generated rapid, continuous progress in the identification and implementation of new approaches to reducing ozone-depleting chemicals. These initial shifts transformed the business environment for firms producing and using ozone-depleting chemicals, suggesting that the 50 percent cuts of 1987 would soon be tightened and possibly extended to other chemicals. For producers, the looming targets imposed grave risks but also carried potential opportunities for the largest and most technically sophisticated producers, because restrictions on CFCs appeared likely to create commercial opportunities in new alternative chemicals. For CFC users, however, agreed and threatened targets posed only risks, whether through losing technologies on which they depended or through requiring them to commit to costlier chemical alternatives of uncertain availability, performance, and regulatory acceptability. This prospect set off a headlong rush to reduce dependence on the threatened chemicals, which various government and industry bodies sought to support by promoting open sharing, exchange, and critical examination of potential alternatives—accelerating further the CFC manufacturers' loss of their former control over technical information about alternatives.

TEAP and its sectoral subbodies played critical roles in promoting innovation and linking it to the evolving negotiations of Protocol targets. These bodies succeeded by exploiting two fundamental differences between technology assessments and scientific assessments. First, technology assessments can to a significant degree change the conditions of technological feasibility on which they are reporting by advancing present technical skill, solving problems, and identifying and removing barriers to product and process development. Second, in accomplishing these tasks, technology assessments are able to jointly provide public and private benefits—the public benefits for which they are established, and the private benefits to participating individuals and their employers that are sufficient to motivate the level of participation and effort the assessment needs to succeed.

Although so many prior attempts at technology assessment had failed, TEAP succeeded by motivating top industry experts to participate and provide their best and most honest efforts and judgments—with little regard for the policy positions or immediate commercial interests of their employers. TEAP organizers were able to accomplish this by exploiting the crisis user firms faced from looming CFC controls and their resultant need to reduce their reliance on ozone-depleting chemicals as rapidly as possible. TEAP's panels offered user firms unique opportunities to solve the technical problems of achieving such rapid reductions by bringing together critical masses of the most respected experts in each sector, both from user firms and from firms developing diverse alternative technologies. The processes of critically examining and evaluating technical alternatives and solving application problems provided a highly rewarding professional challenge and the best chance to reduce the business risk imposed by CFC reductions, thereby increasing both individuals' interest in participating and their firms' willingness to send them. The same activities of gathering data, deliberating, and solving problems that served the needs of participating firms also served the panel's purpose of giving the parties high-quality technical advice on the extent of feasible reductions.

Moreover, the same activities provided still further benefits to the ozone-reduction regime, which were not among TEAP's official responsibilities but were among its most important contributions. Experts' work on the panels advanced the margin of feasible reductions, not just to meet existing regulatory targets but beyond them. After each assessment round, the aggregate effect of the problems solved and the alternatives identified and refined was to reveal opportunities to reduce use beyond existing targets. With industry's vigorous response to the environmental and regulatory challenge almost always bringing them ahead of existing regulatory obligations, and with further reduction opportunities repeatedly identified, repeated further tightening of the requirements was possible. Moreover, as the panel's work proceeded and further opportunities were identified, individual participants increasingly worked to confirm the accuracy of the assessment and spread information about the opportunities it identified among their industry peers, helping to advance the reductions actually achieved.

The Protocol's technology assessment process achieved its success by tying together the provision of public benefits to the ozone regime and private benefits to participating individuals and firms. Participants were attracted by the need to solve their own problems of reducing ozone-depleting chemicals, by prospective opportunities to market alternative technologies and associated services and information (including the expertise gained from participating on the panel), and by the professional challenge, satisfaction, and prestige the process offered (Kuijpers et al. 1998, *172*). In pursuing these private benefits through the assessment process, participants also provided the public benefits of good advice to the parties as to the extent of feasible reductions, and identification of additional opportunities to reduce still more.

In certain key respects, this contribution was achieved by reversing the order of activities in canonical policy choice. Rational policy choice is conventionally viewed as involving the assessment of risks, impacts, and responses prior to deliberations over control measures. But in this case, an initial regulatory target was adopted with little confidence that it could be met at reasonable cost. This fairly stringent target, and the risk of more to come, then set in motion the subsequent processes of technological development, assessment, and strengthening of control measures. In these dynamic processes of adaptation lie some of the most important insights to be drawn from the ozone regime. The problem of ozone depletion was not solved by the 1987 Protocol, but (to the extent that it has been solved) by the subsequent adaptation, refinement, and expansion of the regime.

The technology assessment process for ozone reduction has not succeeded in everything, of course. The assessment process has been most contentious and least effective on those occasions when the greatest individual competitive advantages were at stake in the outcome (Kuijpers et al. 1998, *170*). The assessments have also overreached on a few occasions, particularly on occasions when parties were considering broadening controls to include new chemicals involving new firms. Each time such an expansion was considered, the usual means of eliciting industry input in assessments was unavailable. As long as the relevant firms thought they could block controls, their preferred strategy was to obstruct technology assessments and claim that significant reductions were not feasible, as for CFCs before 1987. Absent the serious engagement of industry experts in these cases, the assessment panel's judgments were weakly founded, contested, and at heightened risk of error. More recently, TEAP's effectiveness has increasingly been challenged by parties' responses to its effectiveness, as they have asked it to answer questions that embed too much policy to be resolved by technical deliberations or tried to assert greater control over specific aspects of the assessment process and conclusions as they have seen its influence grow.

Applying the Lessons

The successful adaptation of the Montreal Protocol was principally driven by interactions between regulatory targets, technology assessment, and industry response that promoted a rapid process of innovation to reduce the use of ozone-depleting chemicals. Although no similar system has yet developed on any other issue, many aspects of its operation are likely to be applicable to other issues, if the required conditions are in place.

Setting this interaction in motion depended on the initial regulatory targets and the risk of more to come. The targets posed a strong enough threat and opportunity to elicit strong efforts to reduce controlled substances by targeted industry sectors and to develop new alternatives by potential market entrants. In addition, the regulations required a system for technology assessment to facilitate and channel these efforts, thus exploiting the effect of the target to harness private interests to public purposes. The technology assessment system effectively engaged the energetic, honest efforts of top industry experts by linking their private interests in solving their existing and anticipated reduction problems or by profiting from alternatives, to the public interests of informing and advancing the regime's control measures. The effective operation of this system depended on many practical design details, such as structuring working groups by specific industries—participants' problems were similar enough that they could all benefit from the common effort.

The strategy of coupling private and public benefits also carried the risk of capture by particular participants' interests. While the existence of regulatory targets diminished this risk by posing immediate priority challenges to participants, organizers also sought to defend against it through the organization and operations of the panel's working bodies. Most importantly, participation in work groups was balanced to include advocates of multiple alternatives and a broad range of material interests. In addition, the stature and closely overlapping expertise of the participants promoted a critical, nondeferential working environment in which implausible or weakly supported claims were vigorously questioned. In most cases, these factors sufficed to ensure the technical quality and perceived impartiality of the proceedings. Achieving this did, however, depend on specific interests of individual participants (e.g., shifting competitive advantages due to the body's evaluation of a proprietary technology) being less prominent than their shared interests in solving common technical problems. The process was least effective when these conditions could not be met.

The prospects for designing similarly structured assessment bodies on other issues appear promising, as long as the necessary conditions are met (e.g., that the relevant technological problems and the expertise most needed to solve them are sufficiently widely shared within some industry subsectors). The Montreal Protocol's model of technical assessment has already been applied once to technical assessment for climate change, for the high global warming potential gases that are implicated in both the ozone and climate issues, with promising results. This collaboration, however, has also highlighted several broad limitations that are necessary costs of this assessment strategy. The ozone assessments have consistently declined to estimate costs quantitatively or even to specify clearly what they mean by economic, as opposed to technical, feasibility. While these omissions have drawn strong criticism, they have conferred clear advantages on the assessments, and TEAP leaders argue that quantitative cost estimates are of little value in a context of rapid technical change and moving regulatory targets. Such estimates would also be vulnerable to attack for their methods and details, as the present opaque process of evaluating options and adding up feasible potential is not. A second limitation of the process is that it cannot be transparent, because it relies on closed deliberations to allow participating experts to make independent judgments, without regard for the positions of their employers. Such secret deliberations are unproblematic if they are overwhelmingly technical in content, but they can represent significant loss of accountability if they move into trade-offs over political and social values. Finally, the process has no provision for independent review comparable to scientific peer review. Rather, it relies on the participating experts and panel leadership to police each other's work both for technical quality and for bias. The lack of transparency and outside review distinguishes this technology assessment process sharply from that conducted for climate change under the Intergovernmental Panel on Climate Change, and it has led to substantial tensions when the two bodies have collaborated. The risks of the Montreal Protocol's technology assessment approach are many, but it may still represent the best review attainable, if—as the Protocol's process assumed—the best technical information is privately held and likely to be unknown to independent reviewers.

Notes

1. This study looked at dozens of separate option assessments in a cross-national and international context.

2. This historical summary of policy and assessments for stratospheric ozone is drawn principally from Parson (2003).

3. The maximum price considered was $2 per pound, a six-fold increase over the market price. An update of this analysis two years later increased the maximum technically feasible reduction to one-third (Palmer et al. 1980, 14; Mooz et al. 1982).

4. The proposed regulation also included two important innovations: the first proposal for a tradable-permit system and the first proposal for joint control of multiple chemicals according to a common metric of their environmental harm (*International Environment Reporter* 3:8, August 13, 1980, 337; Shapiro and Warhit 1983).

5. The major user industries included manufacturers of cooling equipment, foam products, aerosol products, and fire extinguishing equipment, as well as diverse large-scale users of halogenated solvents, particularly in the electronics, computer, and aerospace industries.

6. Although parties nominate individual expert participants, the chairs may—and sometimes do—reject nominees they judge to be unqualified. The chairs can also find parties willing to nominate experts upon their request (Parson 2003).

References

Alliance for Responsible CFC Policy. 1986. Remarks of Donald Strobach (Science Advisor), EPA Workshop on Demand and Control Technologies, March 7, 1986, Washington, DC.

Clark, William C., Josee van Eijndhoven, Nancy M. Dickson, et al. (11 other authors). 2001. Option Assessment in the Management of Global Environmental Risks. In *Learning to Manage Global Environmental Risks, Vol. 2*, edited by Social Learning Group. Cambridge, MA: MIT Press, 49–85.

Dupont. 1980. Fluorocarbon/Ozone Update. Fluorocarbon Products Division. *DuPont Newsletter*, June.

ENDS (Environmental Data Services). 1989. *ENDS Report* 177: 5.

———. 1996. *ENDS Report* 263: 36.

International Environment Reporter. 1980. Washington, DC: Bureau of National Affairs. 3:9, October 9, 1980, 401.

————. 1981. Washington, DC: Bureau of National Affairs. 4:5, May 13, 1981, 826.

————. 1988. Washington, DC: Bureau of National Affairs. April 13, 1988, 210.

Kuijpers, Lambert, Helen Tope, Jonathan Banks, Walter Brunner, and Ashley Woodcock. 1998. Scientific Objectivity, Industrial Integrity, and the TEAP Process. In *Protecting the Ozone Layer: Lessons, Models, and Prospects,* edited by P. LePrestre, J. Reid, and E.T. Morehouse. Boston: Kluwer Academic Publishers, 167–172.

Mauzerall, Denise L. 1990. Protecting the Ozone Layer: Phasing Out Halon by 2000. *Fire Journal* September/October: 11–13.

Molina, Mario, and F. Sherwood Rowland. 1974. Stratospheric Sink for Chlorofluoromethanes: Chlorine Atom Catalysed Destruction of Ozone. *Nature* 249: 810–812.

Mooz, W.E., S.H. Dole, D.L. Jaquette, W.E. Krase, P.F. Morrison, S.L. Salem, R.G. Salter, and K.A. Wolf. 1982. *Technical Options for Reducing Chlorofluorocarbon Emissions.* Prepared for U.S. Environmental Protection Agency. R-2879-EPA, March. Santa Monica, CA: Rand Corporation.

OECD (Organisation for Economic Co-operation and Development). 1981. *Report on Chlorofluorocarbons.* Environment Committee, March 6 (1st revision). ENV(80)32.

Palmer, Adele R, W.E. Mooz, T.H. Quinn, and K.A. Wolf. 1980. *Economic Implications of Regulating Chlorofluorocarbon Emissions from Nonaerosol Applications.* Prepared for U.S. Environmental Protection Agency. R-2524-EPA, June. Santa Monica, CA: Rand Corporation.

Parson, Edward A. 2003. *Protecting the Ozone Layer: Science and Strategy.* New York: Oxford University Press.

Shapiro, Michael, and Ellen Warhit. 1983. Marketable Permits: The Case of Chlorofluorocarbons. *Natural Resources Journal* 23(July 3): 577–591.

UNEP (United Nations Environment Programme). 1981. An Environmental Assessment of Ozone Layer Depletion and Its Impacts. Nairobi, Kenya: UNEP Co-ordinating Committee on the Ozone Layer.

————. 1982. Paper Prepared by the Secretariat. UNEP/WG.78/5, July 9. Nairobi, Kenya: UNEP.

————. 1986. Report of the First Part of the Workshop on the Control of Chlorofluorocarbons. Rome, October 15, 1986. Nairobi, Kenya: UNEP.

————. 1989a. *Report of the Halon Technical Options Committee.* Nairobi, Kenya: UNEP.

————. 1989b. *Report of the Technology Assessment Panel.* Nairobi, Kenya: UNEP.

————. 1989c. Synthesis Report, Assessment Panels UNEP/OzL.Pro.WG.II(1)/4. Nairobi, Kenya: UNEP.

————. 1991. *Report of the Technology and Economics Assessment Panel.* Nairobi, Kenya: UNEP.

————. 1994. *1994 Report of the Technology and Economics Assessment Panel: 1995 Assessment.* Nairobi, Kenya: UNEP.

————. 1997. Report of the 9th Meeting of the Parties to the Montreal Protocol. UNEP/OzL.Pro/9/12. September 25, 1997. Nairobi, Kenya: UNEP.

U.S. EPA (Environmental Protection Agency). 1982. *Report to Congress: Progress of Regulations to Protect Stratospheric Ozone.* Washington, DC: U.S. EPA Office of Toxic Substances.

CHAPTER 12

Global Hazards and Catastrophic Risk

Assessments in the Reinsurance Industry

Mojdeh Keykhah

*H*URRICANE ANDREW RAVAGED the southern coast of Florida and states abut-ting the Gulf of Mexico in September 1992. It inflicted enormous social and economic losses, prompting the federal government to declare the hurri-cane's path a disaster zone. The hurricane also severely impacted the insurance sector, which had never experienced such a great catastrophic loss from a natural disaster. Those few days in September threw the insurance industry in shock, the insured losses having mounted to more than $16 billion. By comparison, the previous major windstorm in the region had been Hurricane Betsy in 1965, which resulted in losses of about $3 billion (1992$). The gap between the indus-try's perception of the probability of catastrophic loss and the reality of Andrew's destruction was a cautionary case of surprise in the insurance system.

The insurance industry responded to the Hurricane Andrew disaster by limit-ing coverage or charging much higher rates in Florida, but the Florida govern-ment reversed these measures. Another option attempted was to transfer more catastrophe exposures to the reinsurance sector, but reinsurers had raised rates considerably after Andrew and had also suffered major losses. A third choice was the construction of stronger houses and reinforcement of existing ones, although this would have required action by state and local governments as well as by the insurance industry. However, this route involved considerable industry–government coordination. The choice most immediately favored by the insurance industry was a fourth one: to invest in assessments of natural hazards.

This chapter aims to evaluate the nature of reinsurance–expertise interactions in two assessment contexts: that of the catastrophe model, an example of assess-ment as product, and that of a joint industry–academic consortium called the Risk Prediction Initiative, an example of assessment as process. The chapter takes as its definition of assessment "the entire social process by which expert knowl-edge related to a policy problem is organized, evaluated, integrated, and pre-sented in documents to inform policy or decisionmaking" (Global Environmen-tal Assessment Project 1997, 53), and applies it to a private sector case. The chapter argues that in order for environmental assessments designed for the private sector to be effective, they must be directly linked to the practitioners' decisionmaking priorities. Moreover, the salience and credibility of the assessment are key to the analysis of assessment receptiveness among private sector practitioners. The case also presents the evolution of participation in an assessment over time reflecting the priorities of practitioners' needs.

The chapter contains four guiding questions:

1. How do reinsurers of catastrophe risk make use of, and debate on a daily basis, assessments provided by academic science, by private catastrophe modelers, and by Intergovernmental Panel on Climate Change (IPCC) reports?

2. How salient and effective are these assessments within the context of competing priorities of the practitioner? Within different temporal and spatial frames?

3. How are the assessment needs and feedback of the practitioner communicated to the scientific community?

4. What contributes to the effective design of assessments for the private sector?

Catastrophe hazard assessments focus on low probability, high consequence natural hazards. To provide the setting in which reinsurers require scientific expertise, this chapter begins with a brief outline of the focus peril: the Atlantic hurricane risk. It examines the role of reinsurance, and the historical context of catastrophe hazard assessments. It then describes two assessment cases in detail within the scale and temporal constraints of the reinsurance practitioner's decisionmaking routines. Finally the chapter evaluates the main factors contributing to the evolution of the design of the assessments and offers some concluding thoughts on the science–commerce interface.

Catastrophes and Insurance

The role of insurance in catastrophe risk is best seen in the compensation of property losses from natural hazards. The premium, or the price of the insurance policy, relies on an estimate of the probability distribution of the risk—a distribution of the frequency and severity of loss. The insurer relies on loss contingencies affecting different contracts at different times, in order to compensate for losses from policyholders with premiums collected from other policyholders in the same pool. Moreover, such losses should generally be estimable through probability and statistics from past occurrences.

Catastrophes represent the tail of the loss probability distribution, for which consequences are high, but the probability is very low. Providing coverage for this tail is problematic for the primary insurer because losses could include all policies at once, with little or no data on past events to base future expectations. Insurers hedge this extreme case by applying for *reinsurance* to cover losses on all property contracts exceeding a certain amount—in essence they are insuring themselves against the ambiguity in the tail of the loss distribution. Without the existence of catastrophe risk coverage from reinsurers, primary insurers would be loath to provide catastrophe coverage to homeowners and businesses.

Reinsurers, unlike their primary insurance counterparts, provide coverage on a global scale. At the same time, they face a similar problem as do insurers: how to participate in the catastrophe market for which the probability of loss is difficult to assess? Reinsurers hedge the possibility of a catastrophe loss in one region by diversifying the locations in which they have catastrophe risk liabilities. A second strategy is to purchase further reinsurance—an often iterated measure. Thus, catastrophe reinsurers offer catastrophe coverage selectively and purchase reinsurance for themselves in a very dynamic and competitive global reinsurance market.

The total "exposure" of providing coverage to a client (a primary insurer or another reinsurer) includes both the probability of a natural catastrophe striking a particular geographic region and the number, value, and vulnerability to damage of the properties being reinsured. The first is referred to as the cause of loss, and the latter as the condition of loss. Surprisingly, for much of the latter half of the twentieth century, reinsurance assessment of catastrophe risk relied on databases of the cause of loss, but neglected to monitor the changing conditions of loss. Moreover, most reinsurance assessment routines involved estimation of future probabilities through an inductive review of past data (however sparse), framed by loss analogies to a few particularly large events, such as Hurricane Betsy.

The Cognition of Catastrophe

The awareness of catastrophe loss potential was heightened in 1986 by the catastrophe risk assessment put forward by the All-Industry Research Advisory Council (AIRAC), a group sponsored by the U.S. property/casualty insurance industry. AIRAC estimated the adverse effects of two hypothetical $7 billion hurricanes on the U.S. insurance industry in the same year (AIRAC 1986) and found that while the first $7 billion event would damage the financial capacity of insurers and reinsurers, the subsequent $7 billion loss would engender major market dislocations. Nonetheless, the assessment did not prove very effective in changing reinsurance strategies, mostly due to the lack of salience of natural catastrophes as an issue for the industry.

In the year after the AIRAC assessment had been published, however, several catastrophic losses heralded a change in the industry's treatment of catastrophes. In 1987, windstorm "87J" hit the southern coast of Britain (an estimated once-in-300-year event, resulting in $4.4 billion in insured losses), followed by Hurricane Hugo in 1989 in the Caribbean ($5.6 billion), a set of winter gales in Europe in 1990 ($5.9 billion), and Typhoon Mireille in Japan in 1991 ($6.9 billion). While these were serious loss events, they did not induce a frame shift in routines of thinking about catastrophes as such. At the same time, the loss trend was considered by some larger reinsurers as a signal of global climate change (Freedman 1997). Global warming was at the time linked by several global environmental assessments to increased storminess and extreme weather events (Dlugolecki 1996). The quotes below illustrate this early perception by reinsurers of hazards as becoming increasingly unpredictable, thereby exposing this private sector coverage provider to an unwelcome level of vulnerability.

Swiss Re, the second largest reinsurer in the world, cautioned in 1990 about greater liability due to climate change:

> There is a significant body of scientific evidence indicating that last year's record of insurance losses from natural catastrophes was not a random occurrence. Instead it may be the result of climatic changes that will enormously expand the liability of the property–casualty industry (Gordes 1996).

Lloyd's of London Deputy Chairman Dick Hazell, contributed a complementary statement at about the same time:

> There is no reason to expect the recent spate of disasters was just bad luck or statistical oddity. The long term impact of global warming on the world's weather patterns and the incidence of disasters due to manmade constructions or industry pollution may both ensure that a significant number of large-scale catastrophes occur somewhere around the world each year (Gordes 1996).

The clarion call to a new threshold of catastrophic loss sounded with Hurricane Andrew in 1992. While not quite a category 5 (indicating the most powerful windstorm) hurricane, its particular geographic track through southern Florida and the Gulf of Mexico states caused more than $19 billion in insured losses, contributing to the total $24 billion in insured catastrophe losses that year. Several insurance and reinsurance companies collapsed, precipitating a market plunge in available capital to cover future risks. One study surmised that if Hurricane Andrew had traveled only 50 miles farther north into metropolitan Miami, insured losses would have mounted to more than $50 billion, sufficient to pierce the solvency of the industry as a whole (Changnon et al. 1997; Doherty 1997).

Two years after Andrew, California fell victim to the Northridge earthquake, producing $14.1 billion in insured losses. One could only imagine Andrew and Northridge occurring in the same year to realize the extent of liability the insurance and reinsurance sectors were potentially facing. The new "cognition of catastrophe" (Meszaros 1997) by insurers moved the boundary posts of disaster for underwriters; it was a far cry from two $7 billion windstorms given in the AIRAC assessment just a decade earlier. One result of these financial shocks was to expose the Achilles' heel of reinsurance expectation routines based solely on past loss levels, disregarding changes in the conditions of loss. Table 12-1 details the largest insured loss events from natural perils (Swiss Re 2000); all have occurred within the past 20 years. As is shown in the table, the losses highly depend on the locations affected and the degree of private insurance protection involved (as opposed to state protection). The industry responded to this unexpected string of losses by turning to scientific expertise. The industry sought to tame the perceived pure uncertainty, or chance, of these events into a loss trend that could be estimated.

Table 12-1. *Ten Most Expensive Natural Catastrophes for Insurers (1999$)*

Event	Region	Date	Insured loss ($ billion)
Hurricane Andrew	USA	1992	19.060
Northridge Earthquake	USA	1994	14.122
Typhoon Mireille	Japan	1991	6.906
Winter Storm Daria	Europe	1990	5.882
Hurricane Hugo	Puerto Rico	1989	5.664
Winter Storm Lothar	Europe	1999	4.500
Autumn Storm	Europe	1987	4.415
Winter Storm Vivian	Europe	1990	4.088
Hurricane George	USA / Caribbean	1998	3.633
Typhoon Bart	Japan	1999	2.980

Source: Sigma/Swiss Re 2000.

The Atlantic Hurricane Risk

Hurricanes and windstorms are among the greatest loss-generating hazards to the insurance industry. While it may seem that the losses are due to an increase in the frequency of hurricanes, the high loss levels are more reflective of the increased societal vulnerability to catastrophes. Home and business settlements along the Eastern Seaboard of the United States in particular have greatly expanded over the past 25 years, with concomitant pressures on land, resources, and technological infrastructure (Burton et al. 1993; Mileti 1997; Pielke and Pielke 1997). In fact, the frequency of windstorms actually decreased in the twentieth century (Smith 1999).

While a more specific discussion of Atlantic hurricane generation and propagation is beyond the scope of this chapter, it is relevant to note a few factors favoring hurricane activity. Many hurricanes that reach the East Coast originate as storms in the tropical waters near northwestern Africa. For a hurricane to form, it must evolve through particular stages generating strength and transportability through temperature, moisture, and pressure gradients.

Some of these factors in hurricane generation in the Atlantic Ocean depend on whether a climate phenomenon termed the El Niño is present. El Niño refers to a reversal of the heat convection current of the Pacific Ocean that results in warmer sea-surface temperatures. It has been found to be associated with global precipitation and temperature anomalies, called teleconnections (Glantz 1996; Philander 1998). During an El Niño year, the conditions leading to Atlantic hurricanes are suppressed. At the same time, conditions are enhanced for hurricane development in the South Pacific. It is mainly for this reason that El Niño and climate studies in general could very well affect reinsurance losses worldwide. The El Niño cycle is often correlated with a change in sea-surface pressure across the Pacific, leading climate scientists to refer to the joint phenomena as the El Niño/Southern Oscillation (ENSO).

While ENSO years have generally witnessed subdued hurricane activity in the Atlantic, it is worth noting that the worst insured loss to date, Hurricane Andrew, occurred during such a year (Trueb 1998). Thus, it is not simply the total number of hurricanes formed, but the number that make landfall in a particular year that is significant for reinsurers. Andrew also testifies to the importance of the geographic track on land in determining total insured losses. Despite great advances in hurricane research over the past 20 years, including greater understanding of the meteorology affecting hurricanes, measurement of hurricane characteristics in situ, and the development of hurricane forecasts, it is very difficult to predict a geographic-specific track over land (Pielke and Pielke 1997). Indeed, some mathematicians and physicists argue that it is fundamentally impossible, due to the chaotic behavior of weather systems (Casti 1997a, 1997b; Woo 1999). On the other hand, numerous studies describe past decadal clustering of hurricanes impacting a particular landmass such as Florida. Hurricane forecaster and Colorado State University meteorologist William Gray argues that based on historical trends over the twentieth century, the Atlantic basin may be due for another clustering of hurricane activity following a lull in the 1970s and 1980s (Gray et al. 1998). However, can the past provide a suitable projection into the future? Climate modelers do not think so.

During the 1990s, general circulation models (GCMs) have projected an almost inevitable rise in the number of storms and extreme events due to climate change (Dlugolecki 1996). On the other hand, meteorologists and hurricane forecasters remain skeptical of the proposed link between Atlantic hurricane activity and climate change (Henderson-Sellers et al. 1998).

It may be useful to note at this point that while greater understanding of hurricane formation and propagation is the vision of many in the scientific community, from the reinsurance practitioner's perspective, a few generalized features may be sufficient. Reinsurance judgment routines center on questions not of activity per se, but of loss. For example, are El Niño years correlated with lower windstorm losses? If Hurricane Andrew is any guide, the answer is ambiguous due to the heavy dependence of loss on geographic track, as described in research by catastrophe expert Guy Carpenter:

> While the ENSO cycle has a definite correlation with the rate of hurricane formation in the Atlantic, there is little apparent correlation between ENSO and the occurrence of hurricanes making landfall on the U.S. coast. The effect of ENSO on insurance losses is also uncertain (Major 1999).

In the meantime, hurricane forecasts and results from climate simulations have been complemented in the reinsurance practitioner's context by the advent of the catastrophe model.

The Rise of the Catastrophe Model

Catastrophe models as assessment tools for underwriting decisionmaking have rapidly proliferated over the past 15 years. Part of this growth is due to demand wrought by the insurance sector's cognition of catastrophe after the watershed losses of Hurricane Andrew (Meszaros 1997). Another factor in the models' commercial success is due to the advances in computing power to combine and visually display results from thousands of simulations. The catastrophe model integrates mini-assessments of the probability of the peril in a particular geographic region, the vulnerability of the properties concerned (materials, construction, environmental conditions) to the peril, levels of intensity and duration of the peril, the insured values in the insurer's or reinsurer's portfolio, and the terms of the insurance or reinsurance policy (such as deductibles, reinstatements, and exclusions, that affect the liability of the reinsurer). The product thus combines hazard, vulnerability, and insurance variables. As the models handle a number of mini-assessments of environmental and societal factors, they are comparable to integrated assessments in global environmental change research (Rotmans 1998).

To provide an output of the probability of losses exceeding a certain level, catastrophe modelers use two different approaches, one mainly deterministic and the other probabilistic (Kozlowski and Mathewson 1997). In *deterministic* modeling, historical data on a particular hazard such as hurricanes serve as a baseline for trends in a region. The model then uses these data to simulate a number of specific (past) events. The technique in *probabilistic* modeling is to run many hypothetical events covering a range of possible outcomes. These two methods allow the modeler to assess the probability and severity of loss and to derive a distribution of exceedence probability (the chance that a loss will exceed a certain level). While the output is technologically sophisticated and visually stunning, catastrophe models contain a number of assumptions and, some may argue, omissions both at the level of global climate variables and local conditions.

For example, acquiring relevant data for the models can be extremely difficult, notably if past weather conditions have not been well recorded, if the reliability and quality of data are questionable, or if data are scattered. Moreover, climate data can be, as one catastrophe modeler exclaimed "fiendishly expensive" (Beatty 1998). When data are not directly available, deductive inferences are sometimes made using contingent data. For example, one catastrophe modeling firm digitized six-hourly pressure maps from the UK Meteorological Office to derive windspeeds at high altitudes. Cynical modelers advise the adage "rubbish in, rubbish out" as applicable to catastrophe modeling as any other modeling enterprise (Beatty 1998). The 1995 IPCC report itself mentions the problem of inhomogeneities in historical data for catastrophe modeling purposes (Dlugolecki 1996).

Underestimating the uncertainties involved is a grave mistake, counsel two catastrophe modelers from Risk Management Solutions, a private risk modeling company (Boissonnade and Collignon 1999). They take the example of their tropical cyclone model and four hurricane forecasts used as its basis, which were provided by William Gray of Colorado State University, James Elsner of Florida State University, Mark Saunders of University College London, and the NOAA-CPC/National Hurricane Center. The modelers remark that the forecasters themselves have spotty success records, and that the accuracy of predicting U.S. landfalling hurricanes is inconsistent. On the other hand, improvements to the resolution of the models could require "an exponential increase in the amount of data, processing power, and money required to build the model" (Beatty 1998). Other sources of uncertainty include assumptions from the raw data—provided in technical estimates of material behavior under stress—to be included in the vulnerability component of the model. For example, the steel frame structures during the Northridge earthquake did not withstand the level of disturbance as had been assumed they would.

Complicating the uncertainties involved in catastrophe models are their proprietary status: they have been termed "black boxes" because their source codes, data, assumptions, and uncertainties are generally kept out of the practitioner's eye. A major foundation for their credibility lies in their assumed ability to predict loss values in particular regions due to different perils. The attractiveness of these models is their capacity to translate scientific data to the temporal and spatial decision frame of the practitioner, including the context of the reinsurer's own portfolio. To evaluate the effectiveness of this kind of assessment as product, a brief description of the temporal and spatial contexts of reinsurance decision-making is provided in the next section.

The Context of the Reinsurance Practitioner

Reinsurers operate within specific temporal frameworks that are common throughout the industry and serve to coordinate key market activities. One important reference is the renewal season for the annual catastrophe reinsurance contract. North American reinsurance contracts, for example, are reviewed and renegotiated each January. Therefore, to have relevance to decisionmaking in reinsurance, environmental assessments must arrive to the reinsurer in advance of the renewal season. The periods are characterized by intense contract negotiations between underwriters and brokers, who mediate risk information from clients.

Spatially, reinsurers seek to diversify accumulation of catastrophe risk in order to avoid being too greatly exposed to loss from a particular region. Reinsurers thus choose to accept risks from particular geographic zones (e.g., exposure to California earthquakes or Florida windstorms) and at the same time choose risks in less disaster prone locations (e.g., continental Europe). However, the recent destructive storms Lothar (in France) and Martin (in Germany) illustrate the perpetual possibility of surprise as to the next major loss for reinsurers. Therefore, a spatial distribution of risks, while limiting potential loss, is not a foolproof method for preventing catastrophic liabilities from reaching the reinsurer.

The Bermudan Context

After the losses from Hurricane Andrew precipitated the collapse of a number of insurance and reinsurance companies, eight specialized reinsurance companies established themselves on Bermuda. They situated themselves on this Atlantic island due to its thriving financial services infrastructure, made more attractive by its tax status. Funded by mostly U.S. and British capital providers, the companies formed an international reinsurance hub dedicated to catastrophe reinsurance. One of the distinguishing characteristics of these small firms was their open search for and incorporation of hazard assessments. Within a few years, these new companies claimed a quarter of the global catastrophe reinsurance market, abetted by the drain in international reinsurance capital after Hurricane Andrew. At the time, demand for catastrophe reinsurance outweighed supply, permitting a large profit base and good returns for shareholders. However, as London and other reinsurance centers began to recover from Andrew, and as new market entrants increased the available capital to underwrite catastrophe risks, rates began to decline, and profitability was more difficult to achieve. By 1999, most of the original eight reinsurance companies had either merged, shut down, or diversified to include other lines of reinsurance.

The salience of hazard assessments reached a peak in the years following Hurricane Andrew. At the same time, catastrophe modeling companies were prepared with an assessment product that communicated directly to the perceived needs of the reinsurance community, particularly the Bermudans. The models as carriers of extended databases of loss were already an improvement, in addition to their exposure monitoring capabilities. As simulators of future catastrophic events, the models also provided substantial scientific expertise that until that time was very unevenly divided among the reinsurance community as a whole (mostly depending on internal technical capacity).

At first, the models' black box proprietary aspects were not considered as much as their compatibility with the numbers they could provide reinsurers. Indeed, in some early cases the models were taken as "truth machines," capable, with the input of teams of scientists in the catastrophe modeling firms, of providing robust probabilities to counter the uncertainty of the future. However, unlike climate models, which have long lead times, catastrophe models can be verified comparatively quickly, for example with the Northridge earthquake, the winter storms in Canada, Hurricane Floyd, Winter Storm Lothar, and Hurricane Hugo. Thus, a learning process took place, not unlike the evolution of expectations about predictability of global circulation models among policymakers during the early 1990s (Henderson-Sellers et al. 1998; Shackley 1997; Shackley and Wynne 1996).

Such experiences, mostly involving verification of the model against catastrophe events, promoted greater internal capacity among reinsurers to become more reflective model users. The three main model providers had different approaches, and loss estimates belonging to certain key catastrophe events (such as a Miami hurricane or a San Francisco earthquake) began to assume an anchoring quality in characterizing the overall quality of a particular model. There is a parallel process of distinguishing among GCMs by policymakers through the models' estimates of future temperature rise (van der Sluijs et al. 1998). The credibility of the catastrophe models, while moving through a learning cycle of expectations of the science, did not diminish the models' placement in the reinsurance tool box. As one reinsurance underwriter remarked, "What else do we have?" In fact, some expressed a modest optimism that as catastrophes continued to occur, the models could only become better calibrated. In addition to purchasing third party models, the Bermudan reinsurers have also created models in house in part to test against the black boxes. While interest remained high among reinsurers for this assessment product, the degree of openness to different sources of expertise began to change. Logistical and technical infrastructures in the firms began to form around catastrophe models and their formalized input requirements. As the models gained greater market presence, conferring substantial informational benefits in a translation of the science to reinsurance decisionmaking frames, other sources of "raw" expertise diminished in applicability. These included rough estimates of return period forecasting, the broker's file on the risk, or rough and ready assessments of hazards as available in the publications available through CRESTA (Catastrophe Risk Evaluating and Standardizing Target Accumulations), an organization jointly managed by Munich Re and Swiss Re. In effect, the models complement and to some extent challenge these older forms of underwriting expertise, leading to a type of "knowledge lock-in" in the specifications of the model.

This process is analogous to Sante Fe Institute economist Brian Arthur's observation of "technological lock-in" (Arthur 1989)—that is, the tendency of economic systems to adopt technologies through historical contingency and of the directions of future technologies to be strongly influenced by what technologies penetrate the market early on. As with technological lock-in, when catastrophe companies adopt a model, the firm congeals a particular entry of hazard assessment to the reinsurer, molding data requirements and evaluation methods to conform to a market-wide standard. Indeed, the visibility and the sophistication of the technology has led to its adoption by many overseas reinsurers. Keeping in mind the numerous uncertainties involved in catastrophe modeling, the standardization of catastrophe risk assessment expertise through the models promotes their transportability among different firms and in different areas of the same firm.

At the same time, the technical capacity of reinsurance practitioners to judge models and use them with appropriate caveats may range markedly from one firm to another (Golnaraghi 1997). This variation is mostly due to the highly technical and knowledge-intensive nature of the models, the lack of transparency of the model design to third parties, and the nonscientific training of most reinsurance underwriters.

Similar to climate modeling outcomes and their roles in international negotiations, catastrophe models and the assessments provided should be viewed within the caveats of modeling endeavors generally (Casti 1997a). The simulation by the model is a representation of reality, and it is constructed on the foundations of assumptions of the scientists involved, the modeling translation, the epistemologies and methodologies brought to bear, and the contemporary understanding of the physical systems (Casti 1997b; Morgan and Morrison 1999). Indeed, Munich Re geophysicist Ernst Rauch cautions against taking the models as literal forecasts:

> It is important to understand that there is a big difference between catastrophe prediction and catastrophe modeling. The term "prediction" is very often used in a misleading way, as it is impossible for anybody to predict catastrophes by means of the exact location, time, and intensity of the event. "Catastrophe modeling" allows the insurance industry, based on scientific and statistical methods, to have a technical, sound understanding of loss occurrence probabilities from a long-term perspective (Dowding 1998).

One main role of the models has been to provide a "technical" price for a particular risk, in order to compare to market rates and also sometimes to justify certain corporate strategies. For example, some reinsurers may abstain from accepting risk contracts they consider underpriced, according to model evaluation. At the same time, competition in catastrophe reinsurance pricing limits opportunities for the model recommendations to significantly affect market decisions. Therefore, similar to political priorities vitiating recommendations based on climate models, the outputs of catastrophe models could be ineffective in the face of market pressures.

The importance of continuity, relationships, and market share strategies also plays a large role in reinsurance practices. Much of the catastrophe reinsurance market relies on client commitment to long-term risk sharing. In light of these contextual considerations, it may be said that the models serve a stronger role as exposure monitors and simulation enablers than as market pricers or as handmaidens to decision outcomes. This role is consistent with much of the Global Environmental Assessment Project's findings that past environmental assessments have been most effective in shaping the perception of knowledge needs and the terms of the debate rather than changing policy decisions (Mitchell et al. forthcoming).

The Risk Prediction Initiative

The Risk Prediction Initiative (RPI) is an industry–academic research consortium formed soon after Hurricane Andrew as part of the Bermuda Biological Station for Research. At the time of its founding, interest among reinsurers for catastrophe assessments and market conditions generating substantial returns for reinsurers had both reached a crest. Thus, there existed great momentum in providing a financial infrastructure for this type of expertise–reinsurance interaction. The RPI framed its approach as the "open" alternative to catastrophe models:

> Existing risk models use different proprietary techniques. Because these techniques are proprietary, it is impossible to determine how the assumptions they employ affect their quality. Yet models using different proprietary techniques often generate significantly different losses for the same insurance portfolio. Some consistency or public validation of these risk models would help model users justify, both to themselves and their clients, that the model-derived decisions they make are valid. Because it is difficult to persuade competing companies to make their models public, it may be expedient to create a public, peer-reviewed wind model. Insurers could use this model as a baseline to compare their various proprietary data sets. Equal access among insurers to a public, peer-reviewed model would improve confidence in existing models (Malmquist 1997, 36).

RPI serves as both an interface and a pooling body: it collects funding streams from member reinsurance companies and several public sources, and it contracts work to various scientific researchers at the Bermuda Biological Station for Research as well as researchers in the United States. The research focus has been on improving the ability to forecast tropical cyclones. Part of the RPI research

Table 12-2. *Areas of Research for the Risk Prediction Initiative*

1.	Improve basic understanding of tropical cyclones and their link to climate
2.	Improve the skill of seasonal and multi-year forecasts of tropical cyclone landfall probabilities
3.	Determine fidelity of various proxies for tropical cyclone landfalls
4.	Create global landfall teleconnection maps
5.	Create/compile public data sets related to landfall of tropical cyclones in key cities/regions
6.	Support improvement and comparison of physical models for seasonal to interannual forecasting of global tropical cyclone landfalls
7.	Participate in and sponsor special sessions at scientific meetings

mission is provided in Table 12-2. As the table notes, much of the work is directed at the refinement of the science with little if any formal translation to insurance practices. Because many scientists working on these projects come from academia, there is a significantly different character to this interface than that of the catastrophe modeling firms. From the wish lists of their majority funders (the reinsurers), the scientists negotiated as to which scales of analysis and data sets were most relevant to pursue, while the incorporation of the results in the reinsurers' daily decision routines was discussed less thoroughly. Indeed, researchers at the RPI viewed their commitment to academic science (funded by reinsurers) in part as the production of peer-reviewed articles and attendance at conferences. The reinsurers, on the other hand, were less interested in the polished standard of proof of the academic article. They sought strategic expertise—that is, results that would provide a greater standard of accuracy than already existed. The first "model" of interaction, that of production of academic science for reinsurance, went through a learning cycle in which the RPI realized that the interface itself was as much a concern as the production of the science.

The expertise–reinsurance interactions were usually concentrated in workshops held at the RPI, in which numerous academic hurricane and climate experts would be invited to present their views. Prominent hurricane scientists, such as Bill Gray from Colorado State University, Jim Elsner of Florida State University, Christopher Landsea of the National Hurricane Center, and Kerry Emanuel of Massachusetts Institute of Technology, offered different expert opinions on hurricane dynamics. It was not rare for the reinsurers assembled to witness a face-to-face contested discussion on the science of hurricane formation or the forecasting of future Atlantic hurricane activity. Indeed, the RPI has become a conduit for reinsurers to gain direct access to the pluralism that exists within a scientific field. While the greater openness in participation and sources of expertise in the RPI increases the legitimacy of the assessment, inclusion of multiple views within the sciences could decrease its credibility.

Since its inception in 1992, the RPI has been exhorted by its reinsurance funders toward ever greater specificity to reinsurance needs in its research products, for example from a focus on forecasting hurricane activity in general to forecasting landfalling hurricanes specifically. Moreover, reinsurers have pressed to obtain such forecasts by November of the preceding year to accommodate the reinsurance renewal season for North American contracts. As reinsurers' knowledge needs have changed or become more specific, the RPI has seen participation of scientists change over time. For example, new research initiatives emphasize the vulnerability of different building materials to hurricane intensities, engaging the research of engineers in addition to earth scientists.

The short-term time frame of reinsurance favors rapid research products that suit a very particular decision need. To a certain extent, the perfection of the research result is not as important as its user access. Thus, the level of acceptable uncertainty in academic science and reinsurance differ, with reinsurance tolerant of greater relative uncertainty, and academic research stressing that propositions should be nearly perfect—that is, as certain as possible. Dissent, such as the lack of consensus among hurricane and climate scientists, is treated as a part of the "evolving science" and therefore from the reinsurance perspective, it is bound to have such controversies. It is also important to note that RPI reinsurers are competitors participating in the same pool of knowledge, learning all the while of the complexities involved in scientific assessments for industry. Moreover, participation as RPI members is used as a credibility credential for reinsurers in relationships with brokers, primary insurers, and other reinsurance companies. The effectiveness of the RPI for reinsurers depends on the structuring of the science–practitioner interface and on the integration of the basic research into wind models and other products similar to those reinsurers already use.

Conclusions

The questions asked at the outset of this chapter have been addressed with respect to the salience and credibility of two forms of assessments, the catastrophe model and the Risk Prediction Initiative. The assessments have proved salient and credible among users because they were shaped by practitioner interest, openness, and capacity. Reinsurance interest in strategic hazard expertise propelled the development of the assessments, and the practitioners themselves were open to different sources and opinions among scientists. Moreover, capacity to evaluate and integrate the assessment products has greatly improved since Hurricane Andrew and has led to the development of the RPI. Both types of assessments are also molded by the procedures of participation and the handling of uncertainty. While the RPI's changing funding priorities for landfalling hurricanes shifted the selection of research scientists funded, the participation of expertise in the catastrophe modeling firm has come to include more former reinsurance practitioners. As for the uncertainties involved, reinsurers compare the scientific uncertainties in their assessments of the Atlantic hurricane risk with the other domains of uncertainty in which they operate, not as ends in themselves; therefore, practitioner tolerance of uncertainty differs substantially from that of scientists providing input to the assessment.

Of the different forms of scientific expertise available to reinsurers, the greatest direct incorporation in reinsurance decisionmaking has been through the catastrophe model, which functions both as a simulator and a monitor of reinsurance exposure worldwide. Its effectiveness in the market is mainly due to the ease with which scientific expertise is translated into the insurance industry's languages of finance and statistics (Elsner and Kara 1999). At the same time, cycles of capital capacity in the market directly influence the degree to which the rein-surance underwriter can take scientific expertise into account. In other words, capital movements seem to influence the effectiveness of private environmental assessments. In contrast, information provided by IPCC assessments were not used by the reinsurers interviewed, very likely due to the differing temporal and spatial resolution of global circulation models and the fact that such models have come under increasing skepticism since changes have been made in estimates of storminess over several IPCC reports.

The case of the catastrophe modeling company provides other insights. The companies employ Ph.D. scientists to contribute to the modeling of hazard probability and the vulnerability of the properties concerned. The underwriters take the models "with a grain of salt" but at the same time, the dissemination of the models across the industry serves to keep the salience of catastrophe risk in the minds of practitioners. There is a certain recognition that the industry as a whole is operating with the same tools, the same uncertainties, and the same omissions. If the models do not converge, or if they are otherwise inconsistent with respect to the input data provided by reinsurers, then the general perspec-tive is that the models will improve over time.

The RPI represents a different kind of assessment; that of assessment as process. The first joint industry–academic workshops of the RPI expressed opti-mism toward the incorporation of science for reinsurers. Workshop participation of prominent academic scientists was outstanding, with reinsurers coming into direct contact with a range of scientific approaches to hurricanes and climate. The workshop also highlighted the contrast between the short term and very directed focus of reinsurers with the long-term and knowledge-building approach of the academic scientists. However, the public nature of the RPI has improved the capacity of the reinsurers to know the range of expertise in exis-tence, and to evaluate expertise as it is given within this range. For example, through the debates on hurricane science they have witnessed in workshops, the reinsurers now have reference points of academic knowledge to compare and contrast in the formation of their recognition of "the science." On the other hand "the science" is broadly understood to be developing, with increased pre-dictability as its goal.

In terms of affecting pricing decisions (i.e., change in decisionmaking out-comes), the two types of assessments are limited by the severity of market compe-tition. The dominance of relationships to brokers and clients and the loss-driven mentality of reinsurers does little to promote effectiveness toward pricing deci-sions in the market as a whole. However, assessments do influence the perception of knowledge needs by the reinsurers and the terms of the strategic expertise debate. And assessments can lead to the use of financial risk mechanisms such as securitization, which is based on the indexing of natural catastrophes.

Reinsurers use a number of communication processes to provide feedback to the assessment providers. In fact, reinsurance practitioners are sometimes hired full time by catastrophe modeling firms, promoting the integration of knowledge from the practitioner's perspective into the scientific models. On the other hand, feedback from practitioners themselves is not as frequent or detailed as in the RPI process, relative to the assessment processes. Through the generation of "wish lists" from reinsurers, the RPI is able to provide a flexible baseline for reinsurance needs for specific expertise. RPI has also obtained information through conferences in which scientific societies communicate their expertise to reinsurers. Many such exchanges also happen informally in RPI-sponsored workshops and at national insurance gatherings.

In response to the four questions raised at the beginning of this chapter, it is worthwhile to regard the reinsurer's approach toward expertise as a combination of familiarity with particular people and a resonance with the practitioners' frames of reference. Most, if not all, the reinsurers interviewed for this chapter were familiar with the prominent hurricane scientists who had participated in RPI workshops and conferences. Many reinsurers had met these experts in person. Their status as hurricane gurus is not established among the reinsurers because of expert prediction, but because they are seen as scientists with differing "expert opinion." Thus in the RPI assessment context, the participation of the hurricane forecasters has become in a sense not simply to provide external perspectives, but also to be a part of the dynamics of the assessment itself.

This study of reinsurance practitioners underscores the point that to reach maximum effectiveness, the assessment design must meet the practitioners' decision needs, frames of reference, and scales of analysis. For example, the first priority of the reinsurance company is not to understand the climate or the generation of a hurricane. It is to produce end-of-year results that outperform the competition and to maintain client continuity and pricing stability. Thus, hurricane *losses,* not numbers, are most relevant. In addition, care should be taken to ensure that the interface is sensitive to the evolving needs of the user community. Learning takes place within the process of an assessment through dynamic interaction among the participants, and thus the interface needs to be flexible.

References

AIRAC (All-Industry Research Advisory Council). 1986. *Catastrophic Losses: How the Insurance System Would Handle Two $7 Billion Hurricanes.* Oak Brook, Illinois: All-Industry Research Advisory Council.

Arthur, W.B. 1989. Competing Technologies, Increasing Returns, and Lock-in by Historical Events. *The Economic Journal* 99(394): 116–131.

Beatty, Alex. 1998. Can Modellers Create Virtual Acts of God? *Reinsurance* March.

Boissonnade, Auguste, and Oliver Collignon. 1999. Insurers Braced for Ill Winds. *Reactions* July.

Burton, Ian, Robert Kates, and Gilbert White. 1993. *The Environment as Hazard.* New York: Guilford Press.

Casti, John. 1997a. *Would Be Worlds: How Simulation Is Changing the Frontiers of Science.* New York: Wiley.

———. 1997b. Complexities of Global Warming: What Scientists Don't Know and Why They Don't Know It. *The Washington Post,* November 30.

Changnon, Stanley A., David Changnon, E. Ray Fosse, Donald C. Hoganson, Richard J. Roth Sr., and James M. Totsch. 1997. Effects of Recent Weather Extremes on the Insurance Industry: Major Implications for the Atmospheric Sciences. *Bulletin of the American Meteorological Society* 78(3): 425–435.

Dlugolecki, Andrew. 1996. Financial Services. In *Climate Change 1995: Impacts, Adaptations and Mitigation of Climate Change: Scientific–Technical Analysis,* edited by Robert T. Watson,

Marufu C. Zinyowera, and Richard H. Moss. Cambridge, United Kingdom: Cambridge University Press.

Doherty, Neil. 1997. Insurance Markets and Climate Change. *Geneva Papers on Risk and Insurance* 83.

Dowding, Tony. 1998. Preparing for the Worst. *Reinsurance* April.

Elsner, James B., and A. Birol Kara. 1999. *Hurricanes of the North Atlantic.* New York: Oxford University Press.

Freedman, Susan A. 1997. *Global Climate Change and Its Policy Implications for the U.S. Insurance Sector.* Master's thesis. Newark, DE: Center for Energy and Environmental Policy, University of Delaware.

Glantz, M.H. 1996. *Currents of Change: El Niño's Impact on Climate and Society.* Cambridge, United Kingdom: Cambridge University Press.

Global Environmental Assessment Project. 1997. *A Critical Evaluation of Global Environmental Assessments: The Climate Experience.* Calverton, MD: Center for the Application of Research on the Environment (CARE).

Golnaraghi, M. 1997. *Applications of Seasonal to Interannual Climate Forecasts in Five U.S. Industries: A Report to the NOAA Office of Global Programs.* Brookline, MA: Climate Risk Solutions.

Gordes, Joel N. 1996. *Climate Change and the Insurance Industry.* Riverton, CT: Environmental Energy Solutions. http://home.earthlink.net/~jgordes/CLIMATE-earth.PDF (accessed January 31, 2005).

Gray, William M., Christopher W. Landsea, John A. Knaff, Paul W. Mielke Jr., and Kenneth J. Berry. 1998. Summary of 1997 Atlantic Tropical Cyclone Activity and Verification of Authors' Seasonal Prediction. Working paper. Fort Collins, CO: Department of Atmospheric Sciences, Colorado State University.

Henderson-Sellers, A., H. Zhang, G. Berz, K. Emanuel, W. Gray, C. Landsea, G. Holland, J. Lighthill, S.-L. Shieh, P. Webster, and K. McGuffie. 1998. Tropical Cyclones and Global Climate Change: A Post-IPCC Assessment. *Bulletin of the American Meteorological Society* 79(1): 19–38.

Jäger, Jill. 1998. Current Thinking on Using Scientific Findings in Environmental Policy Making. *Environmental Modeling and Assessment* 3(3): 143–153.

Kozlowski, Ronald T., and Stuart B. Mathewson. 1997. A Primer on Catastrophe Modeling. *Journal of Insurance Regulation* 15(Spring): 322–341.

Major, John. 1999. The Uncertain Nature of Catastrophe Modelling. In *Natural Disaster Management,* edited by Jon Ingleton. London, United Kingdom: Tudor Rose Holdings. http://www.ndm.co.uk/ (accessed January 31, 2005).

Malmquist, David L. (ed.). 1997. *Tropical Cyclones and Climate Variability: A Research Agenda for the Next Century.* Los Angeles: Risk Prediction Initiative.

Meszaros, Jacqueline R. 1997. The Cognition of Catastrophe: Preliminary Examination of an Industry in Transition. Working paper. Philadelphia: Wharton Risk Management and Decision Processes Center.

Mileti, Dennis S. 1997. *Disasters by Design.* Washington, DC: Joseph Henry Press.

Mitchell, Ronald, William C. Clark, David Cash, and Nancy Dickson. Forthcoming. *Global Environmental Assessments: Information and Influence.* Cambridge, MA: MIT Press.

Morgan, Mary S., and Margaret Morrison (eds.). 1999. *Models as Mediators* Cambridge, United Kingdom: Cambridge University Press.

Philander, George. 1998. *Is the Temperature Rising?* Princeton, NJ: Princeton University Press.

Pielke, Roger A. Jr., and Roger A. Pielke. 1997. *Hurricanes: Their Nature and Impacts on Society* New York: Wiley.

Rotmans, Jan. 1998. Methods for IA: The Challenges and Opportunities Ahead. *Environmental Modeling and Assessment* 3(3): 155–179.

Shackley, Simon. 1997. Trust in Models? The Mediating and Transformative Role of Computer Models in Environmental Discourse. In *International Handbook of Environmental Sociology,* edited by Michael Redclift and Graham Woodgate. Northampton, MA: Edward Elgar.

Shackley, Simon, and Brian Wynne. 1996. Representing Uncertainty in Global Climate

Change Science and Policy: Boundary—Ordering Devices and Authority. *Science, Technology and Human Values* 21(3): 275–302.

Smith, Eddie. 1999. Atlantic and East Coast Hurricanes 1900–98: A Frequency and Intensity Study for the Twenty-First Century. *Bulletin of the American Meteorological Society* 80(12): 2717–2720.

Swiss Re. 2000. Natural Catastrophes and Man-Made Disasters in 1999. *Sigma* 2/2000. Zurich, Switzerland: Swiss Re.

Trueb, Juerg. 1998. *El Niño 1997/98: On the Phenomenon's Trail.* Zurich, Switzerland: Swiss Re.

van der Sluijs, Jeroen, Josee van Eijndhoven, Simon Shackley, and Brian Wynne. 1998. Anchoring Devices in Science for Policy: The Case of Consensus around Climate Sensitivity. *Social Studies of Science* 28(2): 291–323.

Woo, Gordon. 1999. *The Mathematics of Natural Catastrophes.* London, United Kingdom: Imperial College Press.

CHAPTER 13

Making Sustainability Assessments More Useful for Institutional Investors

Bernd Kasemir, Andrea Süess, and Raphael Schaub

PENSION FUNDS CONTROL a significant part of the economies in many developed countries by their stock holdings, and they are thus an essential institution for making large corporations accountable to the public (see, e.g., Hawley and Williams 2000). By definition, pension funds must take a long view with regard to investment strategies, and this has already contributed to the integration of sustainability criteria into the asset management policies of some of these funds. How is this currently done in North America and Europe, and what future trends can be expected? This chapter discusses how sustainability assessments are currently used by institutional investors, and it suggests how future assessments could be improved for their use by investment professionals.

In discussing this question, it is important to consider that today's environmental problems are more complex and more interdependent than the individual problems tackled in the past decades, such as acid rain or stratospheric ozone depletion. Rather, many current individual environmental and social issues are better viewed as multiple dimensions of a common problem. That is the challenge of achieving "sustainable development," in which each generation manages to meet its needs while restoring and nurturing the planet's life support system (Clark 2001). In this unprecedented journey toward sustainability, we have to rely on incomplete information that evolves as we progress with our expedition (National Research Council 1999).

For assessments of the progress toward sustainability to be effective, sustainability researchers have to be aware that modern societies don't have a single central decisionmaker who needs a single type of sustainability information. As Chapter 1 points out, environmental assessments are increasingly important in supporting "decisionmaking for businesses, local and national governments, and international arenas." In discussing the design needs for sustainability assessments by a specific user group within the private sector,[1] namely pension fund investors, this chapter complements other chapters of this volume targeted more at information users in public policy. It also adds to the discussion in Chapter 12, which focuses on private companies as producers of assessment information.

Pension fund investment is an interesting arena that is somewhat between the purely private sector and the public interest domain. As Peter Drucker observed a quarter of a century ago for the case of the United States, employees through their pension funds had already by then assumed a controlling stake in nearly all major companies on the stock market (Drucker 1976). Pension fund managers, with their long term perspective, are an important group in the issue domain of sustainability, as they have a high potential for putting sustainable development on the agenda of large corporations.

Social and environmental issues are already taken into account to a surprising degree in the management of pension assets, and this trend is expected to increase over time. In principle, this is good news for the proponents of sustainable development. The bad news for sustainability researchers is that these investors make little use of academic sustainability assessments. This chapter discusses this lack of attention to assessments and explores how improved designs of future sustainability assessments could make them more useful for institutional investors. Such improved analysis is important, as assessments could help increase the transparency of sustainability information available to the investment community, which is a key element determining the growth potential of sustainable asset management.

The main research questions for the study discussed in this chapter pertain to three linked issues: Why are sustainability criteria integrated into pension asset management today? Which tools and approaches are used for this integration by pension fund managers and by external asset managers working for them? Which elements in the design of future sustainability assessments would make them more useful for application in institutional investment? In discussing these questions, the legitimacy, credibility, and salience of assessments are considered.

The study discussed in this chapter targets the use of sustainability criteria in pension fund investment both in the context of "active" investment, where managers try to optimize the performance of their portfolio by selecting particular high-growth or low-risk shares, and "passive" investment, which aims to mirror the performance of specific stock indices. To capture views of investment professionals on both sides of the Atlantic, interview and questionnaire responses were gathered from pension fund managers and external investment experts in eight research regions throughout Europe and North America (Table 13-1).

The European regions selected include the United Kingdom, the Netherlands, and Switzerland, which hold the highest pension fund assets within Europe (in this order); France was also selected because it is an interesting case on the other extreme in that the private sector plays a very limited role in the occupational pension system. The research regions in the United States and Canada were selected to cover different parts of North America representing diverse cultural traditions with potential influence on practices of responsible asset management.

Interviews were conducted with pension fund managers and external investment experts known to be progressive on this issue, because we were interested in understanding not only current practices but also possible trends for the future. The interviews lasted about one hour each, focused on qualitative issues, and were conducted using a combination of semistructured and narrative inter-

Table 13-1. *Regions Where We Collected Information from Institutional Investment Professionals on their Current Practices and Future Needs*

Europe	North America
UK	Northeastern US
Netherlands	California
Switzerland	Ontario
France	Quebec

view techniques.[2] They were then taped, transcribed, and analyzed for typical patterns appearing across the different interviews. In addition, a questionnaire study was conducted with a wider sample of pension fund managers and external investment experts in the same research regions. These questionnaires addressed motivations, current practices, and future requirements for taking sustainability into account (or not) in institutional investment. In the following empirical sections, points of view of pension fund managers and external investment experts that were found to be typical from the interviews and questionnaire responses are illustrated with interview quotes. To guarantee anonymity for the study participants, we applied the "Chatham House Rule of Confidentiality"—meaning that statements by study participants are quoted in a manner that does not allow the connection of any particular statement to a particular person or organization.

A list of participants who explicitly agreed to be named as interviewees or questionnaire respondents is given in the Acknowledgments section of this chapter. Overall, more than 70 pension fund managers and external investment experts from North America and Europe participated in our study.

Why Are Sustainability Criteria Taken into Account in Pension Fund Investment?

Within the investment field, pension fund investors in particular have reason to be interested in sustainability issues. They must focus on the long term, because they have liabilities for a number of decades ahead, as employees paying contributions today expect pension benefits when they retire. While individual investment decisions are not planned with quite the same time horizon as overall liabilities, pension fund managers have a rather long term investment perspective, for example on the scale of a decade.

> Long term we've got liabilities that might go 50, 60 years ahead. We don't think as far as that ahead. But we do have somewhat of a 10-year plan, which is reviewed every 3 years.
> —Pension fund representative, United Kingdom

A long-term horizon in principle goes together well with sustainability goals of preserving the long-term viability of natural and social systems. Some interviewees from pension funds explicitly told us that preserving living conditions in the long term is part of the value system of their funds, using a language of intergenerational responsibility that seems well suited to the context of pension fund management. However, more often pension fund managers see themselves as having obligations specifically to the members or beneficiaries of their funds, rather than to society or nature as a whole: "by law . . . we're not in the positions to say, well, X investment will benefit mankind. If it's not going to benefit our beneficiaries specifically in terms of making money for them, then we can't do it."

That said, the degree to which sustainability concerns already figure in the management of pension assets is surprising. One reason for this level of concern is that spectacular environmental disasters such as the Exxon-Valdez oil spill have had grave consequences for pension funds, which are major shareholders in large companies. Including sustainability aspects in their asset management criteria helps pension funds and external asset managers working for them to protect their investments from environmental or social risks, and it helps them avoid an image loss not only with regard to external stakeholders but also specifically with regard to the pension plans' members or beneficiaries. In addition, some pension fund managers and external asset managers begin to think that sustainability criteria can help them in selecting high-growth stock by choosing sustainability leaders. Investment managers increasingly believe that environmental and social performance by corporations can be an interesting proxy for selecting companies that are well managed overall.

The consideration of environmental and social criteria in a pension fund's investment decisions is not in all cases prompted by personal preferences of the funds' managers for this. This is illustrated by the example of one of our interviewees, who was initially highly skeptical when his board urged him to include sustainability aspects into the management of his assets. However, working with sustainability criteria, and specifically with sustainability funds provided by external asset managers, taught him a great deal and changed his expectations fundamentally.

> Let me be frank. I have, as I said, such a clause in the investment guidelines, which urges me to occupy myself with [these sustainability criteria]. And then I thought, we'll do this once for a year, and we'll see that it was worthless. That was my opinion. And with this I went and looked at the [sustainability] funds that are around and compared them to each other, bought one or the other, and then surprise, surprise, the performance was quite positive! And above all, the past performance was pretty positive. And then it began to interest me. . . . Of course, this was also a learning process for me and the people around me.
> —Pension fund representative, Switzerland

Concerning the content of sustainability criteria considered in managing pension assets—either by acquiring or selling specific stocks or by engaging with the management of companies the fund owns shares in—social issues are included just as much as environmental questions.

> When a company is . . . imposing, let's say, economically irrational criteria for hiring—that is to say they are not going to hire people who are black

or Hispanic or whatever else—they are cutting themselves off from the services of potentially talented employees. . . . That, we feel, can lead to less efficient business operations. Less efficient business operations means less profit, less profit means less dividend, the stock value can go down; once again we're hurt by that. So when we oppose discrimination on the basis of race here in the United States, . . . all of this activity is related to the bottom line.

—Pension fund representative, Northeastern United States

As in this case, pension funds that include sustainability criteria into their asset management policies usually consider sustainability in investments to be related to the "triple bottom line," which is expressed in the terms environmental/ social/financial, or more poetically, planet/people/profits. This integrated approach includes environmental and social issues on a roughly equal footing, as opposed to consideration of social factors in most of "sustainability science," where these are often considered only insofar as they show a clear interaction with environmental issues.

Different terms are used to describe efforts to integrate environmental and social criteria into pension fund management. Especially in the United States, there is a long tradition of socially responsible investment that today includes both social and environmental considerations. In continental Europe, the term "sustainability" in investment is increasingly used, and it usually also includes social as well as environmental criteria, although sometimes the emphasis leans toward the environmental side. One difference between the connotations of socially responsible investment and sustainability in investment may be that the latter clearly focuses on future developments.

> *Question:* Now some people talk about sustainability or sustainable invest-ment and some others about socially responsible investment. . . . How do you see the differences, or how do these terms relate to each other?
>
> *Response:* I think one is just a progression of the other. If you look at how we've gotten to this point: . . . even in the 1930s, I think a lot of reli-gious and charitable organizations wanted to replicate their ethos or their thinking in their investment policies. . . . So that was *ethical investing.* Then we moved forward to *environmentally responsible investing . . .* so people said, okay we don't want to invest in nuclear energy for instance; or we don't want to invest in South Africa . . . described as *socially responsible investing,* combining the ethical and environmental issues and the social issues. And now, I think the next step from there is the *sustainable* element. . . . In that case you're actually combining the environmental and the social issues but looking forward to future generations, so it's not just a "for now"-type of situation.

—Investment expert, the Netherlands

Overall, our survey indicates that pension fund and asset managers from North America and the United Kingdom often use the term "socially responsi-ble investment" when they take social and environmental criteria into account. In continental Europe, external asset managers seem to use "socially responsible investment" and "sustainability or sustainable investment" rather interchangeably, while pension fund mangers more often refer to "sustainability" or "sustainable investment."

Findings from our survey also indicate that the goal of higher growth investments and risk reduction in investment decisions are to a certain extent reasons for pension fund managers and external asset managers to consider sustainability criteria in their investment decisions, as is their sense of responsibility as long-term investors. However for pension funds, their board of trustees, as well as their members or beneficiaries, seem to be the major driving forces by demanding these investments, while pressure by other stakeholders such as nongovernmental organizations (NGOs) seems to play a minor role. For external asset managers, demands by their pension fund clients are a main driving force for these investment activities.

Of the institutions reporting that they did not apply social and environmental criteria for their investment choices, approximately half believed that such investments were not profitable or not profitable enough, while the other half said that they didn't have sufficient knowledge for engagement in this type of asset management.

How Is Sustainability Integrated into the Management of Pension Assets?

What kinds of approaches and tools do pension funds use for integrating sustainability criteria into the management of their assets? Basically, two approaches can be distinguished. We call the first approach "selective investment," and the second approach pure "engagement."

In the selective investment approach, pension funds actively avoid investments in companies whose actions are seen as environmentally or socially problematic (sometimes called "negative screening"). Or they specifically invest in companies that are seen as sustainability leaders well positioned to show continuous growth as they are mastering the challenges of the future (sometimes called "positive screening"). Sometimes, these different kinds of screening are seen as directly connected to the different terms used for these investment styles.

> The definitions are a little bit different from one country to another. In France, traditionally we speak about ethic funds. . . . For us, ethic funds means often exclusive funds.
> *Question:* So negative screening?
> *Response:* Yes, often. And sustainability funds means more positive [screening]. . . . For me, it's a question of approach, and also a question of performance. For me, sustainability funds can do better than the benchmark. And traditionally, ethic funds in France often did less than the benchmark.
>
> —Investment expert, France

For implementing these investment strategies, some pension funds select stocks according to sustainability criteria developed in-house, while others buy corresponding investment products from external asset management providers. For most pension funds that practice selective investment according to sustainability criteria, these sustainability investments currently amount to only a small fraction of the total assets invested. For the example of a Swiss pension fund in our sample, 4–5 percent of its overall stock market holdings were invested in sustainability funds. In a Dutch pension fund, just 0.4 percent of the overall assets were invested in sustainability funds. But even that small percentage corresponded to a sum of €200 million. And many people we spoke with thought that high growth potential existed for sustainability investment, so that "in 10 years' time it will be a general part of the investments."

Pension funds that invest parts of their assets selectively according to sustainability criteria will usually not invest in a company (with these parts of their assets) that is seen as an environmental or social laggard. This exclusion does not mean, however, that the funds would immediately divest their holdings in a company that has taken questionable actions from a sustainability perspective. The reason for this is that after a scandal, it is usually too late to sell stock as its value has already gone down. And as major investors in the stock market, pension funds are not interested in contributing to high volatility in the market.

> We don't go directly for the straight line. We don't immediately sell that company. If it's a big company, that would mean that you would disturb the market and the market price. Obviously, we have a huge stake in that company, and selling is not in everybody's interest at that moment. We would go for dialogue.
>
> —Pension fund representative, the Netherlands

So, one can say that there is a certain "stickiness" to investments and investment policies once they are in place.[3]

The reference to "dialogue" above already leads us to the second approach pension funds use for taking sustainability into account, pure engagement. In contrast to the selective investment approach discussed above, in a pure engagement approach sustainability is not included in investment decisionmaking. Sustainability is, however, a key element in working with the management of the companies once stock has been purchased. For example, in corporate governance activities, pension funds as major owners of corporations can force the management to consider the funds' specific interests as long-term investors, and in cases of conflict the funds can even contribute to changing the management.

> We have more than a billion dollars invested in Exxon Mobil. And you remember that some years ago, they had this oil spill. We, as major investors, got on the company's case after that, and we campaigned for them and got them to agree to put an environmentalist on their board of directors.
>
> —Pension fund representative, Northeastern United States

This type of shareholder activism has so far been stronger in the United States than in Europe, but there is an increasing trend on both sides of the Atlantic for institutional investors to actively protect their interests as shareholders. The power of pension funds to force corporate management to pay more attention to long-term viable strategies is especially large if they team up as groups of pension funds, which is beginning to happen.

While most selective investment strategies include some elements of engagement, engagement is the sole mechanism for including sustainability aspects into the management of pension assets that are held in indexed funds. In these funds, the investment managers track the performance of specific stock indices without taking other active management decisions. This passive investing is viewed with concern by some investment professionals.

> Currently, there is a trend towards increase in passive asset management (i.e., index tracking). Today approximately one-third of all assets in the stock market are passively managed. It can be expected that there is an upper limit beyond which passive management will become dangerous (if around two-thirds of the market were passively managed). Most players in investment would be followers rather than leaders then. This might imply high volatility in stock prices, if there are too few people who lead actively.
> —Investment expert, United Kingdom

However, index tracking also holds advantages for individual institutional clients—mainly lower costs and some form of control against the risk of bad judgments by individual asset managers. Within index-tracking fund management, engagement is an important method for integrating sustainability criteria.

The distinction between selective investment and pure engagement for the integration of environmental and social aspects also holds for products and services offered by external asset managers. An additional differentiation of their services is whether they offer products that are standardized (in the sense that the same service is sold to a number of pension funds), or whether they offer investment products that are customized for each specific client. The differentiation is partly a question of the size of the investments, because customized services have relatively high costs for small investments—where investments of around €10 million are considered "small" from the perspective of some of our respondents.

Customized services offer the chance for the client to define for him- or herself which aspects of sustainability should be considered as especially important, whereas the client's control over standardized services is mostly related to transparency. It can be expected that transparency of sustainability investment services will be a key to unleashing the high growth potential of the market for these services. International benchmarks such as the Dow Jones Sustainability Indexes (a family of indices based on the Dow Jones Index family, which includes sustainability criteria in the selection of the stocks included in the index) play an important role here. It will be essential that such international benchmarks are as transparent as possible. Transparency is also important for the investment guidelines of pension funds themselves. A recent regulation in the United Kingdom could become an important international benchmark: the regulation requires pension funds to disclose whether they have policies on socially and environmentally responsible investment, and, if so, what these policies are. The emphasis on disclosure rather than on requiring specific investment guidelines makes this type of rule highly effective and acceptable to the financial industry.

Overall, results from our survey indicate that pension funds follow a selective investment approach and an engagement approach roughly equally. Nearly all of the asset management companies we surveyed offer active investment services (regardless of whether the investment is socially responsible), while some companies also offer passive investment services (i.e., pure engagement support). We found that standardized and customized services offered by external asset managers are provided roughly to a similar extent, both for selective investment services and for passive investment with engagement services. If we break these options down into four categories of investment services (standardized selective investment, customized selective investment, standardized engagement support, customized engagement support), we find that customized and standardized selective investment services are roughly equally widespread, while customized or standardized pure engagement services are both less common.

Which Design Aspects of Future Sustainability Assessments Are Important to Investors?

Which types of sustainability information are used by pension funds today, and which characteristics would be important for sustainability assessments to be useful in the future? First, one should keep in mind that integrating sustainability criteria into investment policies is usually voluntary for pension funds. That means that sustainability information has to fulfill less stringent standards of proof than would be required, for example, in the case of information supporting binding international agreements or legal proceedings: "we can say . . . 'it was reported in such and such a journal that XYZ is happening; in light of this allegation, we would like the company to appoint a special committee to report on this, for shareholder'. . . . We don't necessarily have to have the kind of proof that you would need . . . when you would go to court."

Second, it is important to note that currently pension funds usually do not consider sustainability assessments and information from academic sources in their decisionmaking processes. They partly rely on their own in-house research, and they partly access sustainability information from commercial investment service providers. The following quote is from an interview with a pension fund that places great value on environmental reporting by corporations they invest in.

> It's pretty superficial what we are doing, and we can't do it ourselves. So we used an agency. They produced a 166-page report analyzing all the main companies by various factors. And they gave nothing more in their report to us than a "blob" against whether that company was reporting its position. Now, it's still pretty superficial, and all we are trying to do at this stage is to find out what the key factors are and get a "blob" against their name. And I'm using the word "blob" to show how unsophisticated it is at the moment. . . .
> —Pension fund representative, United Kingdom

Pension funds rely on contractors for sustainability information not only concerning their engagement activities, but also for the case of selective investment policies, where they buy pooled sustainability funds or customized sustainability portfolios from external asset managers. As with pension funds, commercial contractors—be they investment consultants offering information services or external asset managers specializing in sustainability investment products—partly do assessments of sustainability characteristics of corporations in-house and partly buy such information from third-party research service providers. Assessments done by actors in commercial investment consulting are based on a combination of discussions with corporate management, questionnaires sent to the corporations, media reviews, and contacts with NGOs and public policy makers, and they also include at least informal contacts with academic sources.

> We buy social research from various groups based in the United States that provide us with information on workforce, communities, human rights, environment. . . . It's not great [but] . . . it's useful. . . . Then we have our own internal team, and we do a number of things: (1) we talk directly to companies; (2) we review publicly available information; and (3) we talk to thought-leaders in the public interest arena, in government, in the academic community, to give us perspectives at a sectoral level, at an industry level, and at a company level.
>
> —Investment expert, Northeastern United States

Legitimacy of the process leading to sustainability assessments was not an issue of concern to our questionnaire respondents and interviewees from pension funds. This may again be a consequence of the fact that the use of sustainability criteria in pension fund investment is voluntary. The goal here is not to fulfill binding mandatory commitments, but rather to design investment policies that benefit the specific pension funds in question. Pension fund mangers are slightly more aware of potential issues of *credibility*, which are raised by some groups they are in contact with. However, the question of credibility of sustainability information is not seen by many pension fund representatives as a major problem today: "We trust those managers that we appointed [for sustainability services] to have the expertise and to strive to get good information. So we don't get an audit on, for example, their investigation of certain companies."

External asset management consultants specializing in sustainability issues are more aware of problems of the quality and credibility of the sustainability information available for investment professionals today. One source of problems is that research conducted on a large number of corporations usually necessitates that each assessment be done superficially. That is the case even for commercial service providers who have highly skilled teams, but who by necessity are limited in the person-power they can bring to the task, for example in conducting questionnaire research on corporate sustainability policies.

> We simply have to take what is offered, the best that exists. . . . I don't think it is possible to determine the "sustainability leaders" [among companies in a given sector]. I think, the leaders that appear [in sustainability indices] are those that have the best investor relations departments.
>
> —Investment expert, Switzerland

A few large research providers in this field have considerable resources for conducting assessments, but conducting thorough analyses of sustainability aspects for a large number of corporations is a daunting task even for them. In using information provided by them, it is important to users that this information and the process for generating it are as transparent as possible. This can be a problem in an environment of purely commercial sustainability service providers, as "nobody wants to show their hand. . . . The fear of losing this [proprietary] know-how is really strong."

While credibility of assessments of sustainability aspects in investment is more a concern for external investment experts than for pension funds themselves, relevance or *salience*—especially of academic sustainability assessments and information—is an issue of concern for both groups. Perceived lack of salience is a main reason for the fact that assessments of sustainable development from the academic domain are seldom used by managers within pension funds, although informal contacts exist between pension fund managers and academic sustainability researchers. And while informal contacts with academics are important sources of information for some external investment experts specializing in sustainability investment, formal academic sustainability assessments are not used much by them either. Contacts with academic researchers are important to some as part of an informal network of personal contacts in NGOs and academe combined, rather than as part of a more systematic interface with academic research and assessments.

Shortcomings of relevance for investors of current sustainability assessments would have to be overcome for assessment products from sustainability science to become more systematically used by the investment community. A major problem is that sustainability assessments would have to address more clearly social and environmental topics in a way that can directly be related to business sectors—and at least in some cases, all the way down to risks and opportunities faced by particular companies in those sectors—to be really useful for institutional investors. For example, one of our interviewees said that the importance of climate change issues in his work depended clearly "on what specific information we can get about what companies in our portfolio are doing that may have a harmful impact."

The connection of aggregated sustainability assessments down to value-drivers at the sectoral level, and in some cases even the company level, is at least as important for external investment service providers as it is for pension fund managers. Only if this were achieved could sustainability criteria have a fundamental impact on how institutional investors manage their assets. Including management scientists in interdisciplinary sustainability assessment teams could prove very useful in better connecting these different levels in academic sustainability assessments.

For sustainability assessments to be useful for institutional investment professionals, it is important that the assessment takes local conditions adequately into account.

We expect that companies we invest in respect the law and show the behavior of a good corporate citizen in their country. And it's different, a good corporate citizen in Canada is one thing, but in Malaysia or in

France, it's another one. And we have to be flexible with that, because it's impossible to implement our view or culture in all of these countries.

—Investment expert, Quebec

This dependence on local conditions contributes to the problem that assessments of sustainability impacts of alternative courses of action often pose tough problems of weighting tradeoffs, which is an area where institutional investors would clearly welcome guidance from academic experts.

We think it is terrible that 13-year-old girls are exploited in these factories, so we lay them off. And then consequently they go into prostitution. So what have we accomplished? So, I don't know the answer. . . .

—Pension fund representative, California

While academic sustainability assessments are not used often in institutional investment today, potentially they could become very useful in the future. One strong point about such assessments, appreciated by investment experts, is that in principle academic assessments can bring not only more people but also more disciplines to the table than is feasible for any individual pension fund or commercial investment consulting company. Apart from an interdisciplinary approach, academic sustainability assessments should possess several features if they are to be used by practitioners in this field. One issue already touched upon above is that they should provide information on how different aspects of sustainability could be weighted in comparing different corporate approaches. Such weighting is a tough problem and is always linked to the set of values upon which it is based. This means that sustainability assessments might be better off discussing in a very transparent manner different possibilities for such weightings and exploring their implications, rather than selecting one single possibility. But weighting different sustainability criteria should be explored further and more systematically.

Furthermore, responses from our study participants indicate that increased use of dynamic scenario techniques in sustainability assessments could improve their usefulness as well. Such scenarios should connect information on technologies and innovations to estimates of when such innovations might reach the market place. In this, they should also include an improved understanding of social science aspects of technological change and sustainable development.

Academic sustainability assessments could help to improve the transparency of sustainability information used in the investment domain. Transparency is a key issue in the current setting in which most information is generated by commercial service providers who can't make all their processes and results transparent, because they are part of the companies' trade secrets. A better connection between academic sustainability assessments and commercially available sustainable investment advice could play an important role in increasing transparency in sustainable institutional investment services, which is essential for many users in pension funds.

Question: What do you think about transparency? If you buy a service from people like [example of sustainable investment service provider], how much do you want to know exactly how they make their decisions? How much transparency would you expect from them?

Response: A lot. If we are going to rely on what they are doing or saying, then we want to know the rationales behind that. Every decision we make, we should be able to go back to our trustees and explain why we made it.

—Pension fund representative, Ontario

By allowing a better connection between commercial services and academic sustainability science, an improved design of academic sustainability assessments could substantially contribute to more transparency in, and support high growth of, sustainable asset management, potentially to the point where, as one of our respondents expected, "we won't be talking about sustainable investing anymore, because everybody will just be doing it."

Overall, results from our survey indicate that pension funds usually rely either on in-house research or on information acquired from specialized asset management consultants for the development and implementation of their social and environmental asset management strategies. Both of these approaches are roughly equally widespread, while only a minority of pension funds uses both types of sources, and very few use academic research as an important source of information. External asset managers use a wider spectrum of information sources. While in-house research is the dominant source for external asset managers, it is closely followed by information acquired from other specialized investment strategy consultants and academic research (in that order). Nearly all asset managers surveyed use two of the sources referred to above, while only very few rely on a single source, and some rely on all three.

Asked about the potential usefulness of academic research in specific fields, nearly all respondents reported that academic research has the potential to be useful for supporting the integration of social and environmental criteria in pension asset management in the future. Most respondents saw the potential usefulness of academic assessments mainly as a way to better understand specific social and environmental topics. This sentiment was closely followed by the expectation that academic research can provide useful information on problems and opportunities expected for specific business sectors. Among respondents from pension funds, only a minority commented that academic research also had the potential to provide information on problems and opportunities expected for particular companies. However, among respondents from external asset management companies a majority saw academic research as a potentially useful source of information on particular companies (although this expectation was expressed less often than the hope that academic assessments could yield useful information on social and environmental topics, and on implications for specific business sectors). Almost no respondent from either pension funds or asset management companies expressed the view that academic research has no potential to be relevant for social and environmental asset management in the future.

Conclusions and Outlook

Our discussion about the connection between pension fund investment and sustainability information complements studies of the role of other financial institutions in the sustainability transition published elsewhere, especially concerning banks (Bouma et al. 2001), insurance companies (Knoepfel et al. 1999), and venture capital investors (Kasemir et al. 2003).[4] Our focus has been on current barriers to the use of sustainability assessments by institutional investment professionals. The three main questions we raised at the beginning of the chapter, addressing motivations, tools and approaches, and design aspects of the integration of sustainability information into institutional investment practices, guided the study discussed in this chapter. For this study, we surveyed more than 70 pension fund managers and external investment experts from North America and Europe, using interviews and a web-based questionnaire.

Within the financial community, pension fund investors are important potential actors in the "issue domain" of sustainability, as their liabilities extend for decades into the future, and they thus have an interest in thinking about longer time scales. Indeed, sustainability criteria are increasingly considered in pension fund investment. Our findings indicate that the goals of higher-growth investments and of risk reduction in investment decisions contribute to this change. However, currently the major driving force motivating pension funds to consider sustainability criteria seems to be demands by their boards of trustees and by their members or beneficiaries. For external asset managers, demands by their clients, that is, the pension funds, seem to be key. Representatives of both kinds of institutions often emphasize that they as investors feel responsible for taking social and environmental criteria into account, and the spectrum of environmental and social issues they see as relevant for their investment management policies is very broad. This ranges from costs of oil tanker accidents to missed opportunities in hiring talent by companies that discriminate on the basis of race or sexual orientation. Pension investment professionals usually understand sustainability issues to also comprise social problems that have no direct environmental links, making their issue framing broader than the one used in much of sustainability science.

Integrating sustainability aspects into pension fund management can be achieved with different approaches and tools. Two approaches used by pension funds are "selective investment," in which funds or managers purchase stock in companies (often as parts of pooled funds) that are seen as sustainability leaders; and pure "engagement," in which sustainability criteria are not taken into account in investment decisions but later become the subject of influence between the funds and the management of companies. For external investment experts, two additional strategies can be distinguished. These are to offer "standardized" investment vehicles or information services to a number of clients simultaneously, or to provide "customized" services—for example, portfolios managed particularly for one pension fund according to its particular investment philosophy. In all of these activities, results from sustainability science are rarely used by practitioners. Pension funds currently rely on in-house research or on information acquired from specialized asset management consultants, while only very few of them use academic research as an important source of information. External asset managers use a wider spectrum of information sources, with in-house research being their dominant source, closely followed by information acquired from other specialized investment strategy consultants and academic sources. But the responses by our interviewees indicate that asset managers use academic sources in the sense of an informal network of personal contacts, rather than using results from academic sustainability assessments more systematically.

Formal sustainability assessments generated with input from academic sustainability science are rarely used by pension fund managers or external investment experts with whom they collaborate. However, many of those surveyed considered such sustainability assessments to be potentially very useful for them in the future. Whether this potential is realized will depend to a large extent on design aspects of such sustainability assessments. Legitimacy of procedures to generate sustainability information is not seen as much of a problem by either pension fund managers or external investment service providers. Credibility is seen as a problem at least by some of the external investment professionals we surveyed, a number of whom said that commercially supplied information on sustainability records of a large number of corporations is by necessity rather superficial, and they stress the need to improve transparency of the information. But the main barrier to the use of sustainability assessments by investment professionals is relevance or salience, as current sustainability science often provides insufficient information on issues relevant to institutional investors. Sustainability assessments should possess three characteristics for improved usefulness to the investment community:

- *Weighting.* Assessments of sustainability impacts of alternative actions often pose tough problems of weighting tradeoffs (like limiting child labor on the one hand, while possibly eliminating the only decent sort of income for these children on the other). This is something that institutional investors would welcome guidance on from academic experts. As such, weighting is always linked to the set of values used, and sustainability assessments might strive to discuss different possibilities for such weightings and their implications in a transparent manner.

- *Dynamic scenarios.* An increasing use of dynamic scenario techniques in sustainability assessments could improve their usefulness for investment professionals, ideally linking information on potential benefits of innovations to estimates of when they might reach the market place. Also, integrated teams that include social scientists, natural scientists, and engineers might be able to provide more information on how different cultural contexts could influence the acceptability of particular innovations in different markets.

- *Proper scale.* Most importantly, the scale must be right for sustainability assessments to be relevant for investment professionals. From their point of view, aggregate assessments of sustainability aspects should be connected to value-drivers at the sector level, and in some cases the company level. Including management scientists in interdisciplinary academic sustainability assessment teams could prove very useful in better connecting these different levels.

If future sustainability assessments were to more effectively include these characteristics, they could be very useful in supporting sustainable pension fund investment, which could play a significant role in furthering sustainability goals. Including assessments from sustainability science would allow commercial

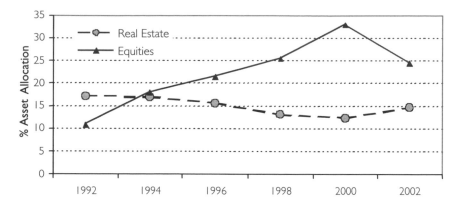

Figure 13-1. *Swiss Pension Assets by Asset Class*
Sources: Values for 1992 to 1996 are from Queisser et al. 2000; values for 2002 are from Swiss Federal Statistical Office 2004.

suppliers of sustainability investment services to improve the *transparency* of sustainability information available to the investment community, which is a key factor for unlocking the potential of high growth in sustainable asset management.

Assessments from sustainability science could also play an important role in addressing two future challenges of sustainable pension asset management. First, a firm maximizing its individual profits may hurt the interests of its institutional shareholders by creating externalities that negatively affect other firms in the shareholders' portfolios (Almaric 2004). This means that pension funds might need tools not only for managing sustainability related risks and opportunities by looking at firms one-by-one, but rather for managing their overall portfolios of share investments, including aspects of interactions between firms. Second, pension funds hold different classes of assets with different exposure to sustainability risks and opportunities. Of particular interest in the context of sustainability are both equity investments and real estate assets. Figure 13-1 shows the relative importance of these two asset classes for the case of Swiss pension assets, where in the 1970s equity investments were only at around 3% of total assets compared with real estate at around 16%, and where in 2002 equity investments increased to roughly 24% of total assets while real estate was still stable at around 15%.

Pension funds as long-term investors could play a very important role in sustainability in the building sector—for example in supporting energy-efficient construction and renovation and adequate housing for all income groups. However, pension fund investors currently lack tools to manage their real estate portfolio from a sustainability perspective, and also to integrate the management policies of their equity portfolio with those of their real estate portfolio.[5] Also, they don't yet have effective procedures for dialogues with stakeholder groups (most importantly with their beneficiaries) on their strategies for managing their assets in a sustainable manner. Insights from sustainability science could be instrumental in helping to address these needs of pension funds in the coming years.

Acknowledgments

The authors are deeply indebted to the pension fund managers and external investment experts who shared their experiences in interviews and questionnaire responses. The following participants, listed below together with their affiliation at the time of our survey, have agreed to be named explicitly in this acknowledgment: Abna Agyeman (Norwich Union Sustainable Future Funds CGNU Life Assurance Ltd.), Philip Ayton (Pavilion Asset Management Ltd.), Rod Balkwill and Judith Lowes (Co-operators Investment Counselling Ltd.), Richard Beaulieu (Gestion de portefeuille Natcan), Erol Bilecen (Bank Sarasin and Cie), Gianni Bottegal and Felix Pfeifer (NEST Sammelstiftung), Paul Burke (Co-operative Insurance Society), Bill Chinery (Barclays Global Investors), Lynn A. Clark (Ontario Municipal Employees Retirement System), Mary Cotrill (CalPERS), Breda Cummins (Robeco), Alan Daxner (McLean Budden Ltd.), Antoine Dehen (ABF Capital Management), Linda Descano (Salomon Smith Barney Inc.), Nico Dijkhuizen (Pensioenfonds Productschappen), Patrick W. Doherty (New York City Comptroller's Office), Peter Dunscombe (BBC Pension Scheme), Michelle Edkins (Hermes Investment Management), Eugene Ellmen (Social Investment Organisation Canada), Jandaan Felderhoff (Lombard Odier Asset Management [Netherlands] N.V.), Richard La Flamme (Desjardins Pension Fund), Georg Furger (Credit Suisse Asset Management), Philippe Gabelier (La caisse de dépôt et placement du Québec), Bernard Gaillard (Caisse de prévoyance du personnel des Etablissements Publics médicaux du Cantone de Genève), James Giuseppi (Henderson Global Investors), Daniel Gloor (Beamtenversicherungskasse des Kantons Zürich), Sharlyn Graham (Laketon Investment Management Ltd.), Paul Grise (Hospitals of Ontario Pension Plan System), Jacques Grubben (Insinger de Beaufort), Elaine Hamilton (Pension Fund, The United Church of Canada), Colin Hay (Lothian Pension Fund City of Edinburgh Council), Karl Herzog (Julius Baer Asset Management Ltd.), Bozena Jankowska (Dresdner RCM Global Investors [United Kingdom] Ltd.), Tamara Jensen (Roxbury Capital Management), David Knight (Knight, Bain, Seath and Holbrook Capital Management), Ivo Knöpfel (SAM Sustainability Group), Laurie Lawson (University of Toronto), André Ludin (NOVARTIS Pension Fund), Colin Melvin (Baillie Gifford and Co.), David Moran (Dow Jones Index), Louis Morisette (SNEGQ), Werner Nussbaum (ISP Innovation Second Pillar), John Owens (Aberdeen Asset Managers Ltd.), Geof Pearson (Sainsbury's Pension Scheme), Anna Pot (ING Investment Management), Robert Preston (AXA Investment Managers), Caroline Schum (ethos, Swiss Investment Foundation for Sustainable Development), Inge Schumacher (UBS), Alexander Schwedeler (Triodos Bank NV), Felix Senn (Pensionskasse des Bundes), Karen Shoffner (Castellum Capital Management Inc.), David Somers (David Somers Clerical Medical Investment Management), Jason Stefanelli (State Street Global Advisors Ltd.), Raj Thamotheram (Universities Superannuation Scheme), David Unitt (FTI–Banque Fiduciary Trust SA), Sylvia van Waveren-Severs (PGGM Zeist), Paolo Wegmüller (Freie Gemeinschaftsbank BCL), Neville White (CCLA Investment Management Ltd.), Thierry

Wiedemann-Goiran (Macif Gestion), and Tom Zimmerman (Nedlloyd Pension Fund).

The authors are also very grateful to Ernst Brugger, William Clark, Nancy Dickson, Hadi Dowlatabadi, Ottmar Edenhofer, Matthew Gardner, Jill Jäger, Carlo Jaeger, Sheila Jasanoff, Don Kennedy, Stephan Lienin, Pam Matson, Steve Rayner, Jan Rotmans, Richard Sandor, Roland Stulz, Marjolein van Asselt, Alexander Wokaun, and Alexander Zehnder for valuable discussions.

This chapter is based on research supported in part by grants from the European Commission (contract ENV4-CT97-0462), the Swiss Federal Office for Education and Science (contract BBW-97.0425), and from the National Science Foundation (Award No. SBR-9521910 for the "Global Environmental Assessment Team"). Bernd Kasemir also acknowledges support from the Belfer Center for Science and International Affairs at Harvard University's Kennedy School of Government, where he was a research fellow while conducting the study described here.

Notes

1. For the important role of private sector participation in a sustainability transition, see Annan (2000).

2. Interviews have been used in research for at least one century (Converse 1987), starting with nonstructured interviews e.g., in psychology. Later, structured interviews with preformulated and preordered questions were used in market research and opinion polling (Denzin and Lincoln 1994). Semistructured interviews, where primary predetermined questions can be complemented by more improvised secondary questions, are particularly suited for fields that are being newly explored. Specific parts of the semistructured interviews in the study reported here were conducted as narrative interviews, for example concerning "how did you personally enter the field of sustainability investment?" Narrative interviews are based on the experience that people constantly tell stories in order to understand the world, which can be encouraged in interviews by refraining from confronting interviewees with the next question before they indicate they have finished their answer (Mishler 1986). Such narratives often indicate respondents' motivations for action (Riessman 1993).

3. For a discussion of the notion of "stickiness" in imperfect markets, see Mankiw and Romer (1991).

4. For a general discussion of financial institutions in sustainable development, see Schmidheiny and Zorraquin (1996).

5. A project on sustainable real estate portfolio management for pension funds, conducted in the context of "novatlantis—sustainability at the ETH domain" (see http://www.novatlantis.ch), is currently addressing these challenges.

References

Almaric, F. 2004. Pension Funds, Corporate Responsibility and Sustainability. Working paper 01/04. Zurich, Switzerland: Center for Corporate Responsibility and Sustainability.

Annan, K.A. 2000. Sustaining the Earth in the New Millenium: The UN Secretary-General Speaks Out. *Environment* 42(8): 20–30.

Bouma, J.J., M. Jeucken, and L. Klinkers (eds.). 2001. *Sustainable Banking.* Sheffield, United Kingdom: Greenleaf Publishing.

Clark, W.C. 2001. America's National Interests in Promoting a Transition to Sustainability. *Environment* 43(1): 18–27.

Converse, J.M. 1987. *Survey Research in the United States: Roots and Emergence 1890–1960.* Los Angeles: University of California Press.

Denzin, N.K., and Y.S. Lincoln. 1994. *Handbook of Qualitative Research.* Thousand Oaks, CA: Sage Publications.

Drucker, P.F. 1976. *The Unseen Revolution. How Pension Fund Socialism Came to America.* London, United Kingdom: Heinemann.

Hawley, J.P., and A.T. Williams. 2000. *The Rise of Fiduciary Capitalism: How Institutional Investors Can Make Corporate America More Democratic.* Philadelphia, PA: University of Pennsylvania Press.

Kasemir, B., F. Toth, and V. Masing. 2003. Venture Capital and Climate Policy. In *Public Participation in Sustainability Science: A Handbook,* edited by B. Kasemir, J. Jäger, C.C. Jaeger, and M.T. Gardner. Cambridge, United Kingdom: Cambridge University Press, 155–175.

Knoepfel, I., J.E. Salt, A. Bode, and W. Jakobi. 1999. *The Kyoto Protocol and Beyond: Potential Implications for the Insurance Industry.* Geneva, Switzerland: United Nations Environment Programme Insurance Industry Initiative.

Mankiw, N.G., and D. Romer (eds.). 1991. *New Keynesian Economics: Imperfect Competition and Sticky Prices.* Cambridge, MA: MIT Press.

Mishler, E.G. 1986. The Analysis of Interview—Narratives. In *Narrative Psychology—The Storied Nature of Human Conduct,* edited by T.R. Sarbin. New York: Praeger.

National Research Council (ed.). 1999. *Our Common Journey: A Transition Toward Sustainability.* B.o.S.D., Policy Division. Washington, DC: National Academic Press.

Queisser, M., and D. Vittas. 2000. *The Swiss Multi-Pillar Pension System: Triumph of Common Sense?* Development Research Group. Washington, DC: The World Bank.

Riessman, C.K. 1993. *Narrative Analysis—Qualitative Research Methods Series 30.* London, United Kingdom: Sage Publications.

Schmidheiny, S., and F.J.L. Zorraquin. 1996. *Financing Change: The Financial Community, Eco-efficiency, and Sustainable Development.* Cambridge, MA: MIT Press.

Swiss Federal Statistical Office. 2004. Die berufliche Vorsorge 2002: Erstmalige Abnahme der Bilanzsumme. Press release 0350-0403-00, June 28, 2004. Neuchâtel, Switzerland: Bundesamt für Statistik.

CHAPTER 14

Improving the Practice of Environmental Assessment

Jill Jäger and Alexander E. Farrell

THE IMPORTANCE OF ENVIRONMENTAL assessments is not disputed. The demands for such assessments have increased over the past 30 years, and their supply has increased correspondingly. The question we had, however, when we started this book, was whether we could learn from an analysis of past experiences and draw some conclusions about the design of assessments. The case studies presented in the book illustrate many of the design decisions that are faced regularly, as well as lessons learned. They also show, however, that there is no blueprint, no "one-size-fits-all" design, for assessments. Still, it is possible to draw some broad conclusions about possible pitfalls and successful strategies that will be useful for those commissioning, designing, carrying out, and using assessments. Confidence in these broad conclusions is based not only on the wide range of assessments addressed in this book but also through cross-reference to other studies and discussions with many of those involved in assessment processes. On the other hand, much also depends on the context within which an assessment is carried out, and it is therefore not possible to provide firm design rules.

It is important to note that many of the chapters in this book provide background and conclusions for previously published parallel studies, such as work edited by Mitchell and colleagues (forthcoming) on information and influence, and a book edited by Jasanoff and Martello (2003) linking local and global levels in environmental politics. Mitchell and colleagues address the community of scholars seeking to understand the interactions of information and institutions in structuring international affairs. Jasanoff and Martello explore case studies that illustrate what it will take to forge robust institutions to address the looming challenges of transnational governance.

As stated in the introduction to this volume, we view assessments as communication processes. An assessment is not just a report—it provides a bridge between the "science realm" and the "policy realm"; it consists of the entire social process by which expert knowledge is organized, evaluated, integrated, and presented to inform policy choices or other decisionmaking. This complexity of goals, processes, audiences, and consequences presents enormous challenges to the design of assessments.

Achieving Effectiveness

The introductory chapter and several chapters in this volume refer to a number of attributes of assessments that play an important role in whether the assessment is effective. Three attributes have been identified:

- Salience: *Does the assessment address questions relevant to decisionmakers?*

- Credibility: *Is the science that is assessed well supported?*

- Legitimacy: *Were various stakeholder interests taken into account fairly during the assessment process?*

The relevance of these attributes to the design of assessments is that design choices that are made can influence how one or more actor groups perceive these attributes. These perceptions play a role in determining the effectiveness of the assessment (see Cash et al. 2003; Eckley 2001). For example, a choice to include only Nobel prize-winners from developed countries in the assessment process would enhance the credibility of the assessment in one sector of the scientific community, but it might lower its credibility in other sectors (e.g., scientists in developing countries). If the participants chose to address questions that are highly interesting from a scientific point of view but do not overlap with the issues high on the agenda in the policy realm, it would lower the perceived legitimacy of the process in developing countries and could also lower the salience of the assessment. Such design choices that influence these attributes have been discussed in several chapters of this book and can be summarized in terms of credibility, salience, and legitimacy.

Credibility

As pointed out above, reliance on "pure science" can enhance the credibility of an assessment process, and there are many cases to exemplify this point. For instance, in assessments that were part of the process of developing international agreements on persistent organic pollutants (Selin, Chapter 4), reliance on the scientific basis enhanced the effectiveness of the negotiations. In addition, the scientific approach adopted in the assessment process can influence the perceived credibility. In the case of impact assessments, discussed by Martello and Iles (Chapter 5), local decisionmakers attached more credibility to scenarios based on field methods than those based on computer simulations.

In the Ozone Transport Assessment Group (OTAG) process (Farrell and Keating, Chapter 3), the policy group meetings were a way to ensure both credibility and legitimacy. However, Farrell and Keating note that people outside the process, such as university researchers, did not recognize the legitimacy of the process (because they were not involved), or the credibility of the process (because it did not use blind peer review). In the Long-Range Transboundary Air Pollution (LRTAP) assessment process, Farrell and Keating conclude that the credibility and legitimacy of the Regional Air Pollution Information and Simulation (RAINS) model increased as the process went on, especially through workshops with policymakers and model developers and a broad range of other European researchers. Furthermore, the multinational character of the institution where the model was developed (International Institute for Applied Systems Analysis [IIASA], Austria) enhanced the credibility of the assessment tool used in the negotiations.

In the case of the assessment in the Severe Sustained Drought study, discussed by Lund (Chapter 7), the drought scenario chosen lacked credibility for at least one stakeholder group, because it was felt to be too far removed from reality. As a result, the study was less effective in leading to policy change in the Colorado River basin. Similarly, the early ozone option assessments (Parson, Chapter 11) relied only on industrial information based on the current state of technology, and the assessments were therefore considered less credible by various actor groups, who thought that possible future developments should also be taken into account.

Credibility also has a different level of importance depending on the stage of development of the issue being considered. For example, in the case of sustainability assessments used for pension fund investments (Kasemir et al., Chapter 13), the credibility of the assessments is not particularly important at the present time but could become much more important when the pressure to invest increases. At the present time, credibility per se is not so important, but trust in the process and participants plays an important role.

As Miller (Chapter 9) points out, perceived credibility also depends on the process and on which conclusions are made public. As shown in many of the assessment processes discussed in this book, different audiences judge credibility using different criteria. Several studies illustrate the tradeoff between credibility and legitimacy, in particular in the case of developing countries (e.g., Ogunseitan, Chapter 10) and countries with economies in transition (e.g., VanDeveer, Chapter 2), pointing out that when issues are addressed in an international forum this tradeoff becomes particularly important but also more difficult.

Finally, there are cases in which innovative solutions have been proposed to deal with enhancing the credibility of assessment processes. One proposal is made by Siebenhüner (Chapter 8), who suggests that credibility of the Intergovernmental Panel on Climate Change (IPCC) assessment process could be increased using an "assessment court," which would judge the external review process. However, Siebenhüner points out that it might be difficult to find appropriate independent (and thereby credible and legitimate) people to be members of such a court.

Salience

As in the case of credibility, decisions that affect salience depend on the stage of issue development. The time-dependence of salience is illustrated, for example, by Keykhah (Chapter 12) and Lund (Chapter 7). In the latter study, the scenarios for the Sustained Severe Drought were not perceived to be salient. If, however, a drought of the magnitude used in the scenario on which the assessment was based had occurred during the time that the assessment was carried out, the salience of the assessment would have increased. In contrast in the Columbia

River basin, the crisis was clear because the decreased salmon population was documented, which enhanced the salience of the assessment.

Several studies also show that salience depends strongly on many factors external to the assessment. For example, Ogunseitan (Chapter 10) shows that the salience of the assessments within the U.S. Country Studies Program depended on political and economic conditions and development goals, in particular, in the countries for which assessments were made. Issue salience varies among stakeholders and participants, and the fact that an assessment is termed "global" does not necessarily guarantee its relevance to every nation state. A reframing in terms of local issues of concern can increase the salience, but possibly at the cost of achieving credibility and legitimacy, especially if the motives for including local concerns are questioned. Similarly, VanDeveer (Chapter 2) concludes that linking assessments to larger political issues or state goals can increase the salience. For example, assessments of long-range transboundary air pollution were linked to the issues of the Cold War and more recently to European Union accession, although this has not necessarily enhanced the effectiveness of the assessments, as shown by the low levels of ratification in Central and Eastern European countries of the LRTAP convention protocols.

Salience also depends on the process chosen for the assessment. In the case of the negotiations of the protocols to the LRTAP convention, the process involving interaction between scientists and decisionmakers using the RAINS model increased the salience of the assessment (Selin, Chapter 4; Farrell and Keating, Chapter 3). The choice of evidence to be included in the assessment also affects the salience. For example, as discussed above for credibility, in the case of climate impact assessments, local decisionmakers considered the use of field methods to be more salient than the use of computer simulations.

Legitimacy

Because we are dealing with global environmental change issues, legitimacy in terms of the process adopted for the assessment and in particular with respect to the selection of participants in the process has often been raised as an important topic. For example, Ogunseitan (Chapter 10) points to the perceptions of legitimacy with regard to the design of the three working groups of the IPCC. Lund (Chapter 7) suggests that legitimacy in the Sustained Severe Drought Study was enhanced by the inclusion of key stakeholders. Selin (Chapter 4) concludes that sufficient vetting of national assessments through international institutions can also increase the legitimacy of the process, because national interests are then seen to be adequately and appropriately included in the process. This inclusion leads to an increased willingness to implement agreements reached internationally at the national level. In other cases, however, legitimacy was not ranked highly. For instance, Kasemir et al. (Chapter 13) find that legitimacy has not been a concern with regard to sustainability assessment production for pension fund investments, although some concerns have been expressed about commercially produced information.

Miller (Chapter 9) raises the important point that perceptions of legitimacy depend on political cultures. What counts as "legitimate science" or a "legitimate process" depends, for example, on the country considered. As Miller points out, the choice of whether to hold open or closed meetings depends on norms expressed in particular national political cultures.

Perceived legitimacy therefore depends on a number of design decisions, but it also depends on external factors. While some will consider that a particular process is legitimate, others will not. Furthermore, the tradeoffs between achieving legitimacy, credibility, and salience in the eyes of one or another group appear to be particularly challenging.

Design Choices

As pointed out in the introduction to this book, as well as in the previous paragraphs, choices in the design of the assessment process can have important implications for the effectiveness of the process. Some of the chapters in this book focus on particular design choices, while others draw more general conclusions. In the latter case, Miller points to the need for flexibility within a process with tight political control and authority. The flexibility is necessary to deal with the rules and expectations of particular national contexts. Design choices are also particularly relevant in the cases of developing countries (e.g., Ogunseitan, Chapter 10) and countries with economies in transition (e.g., VanDeveer, Chapter 2).

Martello and Iles (Chapter 5) point to the need for transparency of design choices to enhance salience, credibility, and legitimacy. On the other hand, in some cases transparency was not essential: Parson (Chapter 11) suggests that the technology assessment panels within the stratospheric ozone regime were not transparent and this was not problematic, because the closed deliberations concerned purely technical issues.

One important consideration in the design of assessments is the treatment of scale issues. Many of the chapters in this book point to the need to take scale issues into consideration. While for the most part this book discusses issues of global environmental change, the effects are also felt at the local and regional levels. While assessment processes might be at the global level, response strategies will also have to be developed and implemented at the regional and local levels. In addition, there are important time scale issues to be considered. The main findings about the treatment of scale issues are presented in Box 14-1.

Initiation and Goals

Design choices at the beginning of the assessment process strongly affect salience, credibility, and legitimacy. Many of the elements discussed in the rest of this section are determined at the beginning of an assessment process; their particular relevance to the entire assessment, however, warrants special attention.

Many of the studies point to the importance of considering the context within which the assessment is to be carried out during the initiation of the process. The context includes both local sociopolitical factors as well as the global situation, trends, and projections. For example, in the case of the Colorado

Box 14-1. Addressing Scale Issues in Assessment Processes

In the impact assessments considered by Martello and Iles (Chapter 5), the use of scenarios based on global circulation models emphasized the long-term issues of climatic change, while local decisionmakers were more worried about shorter-term decisions that had to be taken. Thus the choice of time scale in the assessment process influenced the salience of the assessment. The time scale of the assessment was also mentioned as being important by Keykhah (Chapter 12), who emphasized the short-term focus of the reinsurers.

Scale issues also become important when stakeholders are involved in the process. Thus for example, Kasemir et al. (Chapter 13) point to the need to focus sustainability assessments at the level of company activities in order to bring in that group of stakeholders.

VanDeveer (Chapter 2) points to the challenges that arise because of differences between political scale and natural scale. In the case of air pollution, the political scale of the assessment process was determined by social institutions and not by natural phenomena.

The importance of communication across the different scale levels is pointed out by Ogunseitan (Chapter 10), using the example of the U.S. Country Studies Program, which needed strong communication between the global and local levels. Keykhah points out that information from the Intergovernmental Panel on Climate Change is not used by reinsurers because the level of resolution does not match their needs.

Lund explains that the different effectiveness of the Colorado and Columbia assessments relates to the inclusion of appropriate scale considerations in the assessment design.

River basin assessment (Lund, Chapter 7), the fact that the political and legal context of water management in the Colorado basin were considered in the initiation of the assessment was a critical determinant of the effectiveness of the assessment. In the LRTAP process, the political contexts of the Cold War and EU accession, as well as an awareness of the broader political positions taken by major participants, played an important role in the initiation of the assessment process.

During the initiation of the assessment, a clear identification of the potential threats and opportunities contributes to the determination of the goals and the ultimate outcome of the assessment. This was the case, for example, in the option assessment for the Montreal Protocol on stratospheric ozone depletion (Parson, Chapter 11). Similarly, in the LRTAP negotiations, the determination that critical loads would be the basis of negotiations and that "gap closure" would be used to deal with the time dimension of the negotiation process facilitated compromises during the assessment process (Farrell and Keating, Chapter 3).

Patt (Chapter 6) identifies three broad goals of assessment processes: advisory, advocacy, and agreement. The goal determined at the start of the assessment strongly influences the choices of participation, the way uncertainty is addressed, and which material should be included in the assessment.

The initiation of an assessment process also involves important decisions on the framing of the issue of concern and this, as Ogunseitan (Chapter 10) points out, has important implications for salience and on choices regarding participation in the process. For example, whether the climate change issue is framed as a question of carbon dioxide emissions or of consumption patterns in industrialized countries has an impact on the participation in the assessment as well as on the material to be considered. In the case of impacts of climate change on agriculture and sea level, Martello and Iles (Chapter 5) show that the framing of the issues differed in terms of the emphasis on adaptive capacity and vulnerability of affected populations, and this affected the perceived salience of the assessments. Framing particularly affected the perceived relevance of the findings for decisionmakers, and it also influenced the choice of methodologies used and the participation in the assessments. Patt (Chapter 6) points out that in the case of risks with very low probability, the choices made in framing the risk during the initiation of the assessment has important effects on which response options will be responded to later in the process.

Participation

As Eckley (2001) points out, participation can serve several different functions in assessment processes. Examination of different assessments shows that people and organizations participate in assessments for very different reasons. Some participants engage in assessment processes because they are committed to the development of a particular issue domain. Others participate because they want to enhance the reputation of their organizations. Still others might have an interest in promoting a particular assessment outcome or policy option. Participation can involve stakeholders, interests, and different fields of expertise. How and why participants choose to engage in assessment processes can have a variety of effects, and they often bring credibility, legitimacy, and salience into conflict.

Experience challenges the assumption of many assessment designers that more participation is always better. A workshop at the European Environment Agency suggested that participants (and potential participants) in assessment processes might helpfully be grouped into four separate categories (Eckley 2001):

- **Partners:** The people or groups involved in the production of the assessment

- **Clients:** The assessment's intended audience of users

- **Stakeholders:** Those with an interest in the outcome of an assessment

- **Other users:** Those other than clients who use assessment results (e.g., academics, researchers, and consultants)

Of these groups, partners, clients, and stakeholders may be involved in the design and conduct of the assessment. Their participation can take different forms and can occur at different stages of an assessment process. The effect of participation in assessment processes is highly dependent on the stage the assessment process has reached.

Whether partners, clients, and stakeholders have the capacity to participate as envisioned in environmental assessment is a critical issue. "Capacity" can be scientific (whether a participant with sufficient expertise is available to attend meetings and interpret technical material), administrative (whether sufficient organizational frameworks exist to process information and requests in a timely manner), or financial (whether funds are available for travel costs, salaries, or staff support; see also VanDeveer, Chapter 2).

The choice of who participates in an assessment can significantly influence its credibility to the scientific community (Eckley 2001). One important issue involves whether a scientific assessment is conducted by scientists accountable to governments only, or by scientists participating in their individual capacities. (Of course, these are only two possibilities among many. Scientists could participate in assessments representing industry or nongovernmental organizations as well, and an assessment process could include these participants as well as others.)

When scientists participate in their individual capacities, credibility is likely to benefit, especially among the scientific community. One example is the International Whaling Commission—when this forum was opened to independent scientists, the procedure and agenda changed (Andresen 1998). For particularly contested issues in climate change, the credibility of the IPCC benefited from implementing an exhaustive peer-review process that involved thousands of scientists participating in their individual capacities.

This sort of scientific participation can have significant drawbacks, however. One complication in particular is that it can increase controversy within the assessment process—controversy that may not focus on important issues for policymakers. Increasing credibility in this way can have costs to salience, because questions important to scientists may not be those important to others who are interested in the issue (including the intended users of the assessment). Stakeholders may question whether an assessment conducted by scientists accountable to only their own professional communities took into account their views and circumstances. For example, the first round of the IPCC process ran into similar sorts of legitimacy-based criticisms.

Balancing these tradeoffs requires consideration of the details of the issue at hand. In some cases, such as the LRTAP negotiations, scientists who represent governments (and who often work in regulatory agencies) are credible to both the scientific community and the governments. For some issues, it may be only important to be credible *enough* to the governments for decisions to be taken. These are often the less controversial issues; whereas highly controversial issues, such as climate change, require a higher threshold of credibility. For instance, an LRTAP-like participation system would probably not have been viewed as credible in the climate change area. Similarly, an IPCC-like system would have been too cumbersome—and promoted unnecessary controversy—in an issue area that has been regularly addressed for more than 20 years, as in the case of European air pollution.

In the planning stages of the assessment, encouraging the participation of individuals and groups for whom the assessment is designed to be salient may be helpful. Experience suggests that if participants are engaged in the planning stages of an assessment, the assessment is more likely to ask questions relevant to them. Such participants could include users in the policymaking community or interest groups such as NGOs. Following this strategy often requires close attention to the process used in the assessment to ensure that it retains its credibility. In the case of the OTAG assessment process (Farrell and Keating, Chapter 3), the decision late in the assessment process to add stakeholders' presentations (by groups that had not participated in the process up to then) added very little to the credibility and legitimacy of the process.

Important forms of participation can occur even after an assessment (or an assessment product such as a report) has been completed. For example, users can participate in simulation exercises, query database systems, or use models. One example of an assessment that involves users in this way is IIASA's RAINS model (Selin, Chapter 4). Report-style assessments might promote continued user participation by making authors available for presentations and answering questions after a report is published. New technology is increasingly offering opportunities for such interactivity. The ability of an assessment to respond in a targeted way to specific questions posed by the user clearly has positive implications for the assessment's salience.

The sort of participation required to influence an assessment's salience is quite substantive, however. Experience has shown that users simply "sitting and listening" in an assessment process is not enough for it to become salient to them. Research on the use of IPCC assessments in India shows that these assessments have not succeeded in being useful to Indian decisionmakers (Biermann 2001). In global chemicals negotiations, country representatives were more likely to consider salient those assessments in which they had participated substantively (Eckley 2000).

Many participants do not have the capacity for such substantive engagement. If an assessment is to be salient to them, capacity-building efforts must pay attention to ensuring the ability to participate actively. Merely providing funding to participate in a meeting and covering travel costs is not adequate for this purpose; substantive participation requires training, expertise, and administrative capacity as well as the ability to devote time to the assessment (VanDeveer and Dabelko 2001).

Encouraging such broad-based, substantive participation in assessments to increase salience also has its tradeoffs. Specifically, a process that includes users and stakeholders, who often have clearly defined interests in the assessment's outcome, risks harming its credibility. The process could be perceived as politicized, threatening users' and others' perceptions of its technical quality.

In addition to helping to ensure salience, participation at the beginning of an assessment process can benefit an assessment's legitimacy. Those individuals and groups who participate in the planning stages of an assessment are more likely to perceive the process as fair and one that takes into account their interests and viewpoints.

Research and experience in assessments suggests that the degree of substantive participation required for an assessment to be legitimate to parties is significantly less than that required for salience. In several cases, participation "on paper" seemed to suffice for a process to be considered legitimate to a party—for example, merely being included as an author (without much substantive input), or attending a meeting where an assessment was conducted or approved.

Simple representation, or ensuring that one's voice is heard, seems to have helped a number of assessment processes gain legitimacy. In the IPCC assessment reports, legitimacy to developing countries increased when scientists from developing countries were included as coauthors of all of the chapters—even if some of them did not actively shape the content of their chapters. Many assessments conducted by international organizations achieved legitimacy by being approved by these fully representative institutions. For example, the "blue book" assessments issued in the ozone process under the auspices of the World Meteorological Organization and United Nations Environment Programme (among others) were legitimate to parties involved in the global negotiations, despite the fact that they were conducted by scientists primarily from the United States and the United Kingdom. Because they were reviewed and came out under the authority of these international organizations, they achieved a legitimacy that a report from only the United States or the United Kingdom would not have enjoyed (Clark et al. 1996).

Representation, however, is a tricky concept. It can be difficult to decide who the best person or group is to represent a certain point of view. In international organizations, participation is most often based on country representation, with nongovernmental organizations admitted as observers. It is common practice in U.S. environmental decisionmaking processes to make special efforts to balance the input of industry and nongovernmental organizations in regulatory appraisal. It is also difficult to decide who represents the public at large. Is it nongovernmental organizations that have a broad membership? Elected officials? Because the structure of most assessment processes favors the participation of organized interests, the voice of the "public"—whatever that term may mean—is not often heard in these processes.

Increasing participation to gain legitimacy can also have negative implications for an assessment's credibility, if the process is perceived as too politicized. If a process emphasizes representation by a broad variety of individuals and groups, this wide representation may come at the expense of substantive participation by a smaller subset of participants, risking less salience. This choice can often be made because of resource limitations. Representative, legitimate, broad processes can take a lot of time and money and can often promote significant controversy. Therefore, a process designed to maximize legitimacy may dissolve into adversarial arguments and prevent an assessment from being completed.

Sometimes new groups, with new issue frames, emerge in an assessment process; when participation in the issue area changes, the type of participation in relevant assessments often must change as well. An example of this concept comes from research into the transformation of the climate change issue from a concern by a small group of scientists urging policy attention, to a full-blown international negotiation. Assessments carried out by small groups of experts in the former situation were not legitimate once the issue had been transformed, because countries and interests were brought into the debate that had not been included in previous assessments. The IPCC process developed in the transformed issue area with a more transparent, representative character (Franz 1997).

Another sort of participation change occurs when issues previously assessed or dealt with on a regional level become global concerns. Many issues are pushed onto the global agenda by one or more parties with an interest in the outcome, and early assessments are often carried out by individual nations with particular expertise. In the global ozone negotiations, state-of-the-art science had already been collected by individual parties, and the challenge was to use this information successfully in a global forum; this was achieved by the issuance of reports under international auspices. The addition of parties was also relevant for the issue of persistent organic pollutants—the issue was pushed by Canada and was subject to a regional agreement in Europe and North America before global negotiations began. In this case, participation strategies were able to address concerns of legitimacy in using results of the regional assessment process when additional participants were added, but these assessments were not salient to those who did not participate substantively in their production (Eckley 2000). These cases represent further challenges, but also further support for the linkages detailed above between participation in assessment processes and their credibility, salience, and legitimacy to users.

As shown above, participation in an assessment process is strongly connected to the perceived legitimacy, credibility, and salience of the process. It is important to distinguish between participation in various phases of the process, because it is without doubt possible to have different participants in the initiating, carrying out, and reporting of the process. In particular because these assessments are concerned with global issues, international participation is discussed, with a distinction made between real and nominal participation. Connecting to the previous paragraphs on initiation and goals, VanDeveer (Chapter 2) points out that early patterns of participation and the resulting initial framing of scientific and technical questions are closely connected and can persist during the assessment process.

International participation in assessment processes is undisputedly important in assessment of global environmental change issues. While Selin (Chapter 4) finds that international participation in LRTAP assessment processes increased the legitimacy and credibility, she also points out that the participation of "lead countries" rather than broad international participation also increased effectiveness. On the other hand, within the IPCC and the Subsidiary Body on Science and Technology Assessment (SBSTA) of the United Nations Framework Convention on Climate Change (UNFCCC), participation of developing countries has been frequently discussed (Miller, Chapter 9). Participation of developing countries in the U.S. Country Studies Program also was found to enhance salience of the assessments (Ogunseitan, Chapter 10).

Stakeholder participation is also a key issue. Lund (Chapter 7) emphasizes the importance of the involvement of key stakeholders in the assessment process. Parson (Chapter 11) concludes that participation of industry experts was crucial in getting changed estimates of feasibility of options in the stratospheric ozone regime and also points out that it was important to have shared knowledge on technical problems for which special expertise was required.

The stratospheric ozone technology assessments also illustrate the importance of changing participation over time; because the questions to be addressed changed, the required expertise had to be changed (Parson, Chapter 11). Whereas only user industries and industry associations participated in early assessments, chlorofluorocarbon (CFC) producers participated in later assessment processes. Changing participation led to the creation of an adaptive, flexible regime.

One major issue with regard to participation is capacity. This is closely connected to the question of real versus nominal participation. To participate fully in an assessment process, a basic understanding of the scientific issues and political context is necessary. Several chapters in this book address this issue. Keykhah (Chapter 12) finds that because of a lack of training, there is a gap between the tacit assumptions of those who develop models and those who use them, and this is particularly clear with respect to the understanding and treatment of uncertainties. VanDeveer (Chapter 2) concludes that within the LRTAP assessments, transition country participation was lower than that of the "big player countries." He points out that participation requires a commitment of resources—not just the costs of attending meetings, but material resources and expertise, and data and monitoring capabilities.

The Science–Policy Interface

As discussed in the introduction to this volume and also discussed in detail by Eckley (2001), managing interactions between scientific experts and policymakers (and their representatives) is a challenge that faces most environmental assessment processes. Assessments are carried out in institutions and organizations that have different mandates, institutional structures, and rules. The fit of assessments to institutions depends on the state of the issue at hand (e.g., the level of political contestation about the issue, or the maturity of the science), the history of previous assessments, and the scope of the assessment (especially whether it includes policy recommendations or not). There are many different types of assessment as well—some of which fit better into particular institutions. Assessments, and the institutions within which they are conducted, can be accountable to various groups (e.g., to policymakers, to the scientific community, or to both simultaneously). The institutions in which assessments are conducted, and the interactions between expertise and decisionmaking authority, can influence an assessment's credibility, salience, and legitimacy.

Depending on the state of "issue evolution" (Social Learning Group 2001), the design of the science–policy interface must take different forms. When the issue is on the scientific agenda but has not achieved widespread political attention, as, for example, climate change during the 1970s and early 1980s (Agrawala 1998a, 1998b), the interface can be quite sharp, with the scientific community assessing the state of the art of their knowledge of an issue and communicating this to a distinct policymaking community. When, however, the issue moves higher on the political agenda, institutions that allow closer interaction between scientific experts and policymakers have proven more successful. For example, Selin (Chapter 4) suggests that within the LRTAP negotiation the structure of the communication process between scientists and policymakers created a flexible negotiating process that allowed changes as scientific knowledge changed over time, and this process was able to maintain perceived salience. The institution best suited to conduct an assessment of a problem in the early stages of formulation is most often not the same institution best prepared to conduct a salient assessment of policy options. Such institutional choices must take into account the goals of the assessment, as well as the interests of the intended users.

Iterative assessment processes are often among the most effective, as are those that are structured as continuing, progressively improving assessments of the same issue. In particular, planned iteration can offer the security of knowing that issues can be revisited or decisions can be taken at a later date—when they would otherwise pose serious challenges to short-term or one-time assessments, especially those that aim to produce a consensus report. This advantage of iteration has been the case in LRTAP's assessment process, where areas in which science was not yet fully mature have been noted and addressed in later protocols.

Assessment experience shows that the choice of an assessment's institutional setting matters a lot—not only by shaping the content of an assessment, but also by influencing the way in which an assessment is perceived (Eckley 2001). Organizations that are accountable to both policy users and scientific communities have helped to ensure effective assessments in many of the cases discussed in this volume and by Eckley (2001). Some organizations have developed that seem to straddle the shifting divide between science and policy, helping to maintain scientific credibility while assuring political salience. These "boundary organizations" adhere to distinct lines of accountability to both science and politics, facilitate the transfer of usable knowledge between science and policy, and give both policymakers and scientists the opportunity to construct the boundary between their domains (Guston et al. 2000).

Uncertainty and Dissent

One design element that was perhaps not so obvious when we started to look at lessons in the design of assessment processes is the treatment of uncertainty and dissent. Different assessment processes have different ways of dealing with uncertainty, ranging from explicit recognition and quantification of uncertainty to ignoring uncertainties altogether. Different approaches are taken toward uncertainty, as exemplified by the risk assessment approach taken by the United States and the precautionary approach taken by the European Union.

Stakeholders involved in an assessment process have different levels of tolerance of uncertainty, as illustrated, for example, by Keykhah (Chapter 12). The goal of the assessment also affects the treatment of uncertainty, as illustrated by Patt (Chapter 6), who shows that if the goal of the assessment is to reach consensus on an issue, it is likely to ignore low-probability events. Patt shows that dealing with uncertainty leads to large tradeoffs between salience, credibility, and legitimacy, especially over time.

Farrell and Keating (Chapter 3) point out that uncertainty analysis was ignored in the OTAG assessment process, because it was too time consuming and expensive. This omission in itself was not seen to have damaged the assessment process. For the OTAG process as a whole, decisionmaking was by consensus, while the OTAG Policy Group decisions were taken by majority vote. The consensus decisions were however sometimes speeded up by the leadership forcing temporary decisions in order to help the group move forward. This illustrates the importance of constructive and intelligent leadership within the assessment process and the use of what Selin (Chapter 4) has referred to as "dependable dynamism," that is, participants were willing to accept a temporary consensus with the knowledge that they could return to critical sticking points later in the process. Thus uncertainty was not taken as an excuse to stall the process, but the dynamic design of the assessment process took into account future changes in levels of uncertainty. Miller also discusses this "time dimension," emphasizing the importance of exposing disagreements early in the assessment process.

The Design of Assessments and Future Research Challenges

The chapters in this book suggest that there is indeed quite a lot that can be learned from previous experience and analysis in the design of assessments. Design choices do have an influence on how the assessment process and its results are perceived by those who commission, initiate, carry out, and use the assessment. Design choices shift the weights in the perceived credibility, salience, and legitimacy of the assessment. The previous sections of this chapter detail the four most important elements of design discussed in this book:

- **Initiation of the assessment**—Decisions at the beginning of an assessment on the issues to be addressed (and those not to be included) and the goals of the assessment have important implications. The initiation of an assessment must take into account the social, political, and cultural context within which the assessment is carried out as well as the stage of issue development.

- **Participation**—Although "more participation is not always better," there are interesting examples of the design of participation in an assessment process to enhance credibility, salience, and legitimacy in the eyes of various actor groups.

- **Science–policy interface**—An assessment builds a bridge between the scientific and policymaking communities. The assessments described in this book give many examples of how that bridge can be built. A flexible bridge that can be shaped as the issue evolves is certainly effective. We have noted the importance of a bridge that takes into account the stage of issue evolution, the surrounding political and cultural contexts, and the goals of the assessment.

- **Treatment of uncertainty and dissent**—Assessment processes have failed because they did not deal adequately with uncertainty and dissent within the process. The multitude of actors involved in an assessment and the multiple goals that they have make dealing with uncertainty difficult. Nevertheless, there have been breakthroughs that deserve attention in the future design of assessment processes.

As pointed out in the introduction to this chapter, there is no blueprint design for an assessment process, but experience shows that the above four elements of design are particularly important in dealing with the tradeoffs between the attributes of salience, credibility, and legitimacy to ensure the effectiveness of assessment processes. The complex nature of global environmental risks presents numerous challenges, especially in designing assessments that take the above elements into account. Moreover, of course, these elements are connected. As Van-Deveer shows, for example, patterns of participation within assessment processes and initial framing of the issue are closely interrelated. Dealing with the challenges of assessment design will require serious, worldwide efforts in capacity-building on issues such as issue framing, participation, designing the interaction between the scientific and policymaking communities, and treating uncertainty.

As Siebenhüner points out, however, to absorb the lessons in this book, assessment processes have to include learning mechanisms in their design. It seems fair to assume that advanced learning mechanisms, including adaptive or even more fundamental changes of the value and belief systems of an assessment "organization," would enhance the overall effectiveness of the assessment. Such learning would be enhanced through open communication structures, information processing systems, and supportive shared values among participants and others.

Examples are given in various chapters of how assessment processes have learned over time. For example, Ogunseitan (Chapter 10) points out that the U.S. Country Studies Program built legitimacy into the process through integration with development agendas, prioritized participation in the design of the assessment plan, learned to deal with communication and implementation of the assessment, and developed a flexible process with respect to issue framing, stakeholder participation, and policy relevance.

Our research has highlighted design options in global environmental assessments and shown the importance of design choices. At the same time, it suggests that research in this area still has much to offer. Future research could focus, among other things, on

- how different countries and different stakeholders use expertise and expert information in environmental decisionmaking;

- the role of the public in scientific assessment, in particular in promoting trust in science and governance;

- the roles of political orientations and objectives such as the "precautionary principle" in assessment and decisionmaking processes;

- the role of boundary organizations and of institutions playing boundary-like roles in environmental assessment;

- assessing effectiveness more systematically, and encouraging learning in assessment processes; and

- addressing multiple causes (social, environmental, economic) of global change and multiple outcomes within a broader framework of vulnerability assessment.

In conclusion, much can be learned from the study of assessment processes and the lessons highlighted throughout this book. If these lessons are implemented in the design of future environmental assessments, they can enhance the use of scientific knowledge in decisionmaking. While it is not possible to provide a "recipe book" for assessment processes, this book highlights some of the important ingredients and demonstrates their importance. The environmental challenges faced by the world are large and still growing. For this reason, assessments will remain an important tool for linking knowledge and action, and conscious improvements in the design of assessment processes will contribute to problem-solving.

References

Agrawala, Shardul. 1998a. Context and Early Origins of the Intergovernmental Panel on Climate Change. *Climatic Change* 39: 605.

―――. 1998b. Structural and Process History of the Intergovernmental Panel on Climate Change. *Climatic Change* 39: 621.

Andresen, S. 1998. The Making and Implementation of Whaling Policies: Does Participation Make a Difference? In *The Implemetation and Effectiveness of International Environmental Commitments: Theory and Practice,* edited by D.G. Victor, K. Raustiala, and E. B. Skolnikoff. Cambridge, MA: MIT Press, 431–475.

Biermann, F. 2001. Big Science, Small Impacts—In the South? The Influence of Global Environmental Assessments on Expert Communities in India. *Global Environmental Change* 11(4): 297–309.

Cash, D.W., W.C. Clark, F. Alcock, N.M. Dickson, N. Eckley, D.H. Guston, J.Jäger, and R.B. Mitchell. 2003. Knowledge Systems for Sustainable Development. *Proceedings of the National Academy of Sciences* 100(14): 8086–8091.

Clark, W.C., E. Parson, N. Dickson, and M. Ferenz. 1996. *Connecting Science and Policy for Global Environmental Issues: A Critical Appraisal of the "Atmospheric Ozone 1985" Blue Books.* Paper presented at the American Association for the Advancement of Science Annual Meeting and Science Innovation Exposition. February 1996, Baltimore, MD.

Eckley, N. 2000. From Regional to Global Assessment: Learning from Persistent Organic Pollutants. Belfer Center for Science and International Affairs discussion paper, October 23. Environment and Natural Resources Program. Cambridge, MA: Kennedy School of Government, Harvard University.

―――. 2001. *Designing Effective Assessments: The Role of Participation, Science and Governance, and Focus.* Environmental Issue Report no 26. Report from a workshop co-organized by the Global Environmental Assessment Project and the European Environment Agency, March 1–3. Gare, Luxembourg: Office for Official Publications of the European Communities.

Franz, W.E. 1997. The Development of an International Agenda for Climate Change: Connecting Science to Policy. Belfer Center for Science and International Affairs discussion paper E-97-07. Cambridge, MA: Kennedy School of Government, Harvard University. Also International Institute for Applied Systems Analysis interim report IR-97-034/August. Environment and Natural Resources Program.

Guston, D.H., W. Clark, T. Keating, D. Cash, S. Moser, C. Miller, and C. Powers. 2000. *Report of the Workshop on Boundary Organizations in Environmental Policy and Science.* Belfer Center for Science and International Affairs discussion paper 2000-32. Cambridge, MA: Kennedy School of Government, Harvard University. Also Environmental and Occupational Health Sciences Institute, Rutgers University and University of Medicine and Dentistry of New Jersey – Robert Wood Johnson Medical School; and the Global Environmental Assessment Project, Environment and Natural Resources Program.

Jasanoff, S., and M.L. Martello. 2004. *Earthly Politics: Local and Global in Environmental Politics.* Cambridge, MA: MIT Press.

Mitchell, R., W.C. Clark, D. Cash, and N. Dickson. Forthcoming. *Global Environmental Assessments: Information and Influence.* Cambridge, MA: MIT Press.

Social Learning Group. 2001. *Learning to Manage Global Environmental Risks.* Cambridge, MA: MIT Press.

VanDeveer, S., and G. Dabelko. 2001. It's Capacity, Stupid: International Assistance and National Implementation. *Global Environmental Politics* 1(2): 18–29.

Index